TRANSPORT OF MOLECULES ACROSS MICROBIAL MEMBRANES

This volume considers the transport of molecules, large and small, across the membranes of prokaryotic and eukaryotic microbial cells. A diverse range of related phenomena are covered, but the unifying themes are the signal peptides that target proteins to particular destinations, and the role of chaperonins. Topics covered include: secretion of proteins out of the bacterial cell by Type I, II and III mechanisms, including the newly recognized bacterial signal recognition pathway in Type II; passage across internal membranes of eukaryotic proteins, whether destined for secretion or en route to internal organelles such as chloroplasts and peroxisomes; how bacteria obtain the energy required for solute uptake, the role of phosphorylation, and evolutionary relationships of some of the proteins involved; and efflux pumps for toxic substances in bacterial, animal and plant cells.

JENNY BROOME-SMITH is a Senior Lecturer in Molecular Genetics in the Biochemistry Group at the University of Sussex.

SIMON BAUMBERG is Professor of Bacterial Genetics in the School of Biological Sciences at the University of Leeds.

COLIN STIRLING is Professor of Genetics in the School of Biological Sciences at the University of Manchester.

BRUCE WARD is a Lecturer in Microbiology and Senior Director of Studies in the Institute of Cell and Molecular Biology at the University of Edinburgh.

SYMPOSIA OF THE
SOCIETY FOR GENERAL MICROBIOLOGY

Series editor (1996–2001): Dr D. Roberts, Zoology Department, The Natural History Museum, London
Volumes currently available:

TRANSPORT OF MOLECULES ACROSS MICROBIAL MEMBRANES

EDITED BY

J. K. BROOME-SMITH, S. BAUMBERG, C. J. STIRLING AND F. B. WARD

FIFTY-EIGHTH SYMPOSIUM OF THE
SOCIETY FOR GENERAL MICROBIOLOGY
HELD AT THE UNIVERSITY OF LEEDS
SEPTEMBER 1999

Published for the Society for General Microbiology

CAMBRIDGE UNIVERSITY PRESS

PUBLISHED BY THE PRESS SYNDICATE OF THE UNIVERSITY OF CAMBRIDGE
The Pitt Building, Trumpington Street, Cambridge, United Kingdom

CAMBRIDGE UNIVERSITY PRESS
The Edinburgh Building, Cambridge CB2 2RU, UK www.cup.cam.ac.uk
40 West 20th Street, New York, NY 10011-4211, USA www.cup.org
10 Stamford Road, Oakleigh, Melbourne 3166, Australia
Ruiz de Alarcón 13, 28014 Madrid, Spain

First published 1999

Printed in the United Kingdom at the University Press, Cambridge

Typeset in Monotype Times 10/12pt, in 3B2 [PN]

A catalogue record for this book is available from the British Library

ISBN 0 521 77270 2 hardback

CONTENTS

CONTRIBUTORS

ANDERSON, D. M. Department of Microbiology & Immunology, UCLA School of Medicine, 10833 Le Conte Avenue, Los Angeles, CA 90095, USA

BAKER, A. Centre for Plant Sciences, Leeds Institute for Plant Biotechnology and Agriculture, University of Leeds, Leeds LS2 9JT, UK

BHATTACHARJEE, H. Department of Biochemistry and Molecular Biology, Wayne State University, School of Medicine, Detroit, MI 48201, USA

BROOME-SMITH, J. K. Biochemistry Group, School of Biological Sciences, University of Sussex, Falmer, Brighton BN1 9QG, UK

BROWNSWORD, J. K. School of Biological Sciences, University of Manchester, Oxford Road, Manchester M13 9PT, UK

CHENG, L. W. Department of Microbiology & Immunology, UCLA School of Medicine, 10833 Le Conte Avenue, Los Angeles, CA 90095, USA

EDWARDS, W. Department of Biological Sciences, University of Warwick, Coventry CV4 7AL, UK

FILLOUX, A. Laboratoire d'Ingéniérie des Systèmes Macromoléculaires, CNRS-IBSM-UPR9027, 31 Chemin Joseph Aiguier, 13402 Marseille Cedex 20, France

GHOSH, M. Department of Biochemistry and Molecular Biology, Wayne State University, School of Medicine, Detroit, MI 48201, USA

HIGH, S. School of Biological Sciences, University of Manchester, 2.205 Stopford Building, Oxford Road, Manchester M13 9PT, UK

HULTGREN, S. J. Department of Molecular Microbiology, Washington University School of Medicine, Saint Louis, MO 63110, USA

HYNDS, P. J. Department of Biological Sciences, University of Warwick, Coventry CV4 7AL, UK

LACKS, S. A. Biology Department, Brookhaven National Laboratory, Upton, NY 11973, USA

LEE, V. T. Department of Microbiology & Immunology, UCLA School of Medicine, 10833 Le Conte Avenue, Los Angeles, CA 90095, USA

LEWIS, K. Tufts University Biotechnology Center, Medford, MA 02155, USA

LOPEZ-HUERTAS, E. Centre for Plant Sciences, Leeds Institute for Plant Biotechnology and Agriculture, University of Leeds, Leeds LS2 9JT, UK

LUIRINK, J. Department of Microbiology, Institute of Molecular Biological Sciences, Biocentrum Amsterdam, De Boelelaan 1087, 1081 HV Amsterdam, The Netherlands

MASMANIAN, S. Department of Microbiology & Immunology, UCLA School of Medicine, 10833 Le Conte Avenue, Los Angeles, CA 90095, USA

MITSOPOULOS, C. Biochemistry Group, School of Biological Sciences, University of Sussex, Falmer, Brighton BN1 9QG, UK

MUKHOPADHYAY, R. Department of Biochemistry and Molecular Biology, Wayne State University, School of Medicine, Detroit, MI 48201, USA

O'GORMAN, E. N. S. School of Biological Sciences, University of Manchester, 2.205 Stopford Building, Oxford Road, Manchester M13 9PT, UK

POOLMAN, B. Department of Microbiology, Groningen Biomolecular Sciences and Biotechnology Institute, University of Groningen, Kerklaan 30, 9751 NN Haren, The Netherlands

RAMAMURTHI, K. Department of Microbiology & Immunology, UCLA School of Medicine, 10833 Le Conte Avenue, Los Angeles, CA 90095, USA

ROBINSON, C. Department of Biological Sciences, University of Warwick, Coventry CV4 7AL, UK

ROSEN, B. P. Department of Biochemistry and Molecular Biology, Wayne State University, School of Medicine, Detroit, MI 48201, USA

SAIER, JR, M. H. Department of Biology, University of California at San Diego, La Jolla, CA 92093-0116, USA

SCHNEEWIND, O. Department of Microbiology & Immunology, UCLA School of Medicine, 10833 Le Conte Avenue, Los Angeles, CA 90095, USA

SCOTTI, P. Department of Microbiology, Institute of Molecular Biological Sciences, Biocentrum Amsterdam, De Boelelaan 1087, 1081 HV Amsterdam, The Netherlands

SOTO, G. E. Department of Molecular Microbiology, Washington University School of Medicine, Saint Louis, MO 63110, USA

STIRLING, C. J. School of Biological Sciences, University of Manchester, Oxford Road, Manchester M13 9PT, UK

TAM, C. Department of Microbiology & Immunology, UCLA School of Medicine, 10833 Le Conte Avenue, Los Angeles, CA 90095, USA

TISSIER, C. Department of Biological Sciences, University of Warwick, Coventry CV4 7AL, UK

TSENG, T.-T. Department of Biology, University of California at San Diego, La Jolla, CA 92093-0116, USA

VALENT, Q. A. Department of Microbiology, Institute of Molecular Biological Sciences, Biocentrum Amsterdam, De Boelelaan 1087, 1081 HV Amsterdam, The Netherlands

YOUNG, B. P. School of Biological Sciences, University of Manchester, Oxford Road, Manchester M13 9PT, UK

OVERVIEW: TRANSPORT OF MOLECULES ACROSS MICROBIAL MEMBRANES – A STICKY BUSINESS TO GET TO GRIPS WITH

JENNY K. BROOME-SMITH AND COSTAS MITSOPOULOS

Biochemistry Group, School of Biological Sciences, University of Sussex, Falmer, Brighton BN1 9QG, UK

INTRODUCTION

Our understanding of how molecules are transported across microbial membranes has lagged far behind our understanding of processes that occur within the aqueous compartments of these cells. There is little doubt that this is because it is so difficult to analyse the structures of the membrane proteins that mediate, or play central roles in, these processes. Membrane proteins are inherently difficult to purify and crystallize in (active) forms suitable for high-resolution analysis, because they are amphipathic molecules. The problem is exacerbated by the fact that most are non-abundant, and cannot be successfully overproduced without aggregating within, or even killing, the producing cell. Indeed, it was not until 1985 that Michel's group, applying a novel amphiphile-coating approach, which rendered the surfaces of photosynthetic reaction centre molecules uniformly polar, provided us with the first atomic resolution structure of a membrane protein (Deisenhoffer *et al.*, 1985). Even now, with the structures of soluble proteins being solved at the rate of one or more a day, the number of membrane proteins whose structures have been solved is only just into double figures. In each case ingenious strategies have had to be deployed to get crystals that are suitable for high-resolution analysis – the membrane proteins have been coated with amphiphiles and had their polar surfaces expanded with monoclonal antibodies, or crystallized in two-dimensional lattices (within phospholipid bilayers) or within custom-built three-dimensional lattices (reviewed by Ostermeier & Michel, 1997).

Against this background it is worth reflecting on the considerable importance of membrane transport processes. Eukaryotic microbes have numerous different subcellular compartments, and the proteins they synthesize must be efficiently transported to their correct subcellular destinations. Small molecules (nutrients, ions, drugs, metabolites) are transported into or out of the cell and its organelles, and specialized protein complexes within the

membranes mediate energy transduction and transmembrane signal transduction processes. Even in the relatively simple bacterial microbes a substantial proportion of the proteins synthesized in the cytoplasm (around 25–
30%) are destined for extracytoplasmic locations. In the Gram-negative
bacteria, which have an extra, outer, membrane surrounding the plasma
membrane, extracytoplasmic proteins must be correctly localized to one of
four compartments – the inner membrane, the periplasm, the outer membrane or the exterior. One major question that several articles in this
symposium address is: how do large hydrophilic polypeptide substrates pass
through hydrophobic membranes? Another recurring question is: how are
polypeptide substrates recognized as being destined for different subcellular
locations and correctly targeted to them? Many of the micro-organisms that
have been most intensively studied are human, animal or plant pathogens.
They make contact with their hosts via their external surfaces and appendages. Protein secretion is often of special importance for delivering virulence
factors into the host cell. Finally, we are now in the age of genomics, and it is
clear that amino acid sequence similarity comparisons are hugely impacting
on our insight into protein evolution and biological processes. Such comparisons are of special value where membrane proteins are concerned, since
structural studies lag so far behind those on soluble proteins.

TRANSPORT PROCESSES

Membrane proteins fulfil a variety of crucial cellular functions, and as Saier
& Tseng remind us (this volume): 'These transporters are essential for
virtually all aspects of life as we know it on Earth.' Thus, whilst it has so far
proved impossible to purify, crystallize and obtain high-resolution structural
data for all but a few membrane proteins, there is a very strong impetus to
continue to explore and develop novel approaches that may help shed light
on their structure and function. In the first few chapters of this symposium
we are brought up to date on our knowledge of several different classes of
membrane transport proteins. In an article that reads like a good detective
novel, Kim Lewis describes the proteins that cause multidrug resistance by
catalysing drug efflux. The MDR proteins are ubiquitous and occupy four
different superfamilies of membrane proteins. Clinically significant drug
resistance is caused by increased expression of *mdr* genes. Perhaps the most
taxing question here is: how can MDRs bind and extrude a wide variety of
different substrates? In fact, amino acid sequence comparisons reveal that
MDRs have evolved multiple times from efflux proteins of much narrower
substrate specificity. (Amino acid substitutions in the ancestral proteins have
caused the switch to a broader substrate specificity.) Moreover, although
MDRs extrude a variety of unrelated compounds, their preferred artificial
substrates are almost invariably amphipathic cations. As these substances are
able to partition into the membrane, the possibility that MDRs only

'consider' substances within the membrane as their ligands has been raised. It is now clear that LmrA, a functional bacterial homologue of mammalian P-glycoprotein, can pump ligands from the inner leaflet of the membrane to the exterior. Maybe mammalian P-glycoprotein has evolved its exceptional ability to flip drugs from the inner to the outer leaflet of the plasma membrane, because here they can then be detoxified, whereas extrusion would simply be followed by their re-entry into the cell. As it seemed likely that MDRs could have evolved to protect microbes from the potentially damaging effects of amphipathic cations, Lewis and colleagues searched for natural compounds of this type. They found that a group of plant alkaloids – the isoquinoline alkaloids, such as berberine and palmatine – fitted the bill, and that these had potent antimicrobial activity in the presence of MDR inhibitors. Moreover, they established that a berberine-producing plant also made two different MDR inhibitors. Multidrug resistance is a severe clinical problem, so there is real hope that these natural MDR inhibitors can be used in conjunction with conventional antimicrobials to overcome it.

Arsenic resistance genes are found in nearly all organisms, perhaps because the primordial soup was rich in dissolved metals, and therefore resistance to toxic metals was important to all early life forms. In the article by Bhattacharjee *et al.* we learn that membrane proteins with the ability to extrude arsenicals have evolved at least three times. In bacteria ArsB acts as a secondary transporter, catalysing the extrusion of arsenite coupled to the membrane potential. However, in some organisms the ArsA ATPase is also produced and it binds to ArsB, converting it to a primary transporter that extrudes arsenite at the expense of ATP hydrolysis. Interestingly, the ArsB membrane protein has a topological arrangement [N-in C-in with 12 membrane-spanning segments (MSSs)] that is more reminiscent of secondary rather than primary transporters. (ArsA homologues are found in bacteria through to man, but so far the physiological function of the eukaryotic ArsA homologues remains unknown.) Recently another family of membrane proteins that confer arsenite resistance has been identified in both bacteria and yeasts. One of these 10 MSS proteins, Acr3p of *Saccharomyces cerevisiae*, has now been shown to be a plasma membrane arsenite efflux protein. However, *Sacch. cerevisiae* also harbours the protein Ycf1p, a vacuolar membrane ABC transporter, which is known to confer cadmium resistance by pumping $Cd(GS)_2$ conjugates into the yeast vacuole. Recently it has become clear that Ycf1p also pumps arsenite into the vacuole. Homologues of Ycf1p and Acr3p are likely to exist in all eukaryotes.

Poolman highlights the fact that transporters do not accumulate solutes to such high levels as are predicted from the driving forces for these processes. In fact, leak pathways rarely make a significant contribution, at least in primary (ATP-driven) transport processes, and product inhibition is a major player. When cells are starved of energy and the ion motive force drops, then solutes would be expected to leak out via their secondary transporters. However, in some microbes the solutes are retained because the transporters themselves are highly sensitive to changes in the internal pH, and as the pH value falls below the physiological level they lose activity. Other mechanisms such as inducer exclusion in Gram-negative bacteria, osmosensing and catabolite repression all act to regulate transport activity. This article serves as a salutary reminder that transporters are sophisticated devices, and even when we understand their basic mode of action, we can only meaningfully relate this to actual cellular physiology if we take into account mechanisms for modulating their activity to prevent catastrophically high solute accumulation.

Given the dearth of high-resolution structural information on membrane proteins, and the current explosion in genomic sequencing, molecular archaeological studies are particularly pertinent to the analysis of trans-membrane transport systems (see Saier & Tseng, this volume). The considerable effort of Saier and co-workers has led to the identification of over 200 different families of transporters. These studies reveal that transporter families have arisen continuously over the last 4 billion years and some, for example the major facilitator superfamily, are ancient and ubiquitous, whilst others, for example the mitochondrial carrier family of anion exchangers, arose much later and are confined to particular eukaryotic organelles. We also learn that many permeases arose by tandem intragenic duplication and that a 6 TMS module is, for currently unknown reasons, particularly popular. Phylogenetic analysis is now sufficiently refined that virtually every newly sequenced transporter can be classified with respect to its structure, function and mechanism just by considering how similar it is in amino acid sequence to previously identified transporters.

The other contributions to this symposium are concerned specifically with the translocation of polypeptides across microbial membranes. No one chapter deals exclusively with the process by which polypeptides are translocated across the bacterial cytoplasmic membrane using the Sec machinery. This process is, however, briefly described by Filloux and alluded to by Soto & Hultgren, in their descriptions of two different pathways for the translocation of polypeptides from the periplasm to the exterior of Gram-negative bacteria, the substrates for which are Sec-dependent periplasmic proteins. However, Young *et al.* review our current knowledge of protein translocation across the endoplasmic reticulum (ER) membrane, and it is clear that the translocon – the proteinaceous membrane channel through which the polypeptide exits the cytosol – as well as various features of the translocation process are fundamentally similar in bacterial and eukaryotic microbes. In recent years it has proved possible to complement the elegant genetic analysis of protein export in yeast with sophisticated *in vitro* studies, most notably involving the identification of cross-linking partners of translocating polypeptides, and fluorescence quenching studies.

Such studies are either impossible or extremely difficult to conduct on bacteria, largely because of the technical complications that result from having to turn the membrane vesicles derived from the bacterial cells inside-out in order to bring the cytoplasmic contents to the outside. Just as we had settled into thinking of the translocon as an environment for the one-way transport of unfolded polypeptides, the application of this barrage of elegant techniques has yielded some big surprises. These recent studies have revealed that the translocon is wider than required for linear extrusion of poly-peptides, so have led us to consider that maybe polypeptides start to fold even within the translocon. We have also learnt that translocation will apparently run in reverse if the polypeptide is not properly modified or fails to fold, enabling its degradation via the cytosolic ubiquitin-proteasome pathway. There is a growing awareness of the importance of the gating at both ends of the translocon. The ribosome makes intimate contacts with the translocon and it has been suggested very recently that the ribosome controls translocon gating by a conformational mechanism. During translation, the ribosome undergoes conformational changes, which then induce conforma-tional changes in the translocon to control gating.

Proteins destined for translocation across the bacterial cytoplasmic mem-brane or the eukaryotic ER membrane are made with hydrophobic N-terminal signal peptides that are essential for translocation, and, in the case of soluble proteins, are eventually proteolytically cleaved from the translo-cated protein. It has been known for over two decades that higher eukaryotes contain a ribonucleoprotein particle, termed signal recognition particle, or SRP, that recognizes signal peptides and binds and delivers nascent pre-proteins to the ER membrane, by docking with the SRP receptor. Although genetic screens failed to reveal a bacterial SRP, sequence comparisons eventually revealed that bacteria do contain an SRP, albeit of a rather more primitive form than in higher eukaryotes. For a long time no role in protein targeting could be positively ascribed to bacterial SRP, and it was argued that bacterial SRP could have a different function to mammalian SRP. Valent *et al.* provide us with a historical perspective on the discovery of bacterial SRP and the eventual acceptance of a role for it in targeting membrane proteins, in particular, to the cytoplasmic membrane. Since signal peptides differ considerably in amino acid sequence, a key question concerning the targeting of signal-peptide-containing proteins is: how can such diverse ligands be recognized by a single receptor (SRP)? The structure of the signal-peptide-binding domain of the P48 SRP component of *Thermus aquaticus* reveals that, as predicted more than 10 years ago, this highly hydrophobic methionine-rich domain forms a hydrophobic groove that is lined with flexible amino acid side chains. It is thus sufficiently large and pliable to be able to accommodate signal peptides of different shapes and sizes. Finally, SRP is proving to be ubiquitous – it is present in all bacteria and eukaryotes so far examined, and it is found in the stroma of chloroplasts as well as the cytosol.

In the Gram-negative bacteria secretion of proteins to the medium can occur in two stages, with proteins being exported in a Sec-dependent fashion to the periplasm, and then being translocated across the outer membrane. Alternatively it can occur in a single step, with the exoproteins being transported from the cytoplasm across both the inner and outer membranes, without the involvement of the Sec machinery and a periplasmic intermediate. Type I secretion systems are the simplest and, perhaps for this reason, currently the best understood systems for the direct secretion of exoproteins from the cytoplasm to the exterior of Gram-negative bacteria. Most type I systems are responsible for the secretion of just one or a few closely related exoprotein substrates, belonging to the toxin, protease or lipase families. The first type I secretion system to be characterized, and the most extensively studied, is the system responsible for the secretion of α-haemolysin (HlyA) by haemolytic *Escherichia coli*. However, related systems have since been found in a wide variety of bacteria. They are responsible for the secretion of metalloproteases (*Erwinia chrysanthemi*), lipases (*Pseudomonas fluorescens*), S-layer proteins (*Campylobacter fetus* and *Caulobacter crescentus*) and, in some bacteria, several unrelated proteins (a metalloprotease, a lipase, a haem-binding protein and an S-layer protein in *Serratia marcescens*, and glycanases and a nodulation protein in *Rhizobium leguminosarum*) (Binet *et al.*, 1997; Awram & Smit, 1998; Thompson *et al.*, 1998; Kawai *et al.*, 1998; Finnie *et al.*, 1998). Type I secretion systems are relatively simple. Just three proteins form the substrate-specific channel and drive exoprotein transport through it to the exterior. As shown in Fig. 1(a), they are an ATP-binding cassette (ABC) protein exporter (e.g. HlyB), a membrane-fusion protein, or MFP (e.g. HlyD), and an outer membrane protein, or OMP (e.g. TolC). The ABC protein exporter is a polytopic inner membrane protein which recognizes the exoprotein substrate(s) and which binds and hydrolyses ATP. The MFP is an N-in C-out inner membrane protein. It interacts both with the ABC protein exporter and, via its extended C-terminal domain, with the periplasmic domain of the β-barrel OMP. Usually the three genes encoding the 'ABC exporter' are linked to those encoding the exoprotein substrates. However, NodO, one of four or more substrates for the chromosomally encoded type I exporter of *R. leguminosarum*, is plasmid-encoded (Finnie *et al.*, 1997). Likewise, the gene encoding TolC, the OMP of the α-haemolysin secretion system, is unlinked to *hlyABD*. But TolC is also used by another ABC transporter, the colicin V transporter, and it has additional roles in colicin E1 permeation and chromosome segregation.

The exoprotein substrates do not have N-terminal signal peptides but instead they contain short C-terminal secretion signals. Their other striking characteristic is that many contain glycine-rich repeated motifs that are

(a)

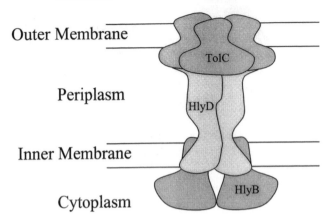

Exterior

Outer Membrane

Periplasm

Inner Membrane

Cytoplasm

(b)

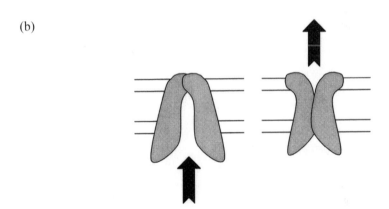

Fig. 1. Type I HlyA secretion system. (a) The three protein components of the secretion machinery are depicted. The polytopic inner membrane ABC protein exporter, HlyB, has a cytoplasmic ATPase domain. The MFP, HlyD, is a bitopic inner membrane protein with an extended C-terminal periplasmic domain. Its N-terminus contacts HlyB and its C-terminus interacts with the periplasmic domain of the outer membrane pore (the OMP), TolC. (b) The outward movement of HlyA is depicted by the filled arrow. According to current data, ATP and HlyA binding are believed to promote opening of the channel entrance (left-hand panel), whereas ATP hydrolysis is required to close the channel entrance and open the channel exit.

implicated in Ca^{2+}-binding, and, hence, rapid and stable folding of the proteins following their secretion. The precise nature of the C-terminal secretion signals currently remains elusive. Often the extreme C-terminus of the exoprotein consists of a negatively charged amino acid followed by several hydrophobic amino acids, and in some exoproteins an α-helical structure is believed to exist just N-terminal to this motif. In some exoproteins the glycine-rich repeats may play an additional role in helping to keep the targeting signal exposed on the surface of the protein, and hence visible to its receptor, the ABC protein. Signal recognition by the ABC protein is usually limited to exoproteins of the same type. But foreign proteins can often be recognized and secreted by an ABC exporter if they are fused to a cognate C-terminal signal. For example, when a C-terminal portion of HlyA was fused to β-lactamase (minus its N-terminal signal peptide) this normally periplasmic enzyme was efficiently and specifically secreted by *E. coli* in an HlyB- and HlyD-dependent fashion (Chervaux *et al.*, 1995). It has been noted that the *Caul. crescentus* S-layer protein is particularly abundant for a type I secretion product (accounting for 10–12% of the total cell protein) and therefore the possibility of using this ABC exporter to secrete foreign proteins looks particularly attractive.

Very recently, elegant studies by Thanabalu *et al.* (1998) have revealed some details of the dynamics of α-haemolysin export. Their strategy was to express HlyA, B, D and TolC in different combinations in *E. coli* and to analyse the complexes that formed (the components that could be cross-linked to one another) *in vivo*. They found that the ABC protein and the MFP formed a complex to which the OMP was recruited only when HlyA engaged the complex, and from which it separated after HlyA had been secreted. TolC was previously found to be a trimeric pore and in this study HlyD was also found to be trimeric and to form the primary inner membrane–outer membrane bridge. Intriguingly, ATP binding and substrate binding both promoted opening of the channel entrance, but ATP hydrolysis was required for HlyA to exit the channel. It is tempting to speculate that ATP hydrolysis is required to close the channel entrance and open the channel exit, thus ensuring gating of the channel, which is presumably necessary if leakage of cytoplasmic proteins to the exterior is to be prevented (see Fig. 1b). Finally, other studies, and in particular the work reported by Delepelaire & Wandersman (1998), highlight the possibility that some, perhaps all, exoproteins may have to be prevented from folding, or even actively unfolded, in order for them to be efficiently secreted by type I systems.

In comparison to type I export, the type III export process, which is responsible for the delivery of Yops (*Yersinia* outer proteins) from the cytosol of pathogenic *Yersinia* species to its outer surface, to the external medium, and into the cytosol of the eukaryotic host cell, is poorly under-stood. As discussed by Anderson *et al.*, some 25 genes are involved in specifying the type III machinery. Contact with eukaryotic cells at 37 °C induces the type III machinery and the programmed secretion of some 14 different Yops to their specific extracellular destinations. Intriguingly, the secretion signals of Yops, which lie within the first 15 or so codons of the *yop* genes, are of a distinctly different nature to all other targeting signals, in that they are tolerant of frame-shift mutations. Presumably these nucleotide-encoded signals ensure that *yop* mRNAs are only translated when the ribosomes attached to them have docked onto the type III machinery. For some Yops, cytoplasmic chaperones are additionally required for their successful secretion. A comparison of the genes required for type III secretion in other Gram-negative pathogens reveals that homologues of nine proteins are found in all known type III machines, and that eight of these are homologous to products needed for the assembly of the flagellar basal body hook complex. The ninth is a multimeric outer membrane 'secretin' protein. Secretins form gated channels in the outer membrane and function in the translocation of proteins and bacteriophage across this membrane. Yops that are injected into eukaryotic cells must cross three membranes. It has been proposed that, for these Yops, the type III machine forms an injection device extending from the bacterial to the eukaryotic cytoplasm.

The main terminal branch of the general secretory pathway in Gram-negative bacteria, or the type II secretory pathway, is used by a wide variety of bacteria to transport exoproteins from the periplasm to the exterior, following their Sec-dependent translocation across the cytoplasmic membrane (Filloux, this volume). Some 14 or so products of linked genes, moderately to highly conserved in all the bacteria in which they have been found, form the export machinery. The clue to why the type II machinery should be so complex comes from the finding that the components include four polypeptides with N-termini resembling those of pilin subunits and a prepilin peptidase. The proteins they resemble are crucial components in the formation of type IV pili (long cell surface appendages at the poles of the producing bacteria). The prepilin peptidase is required for the processing of these 'pseudopilins', and, based on their strange fractionation (when over-produced they fractionate with the outer membrane), these subunits have been proposed to form a 'pseudopilus' – a rudimentary structure spanning the periplasm and connecting the inner and outer membranes. Other components of the type II machinery include a peripheral cytoplasmic membrane ATPase, which might be involved in driving the export of pseudopilins to the periplasm, and an outer membrane secretin, which, in its multimeric form, has a large central pore, some 95 nm wide. Further components are believed to energize gating of/transport through the pore, via a TonB-like energy transduction process. The pseudopilus, assuming it really exists, might either push exoproteins through the pore, or it might act like a cork to keep the pore blocked when not in use. Type II exoproteins do not share regions of amino acid sequence similarity, and molecular genetic analysis has revealed that their secretion signals are 'patch' signals, made up from different portions of the linear amino acid sequence. Conflicting data on the precise constitution of the secretion signal in specific exoproteins have led to the view that either the secretion signal is recognized as the exoprotein folds, or that it comprises a series of signals that are recognized sequentially as the protein is directed outwards via the type II machine towards the exterior.

Gram-negative bacteria make many different kinds of pili and other organelles of attachment that play crucial roles in the early stages of bacterial infection. The chaperone–usher pathway, reviewed by Soto & Hultgren, is responsible for the assembly of many of these structures, and, amongst these, the expression and assembly of P pili and type 1 pili are currently the best understood. Eleven clustered *pap* genes are responsible for the assembly of P pili. The Pap subunits are translocated across the cytoplasmic membrane by the Sec machinery, but are retained on its periplasmic surface by their C-termini. The periplasmic PapD immunoglobulin-like chaperone then binds to a highly conserved C-terminal motif by β-zippering, and the Pap subunit, thus removed from the membrane, folds on the chaperone. The chaperone–subunit complexes are targeted to the outer membrane usher protein PapC. PapC is predicted to have a transmembrane β-barrel structure and a large periplasmic domain for interaction with the chaperone–subunit complexes. It assembles into liposomes as ring-shaped multimeric complexes with central pores of 2–3 nm diameter. The helicoidal pilus rod is, however, some 6–8 nm wide, but pili can be unwound into linear fibres, in which the subunits interact head-to-tail fashion, that are only about 2 nm wide. It has therefore been proposed that the polymerized subunits pass through the usher pore in their extended form, and that their maturation to the final helical form helps drive the pilus assembly process. Intriguingly, recent studies have revealed that attachment of type-1-piliated bacteria to murine host cells is accompanied by an apparent shortening of the pilus, which could reflect its retraction or the cessation of pilus growth. Whatever the molecular basis, presumably there is a consequential build up of excess subunits in the periplasm. It is already established that subunit misfolding or chaperone absence is sensed by the CpxA–CpxR two-component system, and results in the up-regulation of expression of periplasmic folding factors and proteases. Hung & Hultgren (1998) propose that activation of this pathway by host attachment also serves to switch on expression of an array of virulence genes that are needed to establish infection.

E. coli, the laboratory favourite amongst bacteria, must be quite brutally treated in order to make it competent to take up DNA. The bacteria are subjected to either abrupt shifts in divalent cation concentration and temperature or to high-voltage electrical pulses. Both treatments presumably induce transient pores within their membranes, through which the DNA can enter. However, some bacteria are naturally transformable, and Lacks discusses how DNA is taken up by such bacteria. In *Streptococcus pneumoniae*, one of the best studied cases, this process entails DNA binding to external receptors, its conversion to the single-stranded form, and then its unwinding and entry into the cell, 3′ end first, at a rate of about 100 nucleotides s^{-1}. Meanwhile the other strand is degraded and released from the cell surface. Amongst the *Strep. pneumoniae* proteins that are essential for DNA uptake are several related to the type IV pilins, and an energy-transducing protein and a membrane-spanning protein, responsible for their export, as well as a prepilin peptidase. The presumption is that the synthesis of these proteins results in the formation of an external appendage – necessary in some way that is not yet understood – for DNA uptake. Amongst the other proteins essential for uptake is one with multiple hydrophobic stretches which is a good candidate for being the membrane channel through which the DNA enters. Intriguingly, in the case of another naturally transformable bacterium, *Haemophilus influenzae*, the DNA was found to be contained within a membrane vesicle (a transformasome) prior to its uptake. One proposal is that the binding of DNA to the bacterium's surface triggers membrane curvature and vesicularization of the DNA. Fusion of the transformasome with the bacterial membrane has further been proposed to be involved in delivering the DNA to the uptake apparatus. Whether this process is somehow related to membrane vesicle trafficking processes in eukaryotic cells remains to be seen.

Two chapters near the end of this symposium volume deal with various aspects of the import and localization of proteins in two very different kinds of eukaryotic organelles, peroxisomes and chloroplasts. Proteins carrying chloroplast transit peptides are imported post-translationally across the chloroplast double membrane into the stroma. From the stroma some are targeted to and then translocated across, or integrated into, the thylakoid membrane. Robinson *et al.* review the recent rapid progress in our under-standing of these targeting and translocation mechanisms. In keeping with the prokaryotic origin of chloroplasts, proteins with classical N-terminal signal peptides are directed across the thylakoid membrane in a SecA-dependent manner. Moreover, a stromal SRP exists, and many thylakoid membrane proteins require stromal SRP for their delivery to the membrane, after which they are integrated into the membrane in a Sec-dependent fashion. (As in bacteria, SRP acts only on the more hydrophobic secretion targets.) Translocation across the thylakoid membrane can also occur via the ΔpH-dependent pathway, which is unique in that it requires neither soluble proteins nor NTPs. Preproteins that use this pathway have N-terminal signal peptides that appear to differ only subtly from classical signal peptides, most notably in that they contain a twin-arginine motif immediately N-terminal to the hydrophobic core. However, such pre-proteins are not substrates for the Sec machinery, and it is possible that their signal peptides incorporate a 'Sec-avoidance' signal. Moreover, the proteins to which they are attached are inherently difficult for the Sec machinery to translocate, probably because they fold tightly. Remarkably, this ΔpH-dependent pathway has the capacity to translocate fully folded proteins (such as the methotrexate-bound form of DHFR). Homologues of two recently characterized components of this machinery are found in nearly all bacteria examined, and a subset of bacterial preproteins that have the twin-arginine motif in their signal peptides have been identified. Their mature products are periplasmic proteins that bind redox cofactors, almost certainly in the cytoplasm, and thus, like their thylakoid counterparts, are translocated across the membrane in a fully folded state. How fully folded proteins can be translocated across tightly sealed membranes remains to be seen. Certainly thylakoids, unlike bacteria, offer excellent *in vitro* systems for the analysis of this translocation pathway. Finally, some proteins can integrate into the membranes of protease-treated thylakoids, implying that they are able to spontaneously insert into the membrane. Whether this pathway shares fundamental similarities with that followed by the major coat proteins of filamentous bacteriophage as they insert into the bacterial cytoplasmic membrane remains to be seen.

Lopez-Huertas & Baker discuss the biogenesis of peroxisomes. Unlike chloroplasts and mitochondria, peroxisomes contain no genetic material, and different subsets of these organelles have different metabolic functions. Most peroxisomal proteins are post-translationally imported from the cytosol. Most matrix proteins have a C-terminal tripeptide targeting signal consisting of a small neutral amino acid, followed by a basic amino acid and terminating with a hydrophobic amino acid. Others have an N-terminal targeting signal and/or internal targeting signals. The Pex5p and Pex7p receptors recognize the C-terminal (PTS1 type) and N-terminal (PTS2 type) signals, respectively, and probably act by binding to the proteins containing them in the cytoplasm and delivering them to the peroxisomal membrane. (The targeting signals within peroxisomal membrane proteins remain poorly understood.) Remarkably, peroxisomes can import folded and even oligomeric proteins. Only one of the monomers of the dimeric malate dehydrogenase of *Sacch. cerevisiae* needs to contain the peroxisome-targeting signal for import to occur. Elegant genetic selections have been used to obtain mutants defective in peroxisome biogenesis, but our knowledge of the import machinery remains incomplete and the actual import mechanism is currently unknown. Clearly it must be significantly different to chloroplast and mitochondrial import, where precursors are maintained or rendered unfolded prior to import, translocated into these organelles in an extended conformation, and then refolded within them. But as yet there is no evidence either for the existence of any sufficiently large regulated pore (like a nuclear pore) or for any endocytic-like process. On the other hand, fully folded proteins are translocated across the thylakoid membrane and the bacterial cytoplasmic membrane, using the ΔpH-dependent pathway (or its bacterial equivalent). Finally, there is now a wealth of circumstantial evidence that peroxisomes do not receive all their proteins by import from the cytoplasm, but that some peroxisomal membrane proteins and lipids are derived, by vesicle budding and fusion, from the ER.

CONCLUDING REMARKS

This symposium serves to remind us how difficult membrane proteins are to analyse, and how resourceful those working on them are. We are optimistic that, as we move into the next millennium, new approaches will be devised that will provide deeper insights into their functioning, much as advances in crystallization methods, strategies for genetic screening, gene fusion studies, molecular archaeological studies, and elegant *in vitro* studies on reconstructed transport pathways and translocation intermediates have done in the 1980s and 1990s. In this chapter we have discussed that hydrophilic proteins can be transported across hydrophobic membranes by a diversity of transport machineries. At last we are beginning to appreciate the molecular details of how some of these machines function. Recent studies have certainly made us revise our view that polypeptides are necessarily translocated across biological membranes in unfolded states. Understanding the structure and function of the machineries that translocate folded proteins is one of the major goals for the future. Some of the other major items on the agenda will be understanding how channels within membranes are gated, a process that is, presumably, crucially important for maintaining the integrity of membrane-bound cellular compartments, the roles of molecular chaperones, and the processes by which each of the many different types of targeting signals within extracytoplasmic proteins are recognized.

ACKNOWLEDGEMENTS

We gratefully acknowledge funding from The Wellcome Trust.

REFERENCES

Awram, P. & Smit, J. (1998). The *Caulobacter crescentus* paracrystalline S-layer protein is secreted by an ABC transporter (type I) secretion apparatus. *Journal of Bacteriology* **180**, 3062–3069.

Binet, R., Létoffé, S., Ghigo, J. M., Delepaire, P. & Wandersman, C. (1997). Protein secretion by Gram-negative bacterial ABC exporters – a review. *Gene* **192**, 7–11.

Chervaux, C., Sauvonnet, N., Le Clainche, A., Kenny, B., Hunt, A. L., Broome-Smith, J. K. & Holland, I. B. (1995). Secretion of active beta-lactamase to the medium mediated by the *Escherichia coli* hemolysin transport pathway. *Molecular & General Genetics* **249**, 237–245.

Deisenhoffer, J., Epp, O., Miki, K., Huber, R. & Michel, H. (1985). Structure of the protein subunits in the Photosynthetic Reaction Center of *Rhodopseudomonas viridis* at 3Å resolution. *Nature* **318**, 618–624.

Delepelaire, P. & Wandersman, C. (1998). The SecB chaperone is involved in the secretion of the *Serratia marcescens* HasA protein through an ABC transporter. *EMBO Journal* **17**, 936–944.

Finnie, C., Hartley, N. M., Findlay, K. C. & Downie, J. A. (1997). The *Rhizobium leguminosarum prsDE* genes are required for secretion of several proteins, some of which influence nodulation, symbiotic nitrogen fixation and exopolysaccharide modification. *Molecular Microbiology* **25**, 135–146.

Finnie, C., Zorreguieta, A., Hartley, N. M. & Downie, J. A. (1998). Characterization of *Rhizobium leguminosarum* exopolysaccharide glycanases that are secreted via a type I exporter and have a novel heptapeptide repeat motif. *Journal of Bacteriology* **180**, 1691–1699.

Hung, D. L. & Hultgren, S. J. (1998). Pilus biogenesis via the chaperone/usher pathway: an integration of structure and function. *Journal of Structural Biology* **124**, 201–220.

Kawai, E., Akatsuka, H., Idei, A., Shibatani, T. & Omori, K. (1998). *Serratia marcescens* S-layer protein is secreted extracellularly via an ATP-binding cassette exporter, the Lip system. *Molecular Microbiology* **27**, 941–952.

Ostermeier, C. & Michel, H. (1997). Crystallization of membrane proteins. *Current Opinion in Structural Biology* **7**, 697–701.

Thanabalu, T., Koronakis, E., Hughes, C. & Koronakis, V. (1998). Substrate-induced assembly of a contiguous channel for protein export from *E. coli*: reversible bridging of an inner-membrane translocase to an outer membrane exit pore. *EMBO Journal* **17**, 6487–6496.

Thompson, S. A., Shedd, O. L., Ray, K. C., Beins, M. H., Jorgensen, J. P. & Blaser, M. J. (1998). *Campylobacter fetus* surface layer proteins are transported by a type I secretion system. *Journal of Bacteriology* **180**, 6450–6458.

MULTIDRUG RESISTANCE EFFLUX

KIM LEWIS

Tufts University Biotechnology Center, Medford, MA 02155, USA

Among the many mechanisms of drug resistance in micro-organisms, multidrug pumps seem to be the most puzzling and to some of us, also the most fascinating to study. The central issue regarding multidrug resistance pumps (MDRs) is the principal mechanism of substrate recognition – i.e. how does a single protein discriminate between structurally unrelated drugs, and all of the intracellular molecules? This problem is closely related, as we shall see, to the intriguing questions regarding the nature of MDR substrates, the natural function of MDRs, and the mechanism of MDR regulation by multidrug sensors.

MDRs have been found in all organisms studied to date. The putative number in some species is very high – more than 10 in *Escherichia coli* (Blattner *et al.*, 1997; Lewis, 1994) and more than 20 in the yeast *Saccharomyces cerevisiae* (Kolaczkowski & Goffeau, 1997) – which indicates that a similarly large number might be expected in the pathogen *Candida albicans*. MDRs fall into four unrelated families (Fig. 1) and can be relatively simple, like the 12 transmembrane segment (TMS) NorA of the major facilitator (MF) family, or form complex multi-component envelope translocases found in Gram-negative bacteria, such as EmrAB–TolC or AcrAB–TolC of *E. coli* (Fig. 2).

MDR FAMILIES

Major facilitator MDRs

The major facilitator is the largest family of translocases (Marger & Saier, 1993), which comprises pmf-dependent uptake transporters of regular substrates, such as the LacZ lactose/H^+ symporter of *E. coli*, and a subfamily of drug/H^+ antiporters that includes both specific transporters and MDRs (Figs 1 and 2). The MF transporters are likely descendants of a gene duplication event, as proposed originally for TetA where the two halves of the protein share significant homology (Rubin *et al.*, 1990). There is good evidence that TetA is in fact a dimer (McMurry & Levy, 1995). The MDR homologues of TetA might also function as dimers. The N-end half of the MF transporters is better conserved than the C-half, suggesting that the C-half domains harbour the drug-binding site. MF MDRs share important consensus sequences with other MF transporters. A number of

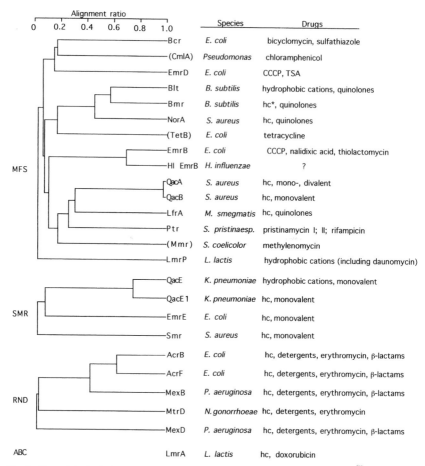

Fig. 1. Bacterial MDR families. The MF family tree shows primarily MDRs and provides some examples of closely related specific drug transporters. EmrB has additional peptides EmrA and TolC. Known RND MDRs have additional membrane fusion peptides and outer membrane proteins (see Fig. 2). LmrA is a member of the large ABC family, but it is so far the only bacterial MDR pump from this group. *hc, Hydrophobic cations. Examples of specific transporters of the MF family are given in parentheses.

sequences are conserved throughout the MF family and are shared by both uptake and extrusion transporters. A particularly interesting motif, GxhyxGPhyhyGGxhy (hy = hydrophobic), is found only in the efflux transporters. It was proposed that this motif forms a kink in membrane domain 5 that changes the ligand/symport coupling of intake transporters into a ligand/antiport pathway of the efflux pumps (Varela *et al.*, 1995). This consensus is very useful in homology analysis, since its presence in a newly sequenced gene clearly indicates that the gene encodes an efflux translocase.

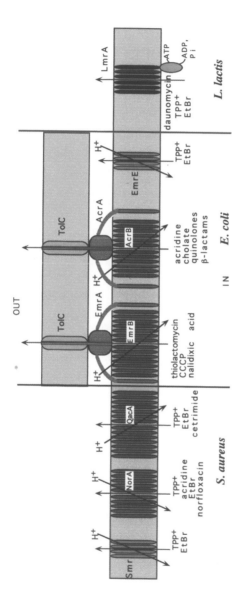

Fig. 2. Schematic model of MDR pumps from different bacterial families. The MF transporters are shown in the same colour, and so are the homologous membrane fusion proteins of the *trans*-envelope EmrAB and AcrAB transporters.

A well-studied member of the MF family is Bmr of *Bacillus subtilis*, which confers resistance to mainly hydrophobic cations, such as TPP$^+$ and rhodamine (Neyfakh *et al.*, 1991). A close homologue of Bmr is NorA, a chromosomally encoded protein of *Staphylococcus aureus* whose over-production, due to promoter mutations, causes clinically significant resistance to quinolones (Kaatz *et al.*, 1993; Neyfakh *et al.*, 1993; Ng *et al.*, 1994). An interesting subgroup of MF MDRs are proteins with a 14 membrane domain structure. QacA, found on plasmids of *S. aureus*, can extrude amphipathic divalent cations (Rouch *et al.*, 1990), a function that might require additional complexity. EmrB of *E. coli* (Lomovskaya & Lewis, 1992) is particularly interesting in that it forms a *trans*-envelope complex with a periplasmically located EmrA and outer membrane TolC. This complex structure allows the pumps to 'take advantage' of the outer membrane barrier, which restricts passage of hydrophobic compounds that readily partition in the cytoplasmic membrane. It seems that the only way to prevent their rapid re-entry into the cytoplasmic membrane is to extrude them across the outer membrane barrier. Using this strategy, the EmrAB and Acr pumps protect the cell from such membrane-active agents as uncouplers of oxidative phosphorylation and detergents like SDS and bile acids. Not surprisingly, MDRs capable of uncoupler extrusion have not been found in Gram-positive species.

The analysis of relatedness within the MF family suggests that MDRs were not derived from some primordial MDR, but rather evolved independently many times in the course of evolution (Lewis, 1994). For example, an *S. aureus* MDR, NorA, is much closer to the tetracycline extrusion pump TetB from *E. coli* than it is to the *Lactococcus lactis* LmrP multidrug resistance pump (Fig. 1). It seems that MDRs were derived from specific drug extrusion pumps (such as Mmr) of bacteria producing amphipathic antibiotics. Selecting for a mutation that allows an enzyme or a translocase to recognize a new substrate usually leads to broadening the spectrum of specificity, rather than to switching specificity from one compound to another. We have proposed that MDRs might have been intermediates in the evolution of one specific drug extrusion pump from another (Lewis, 1994). Some of these intermediates might have taken on a life of their own, with their unspecific amphipathic binding domain providing the cell with the capability to rid itself of a wide range of chemically unrelated amphipathic toxins. Similarly to members of the MF family, resistance–nodulation–division (RND) and ABC MDRs are related to, and might have arisen from, specific translocases of amphipathic solutes.

RND MDRs

The RND pump AcrAB (formerly AcrAE) is responsible for what has been known since the 1960s as the 'Acr phenotype' in *E. coli* (Ma *et al.*, 1993,

1994). A mutation in the pump leads to sensitivity to acridine and other hydrophobic cations, to tetracycline, β-lactams, and detergents like bile acids and SDS. Not surprisingly, the *acrAB* mutation has been originally thought to disrupt the outer membrane. AcrB is a large 113 kDa peptide that is the translocase proper. It has a putative 12 membrane domain structure with two very large periplasmic segments. AcrA belongs to the same membrane fusion family as EmrA. It appears that AcrAB together with TolC form a *trans*-envelope structure (Fralick, 1996). Other RND MDRs with similar composition are found only in *E. coli* and in other Gram-negative bacteria such as *Pseudomonas aeruginosa* and *Neisseria gonorrhoeae* (Poole, 1994; Hagman *et al.*, 1995). All RND MDRs are characterized by a surprisingly broad range of substrate specificity. The closest non-MDR homologues are extrusion pumps for heavy metals.

SMR, small multidrug resistance pumps

The peptides encoding SMRs form only four membrane domains, which makes them the smallest known translocases at about 110 amino acids long (reviewed by Paulsen *et al.*, 1996b). The substrates are invariably hydrophobic cations, and the spectrum is narrower than that of QacA, for example. Representative pumps that have been studied extensively are chromosomally coded EmrE of *E. coli* and plasmid-coded SMR present in some strains of *S. aureus* (see 'Mechanism' for a more detailed description). The only known non-MDR protein of this group is SugE (and its homologues), which suppresses mutations in the GroE chaperone (Greener *et al.*, 1993). It has been suggested that the distantly related tellurite resistance transporter also belongs to this group (Paulsen *et al.*, 1996a).

ABC MDRs

There is a large family of ATP-dependent translocases in bacteria, most commonly involved in the uptake of nutrients, which requires a periplasmic binding protein. The maltose and histidine transporters are pertinent examples. There are also ABC efflux pumps, such as the haemolysin transporter of *E. coli* and ABC transporters of antibiotics. Recently, an ABC MDR, LmrA, has been discovered in *L. lactis* (van Veen *et al.*, 1996). With six transmembrane domains, the pump looks like half of the duplicated mammalian P-glycoprotein MDR. The pump is closely related to the P-glycoprotein, and the homology is over the entire sequence range. The substrate spectrum is similar to that of P-glycoprotein as well (van Veen *et al.*, 1998).

MECHANISM

It is tempting to simplify the analysis of a transport mechanism by separating it into two parts – the energy-dependent transport of a ligand, and the selective binding. MDRs of the MF family, for example, are homologous to specific efflux antiporters and share with them a number of conserved domains thought to be involved in energy transformations and ligand movement (Paulsen *et al.*, 1996a). However, it appears that the path of the ligand in an MDR-driven efflux might be intimately related to the mechanism of discrimination.

It has been noted by Higgins & Gottesman (1992) that drugs with intracellular targets need to be amphipathic in order to cross the cytoplasmic membrane, and by the same token intracellular compounds must be hydrophilic in order to avoid escaping into the external environment. This difference in polarity, rather than in chemical structure, could serve as a basis for drug/self discrimination. Perhaps taking this logic to its limit, the same authors proposed that it is the membrane that acted as a discriminator – any substance partitioned in the internal leaflet of the bilayer would then be picked up by the completely unspecific MDR and flipped to the outer leaflet. This would shift the equilibrium in the direction of efflux, decreasing the intracellular concentration of the drug. This specific model of substrate discrimination proposed for P-glycoprotein does not appear to be correct – purified MDR clearly has binding preferences of its own (Liu & Sharom, 1996; Sharom, 1997) – and amphipathic cations are by far better substrates than neutral molecules or amphipathic anions with similar polarity (review, Sharom, 1997; Ueda *et al.*, 1997). What is truly impressive, however, is that the idea of a flippase mechanism for the P-glycoprotein appears to be correct. It was found that human MDR3, which is highly homologous to P-glycoprotein MDR1, is indeed a phospholipid flippase (Fig. 3), responsible for phosphatidylcholine transport into the bile (Ruetz & Gros, 1994). It was further found that P-glycoprotein can transport (flip) analogues of phospholipids with short fatty acid chains from the inner to the outer leaflet of the membrane of epithelial cells (van Helvoort *et al.*, 1996). This suggests that the membrane does influence the substrate specificity of the MDR through a prescreening process – only substances that partition in the membrane will be 'considered' as ligands by the MDR.

Interestingly, the flippase mechanism appears to be an exception rather than a rule among MDRs. A fairly close homologue of mammalian

---

Fig. 3. Path of the ligand in different MDR pumps. MDR3 is an ATP-dependent phosphatidylcholine flippase of the ABC family; P-glycoprotein is the homologous human MDR that is a flippase of drugs; LmrA of *L. lactis* (ABC transporter) transfers drugs from the inner leaflet to the outside medium; LmrP MDR of *L. lactis*, which belongs to the MF family, transfers drugs from the inner leaflet to the outside medium. RND MDRs of Gram-negative species apparently transfer drugs from the outer leaflet to the outside medium.

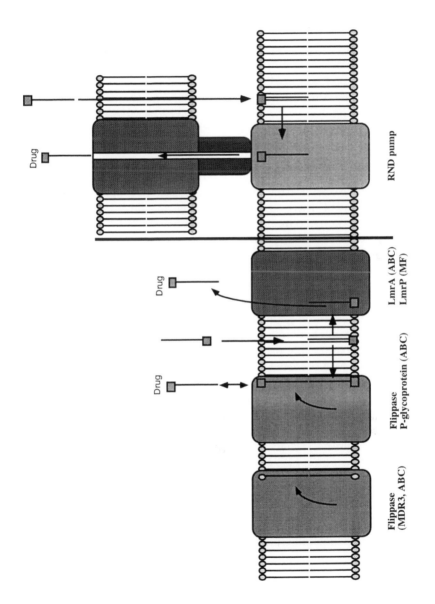

Flippase
(MDR3, ABC)

Flippase
P-glycoprotein (ABC)

LmrA (ABC)
LmrP (MF)

RND pump

Drug

Drug

Drug

Drug

P-glycoprotein, the *L. lactis* LmrA, has been studied by Wil Konings and coworkers (van Veen *et al.*, 1996). LmrA has an interesting structure – it is exactly half of the 12 TMS P-glycoprotein, and apparently forms a dimer. The substrate specificity of LmrA is similar to P-glycoprotein, and LmrA was found to successfully replace P-glycoprotein in mammalian cells (van Veen *et al.*, 1998). However, experiments with amphipathic dyes suggest that LmrA is not a flippase. A hydrophobic probe, 1-[4-(trimethylamino)phenyl]-6-phenylhexa-1,3,5-triene (TMA-DPH), which is fluorescent in the membrane, inserts into the outer leaflet of the membrane and then slowly flips, like a phospholipid, to the inner leaflet. Activation of an MDR by providing an energy source caused rapid efflux of the probe into the external environment. This experiment showed that the MDRs were not acting as flippases, since merely flipping a substance from the inner to the outer leaflet would have little effect on fluorescence. Importantly, the experiments suggest that the binding site is accessible from the membrane (Bolhuis *et al.*, 1996a). This and similar experiments indicate that the path of an LmrA ligand is from the inner leaflet of the cytoplasmic membrane all the way to the external medium (Fig. 3). The ligand pathway was analysed in similar experiments for *L. lactis* LmrP MDR, which belongs to the MF family of translocases. The results showed that LmrP also transports its ligands from the inner leaflet of the bilayer to the outside medium (Bolhuis *et al.*, 1996b). LmrP belongs to the MF subfamily together with specific translocases of antibiotics. Hydrophobic antibiotics partition preferentially into the membrane, and it seems very reasonable for an antibiotic translocase to have the ability to pick its substrate from the membrane.

If microbial MDRs indeed evolved from homologous specific translocases of antibiotics, as discussed in the first section, it would seem that MDRs had to develop a distinctly unique ability to pick up their substrates from the membrane. However, it is quite possible that binding of ligands within the membrane had actually originated in efflux translocases of hydrophobic antibiotics. The polyketide synthase of tetracenomycin is a membrane protein (Gramajo *et al.*, 1991). Tetracenomycin is fairly hydrophobic and is exported by a translocase of the 14 TMS subfamily homologous to EmrB and QacA (Caballero *et al.*, 1991; Guilfoile & Hutchinson, 1992). It is reasonable to assume that the membrane synthase deposits its hydrophobic product into the membrane, where it is picked up by the efflux transporter.

The RND pumps of Gram-negative species apparently have the most unusual pathway for their ligands. In *P. aeruginosa*, MexAB–OprM protects the cell from β-lactams that do not penetrate into the cytoplasm (Li *et al.*, 1994, 1995; Srikumar *et al.*, 1997). The AcrAB pump of *E. coli* protects the cell from β-lactams as well (Ma *et al.*, 1995). It was established that in *Salmonella typhimurium*, transport activity by the AcrAB pump increases with decreasing polarity of β-lactams (Nikaido *et al.*, 1998). It was therefore proposed that the pump picks up the substrate from the outer bilayer (as well as from the inner bilayer) and transports it across the outer membrane (Li *et al.*, 1995) with the aid of a membrane fusion protein and an outer membrane porin, TolC (Fig. 3). It seems that a simpler possibility is for RND pumps to pick up substrates exclusively from the outer leaflet of the bilayer; there is no apparent reason to complicate this model.

The advantage for an RND pump to pick up its substrate at the outer leaflet of the cytoplasmic membrane is evident – this allows for extrusion of fairly hydrophilic substances that will only partially partition in the membrane. The ability of the pump to transport them across the outer membrane makes this a reasonable strategy, especially for substances like β-lactams whose target is the cell wall. The interesting ability of the RND Mtr pump of *N. gonorrhoeae* to protect the cell from mammalian antimicrobial peptides such as protegrin-1 and LL-37 (Shafer *et al.*, 1998) would also be aided by an outer leaflet–outer medium transport pathway. After insertion in the membrane, antimicrobial peptides are known to form oligomeric pore structures that would no longer be subject to export.

It also seems very reasonable for the MDRs of Gram-positive species like LmrA and LmrP to have an inner leaflet–outer medium ligand pathway. Many of the substrates are cations, and accumulate in the cell driven by the membrane potential. The concentration of the substrate will therefore be considerably higher in the inner leaflet equilibrated with the cytoplasm, rather than in the outer leaflet equilibrated with the external medium. Without being able to take advantage of an additional permeability barrier like the outer membrane, it does seem that the optimal strategy for a pump is to take the substrate out of a site where its concentration is highest. This would allow the pump to clear the membrane of compounds present at a very low extracellular concentration – we have found that the *S. aureus* NorA pump protects the cell from nanomolar concentrations of hydrophobic cations (Hsieh *et al.*, 1998).

What appears rather puzzling is the unusual ligand pathway of the P-glycoprotein. A flippase is a natural solution for a lipid translocase – depositing a single lipid molecule in the outer aqueous environment is simply not an option. But it would seem that a drug should be properly cleared from the membrane and deposited in the external aqueous environment. Why would P-glycoprotein be designed to deposit drugs in the outer leaflet of the membrane? One notable difference between P-glycoprotein and the Lmr pumps of Gram-positive *L. lactis* is that an 'external environment' does not really exist for the mammalian MDR. Even deposited away from the cell, the drug will still end up within the body. One possible explanation for the flippase pathway is drug detoxification. The ultimate goal of the organism is to detoxify drugs and clear them from the system. One type of toxins that are especially dangerous to the membrane are hydrophobic radicals and oxidants. The cytoplasmic membrane has a redox chain that begins with an NADH dehydrogenase at the cytoplasmic side and ends with coenzyme Q_{10}

on the outer side of the membrane. The function of this pathway remains an open question, and it does have the ability to reduce a wide range of redox agents, such as ascorbate (Gomez-Diaz *et al.*, 1997). It is tempting to suggest that the main function of P-glycoprotein is to act in tandem with this redox chain in bringing substrates to its site of action. Once reduced and thus converted to a less harmful/more polar reduced form, the product will escape into the aqueous environment outside the cell and will be cleared through the kidney or bowel, the two major sites of high-level MDR expression.

Much less is known about the ligand-binding site itself. Numerous attempts to localize it in the P-glycoprotein, for example, revealed mutations that affect binding affinity and specificity to be scattered around the transmembrane segments (review, Ueda *et al.*, 1997). A similar picture emerged from a more limited mutagenesis analysis of the bacterial BmrR (Ahmed *et al.*, 1993; Klyachko & Neyfakh, 1998; Klyachko *et al.*, 1997). Interesting structural work has been done with EmrE of *E. coli*, the small MDR of the SMR family. This very hydrophobic MDR was solubilized in chloroform/methanol/water and analysed by NMR (Schwaiger *et al.*, 1998). Results indicated the presence of four α-helical bundles in good agreement with computerized predictions. A biochemical study of membrane topology with a closely homologous Smr protein from *S. aureus* using reporter protein fusions also shows a 4 TMS structure (Paulsen *et al.*, 1995). Smr has a single charged residue in the membrane, a glutamate. This residue is conserved in the family and was found to be essential for drug transport (Grinius & Goldberg, 1994). Even a conserved substitution with an aspartate abolished efflux. According to a proposed model, Glu acts to bind a cationic drug, and a proton coming in from the external medium replaces the cation, which then continues on its way out of the cell (Grinius & Goldberg, 1994; Paulsen *et al.*, 1996b). It is important to note that all substrates of Smr are cations, which agrees well with this model of translocation. However, the model does not, and hardly can, address the question of drug recognition.

An overall structure of purified P-glycoprotein was recently obtained by electron microscopy and Fourier projection maps of small two-dimensional crystalline arrays. From above the membrane plane the protein appears to have a diameter of about 10 nm, with a large central pore of about 5 nm in diameter closed at the cytoplasmic side of the membrane. Two 3 nm lobes are exposed at the cytoplasmic face of the membrane, likely corresponding to the nucleotide-binding domains (Higgins *et al.*, 1997; Rosenberg *et al.*, 1997).

With little indication that a crystal structure of an MDR will be obtained any time soon, a number of laboratories have been turning to the study of multidrug sensors, small soluble proteins that control the expression of MDR pumps. Three such proteins have been described so far. In *E. coli*, we identified a regulator of the EmrAB pump, EmrR, an 18 kDa protein that is encoded by the first gene of the *emrRAB* operon (Lomovskaya *et al.*, 1995) and belongs to the MarR family of transcriptional repressors (Miller &

Sulavik, 1996). EmrR binds to such EmrAB substrates as uncouplers of oxidative phosphorylation and binding releases repression and activates transcription. Experiments with purified EmrR show that it directly binds its ligands *in vitro* (Fig. 4). Scatchard analysis of equilibrium dialysis data showed 1 ligand per monomer with good affinity – K_s around 1 μM for FCCP and CCCP, and K_s around 10 μM for the more hydrophilic DNP (A. Brooun and others, unpublished). The central region of the protein is fairly well-conserved within the family and corresponds to a helix–turn–helix motif (Miller & Sulavik, 1996). This would indicate that the ligand-binding site would be in the divergent N- or C-terminal portion of the protein. According to our data, the C-terminal half expressed from a recombinant vector had no ligand-binding activity, and the N-terminal half was insoluble (A. Brooun and others, unpublished). Structural studies will thus have to be done with a full-length protein. Native gel electrophoresis and molecular sieve chromatography show that the protein is a dimer (Brooun *et al.*, 1999).

Efforts to localize a ligand-binding site were successful in the study of the *B. subtilis* BMR multidrug sensor and the crystal structure of this C-terminal domain has recently been resolved (Zheleznova *et al.*, 1997, 1999). BmrR is a transcriptional activator of the MF BMR pump (Ahmed *et al.*, 1994; Markham & Neyfakh, 1996). BmrR binds chemically unrelated hydrophobic cations, such as TPP^+ and rhodamine, substrates of the BMR pump, and activates the BMR transcription. BmrR is a 32 kDa protein that forms a dimer and binds one ligand molecule per monomer. The 18.4 kDa C-terminal ligand-binding domain still forms a dimer, and this peptide was crystallized both with and without the ligand TPP^+. The binding site is rather unusual – it is not obviously present in the apoprotein and is formed in the process of ligand binding. TPP^+ apparently aids unfolding and displacement of an α-helix, which exposes a hydrophobic binding pocket with a buried glutamate residue. Once past the gate, the ligand is bound by stacking and van der Waals interactions with residues of hydrophobic amino acids and by electrostatic interaction with the glutamate. This interesting and unusual structure explains the main features of selectivity. Apparently, hydrophilic molecules will not gain access to the hydrophobic site, which is not even open in the apoprotein; once inside the pocket, the amphipathic ligand will be retained by hydrophobic interactions, and the presence of a strong

---

Fig. 4 (pages 26–27). (a) *In vitro* ligand binding by the EmrR protein. Equilibrium dialysis binding was performed in Pierce Slidalyzers, and the concentration of free ligand was determined spectrophotometrically. Upper and middle panel, ligand binding was measured in cell extracts expressing (+) or not expressing (−) EmrR. B, Bound ligand; F, free ligand; K_{ns}, non-specific binding constant; K_s, specific binding constant; $n[E]_t$, the *x*-axis intercept showing maximal ligand binding, which is used to calculate the ligand/monomer ratio. Lower panel, binding of DNP to purified EmrR. *n*, Ligand/monomer ratio. (b) A model of MDR expression by the EmrR drug sensor.

Specific vs. Non-specific Binding of FCCP to EmrR

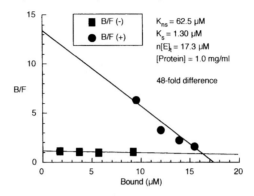

Specific vs. Non-specific Binding of CCCP to EmrR

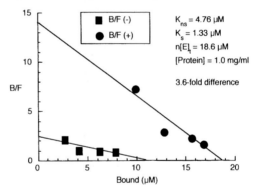

Specific Binding of DNP to EmrR

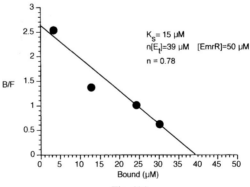

Fig. 4(a).

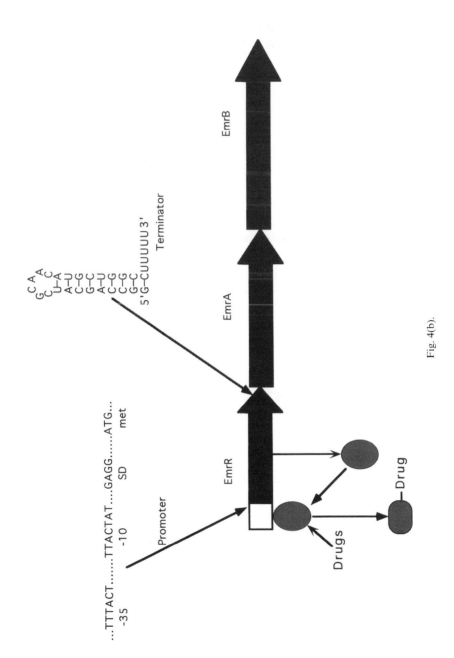

...TTTACT......TTACTAT....GAGG......ATG...
-35   -10   SD   met

Promoter

EmrR

```
        C   A
      G   C  A
        C   C
        U—A
        A—U
        C—G
        G—C
        A—U
        C—G
        C—G
        G—C
    5'G—CUUUUU 3'
      Terminator
```

EmrA

EmrB

Drugs

Drug

Fig. 4(b).

negative charge in this hydrophobic environment will select for cationic species.

A third multidrug sensor has been recently described that regulates expression of the QacA MF MDR found on multidrug resistance plasmids of *S. aureus*. It is encoded by a *qacR* gene located immediately upstream of *qacA* and is divergently transcribed from its own promoter. The QacR repressor binds hydrophobic cations (Grkovic *et al.*, 1998), exclusive substrates of QacA, and belongs to the TetR family of repressors.

There is evidently intrinsic value in understanding the mechanism of drug discrimination by multidrug sensors, but will they also serve as a useful model to understand the more complex MDR pumps? The sensors share a substrate specificity with the pumps, but no homology. However, it seems possible that the general principles of ligand binding/drug discrimination that emerge from the structural studies of drug sensors will be applicable to MDR pumps. At the very least, information gained from the study of the sensors will stimulate the design of experiments to test particular models of drug discrimination by MDR pumps.

FUNCTION AND THE SEARCH FOR NATURAL SUBSTRATES

MDR pumps can be very effective in protecting the microbial cell from toxic compounds. For example, the EmrAB pump of *E. coli* (Lomovskaya & Lewis, 1992) confers a 60-fold increase in resistance to the antibiotic thiolactomycin (Furukawa *et al.*, 1993); the AcrAB pump of *E. coli* confers a 60–100-fold resistance to novobiocin and erythromycin (Ma *et al.*, 1995); and the MexAB pump of *P. aeruginosa* confers a 50–100-fold resistance to some quinolones and hydrophobic antibiotics (Poole, 1994). These levels of resistance rival specific resistance mechanisms such as tetracycline efflux by a dedicated pump, or inactivation of kanamycin by a specific acetyl transferase.

Even though MDR pumps have the potential to protect cells from antimicrobials and do this both in the laboratory and in a clinical setting, this does not necessarily mean that drug resistance is the natural function of MDRs. For example, the Blt pump of *B. subtilis*, which protects cells from a panel of amphipathic cationic antimicrobials, is part of an operon that encodes a putrescine acetyl transferase (Woolridge *et al.*, 1997). The Blt pump extrudes putrescine from the cell, suggesting that this might be its natural function, and drug resistance is a coincidental consequence of the protein being a 'sloppy translocase'.

However, other data indicate that drug resistance is the natural function of MDRs. The QacA MDR pump is found on broad-host-range plasmids that also carry specific gentamicin and trimethoprim resistance genes (Rouch *et al.*, 1990). This context suggests QacA is a dedicated drug resistance component of the plasmid as well. The induction of MDRs by their substrates acting through MDR sensors is a very strong argument for multidrug resistance being the natural function of at least some of these translocases. The BmrR ligand-binding site discussed above seems to have evolved to accommodate a wide range of substrates, which it selects largely on the basis of polarity, a salient feature of MDR pump recognition.

But if there are dedicated MDRs, what are their natural substrates? RND pumps extrude natural antibiotics, but handle artificial substrates just as well or better, and have such a broad substrate spectrum that it is unclear whether antibiotic extrusion plays a leading role outside of the clinical setting. The AcrAB pump was shown to strongly protect *E. coli* from bile acids, and it was proposed that bile acid extrusion might be the natural function of this pump (Thanassi *et al.*, 1997). This is an appealing hypothesis, and bile acids might indeed be among the many natural substrates of the extremely broad spectrum AcrAB pump. Protection from bile salts appears to be one of the functions of the Mtr RND pump of *N. gonorrhoeae* (Hagman *et al.*, 1995) and of VceAB, which is a *Vibrio cholerae* homologue of the *E. coli* EmrAB pump (Colmer *et al.*, 1998). As mentioned above, the Mtr pump of *N. gonorrhoeae* has been reported to protect the cell from small defensin peptides, and this might be part of its natural function. It does not seem, however, that RND pumps, which are widely distributed among Gram-negative bacteria, evolved to cope with bile acids or defensins of animals. Another possible example of a natural substrate has been described recently in a study of an IfeAB RND pump from *Agrobacterium tumefaciens* (Palumbo *et al.*, 1998). Mutation of the pump decreased accumulation of an isoflavonoid, coumestrol, which is produced by its host, the alfalfa plant. Coumestrol induced expression of the pump and the mutant was out-competed by the wild-type in colonizing the plant. However, neither the wild-type nor the mutant was sensitive to coumestrol. As the authors note, coumestrol might be acting as an inducing signal for the pump, rather than its natural substrate.

The issue of natural MDR substrates in most cases remains very much an open question.

Amphipathic cations

Even though MDR pumps extrude structurally unrelated compounds, a general theme emerges if one considers the preferred artificial substrates of most MDRs. These substances with little exception are amphipathic cations (Fig. 5). This observation suggests that amphipathic cations represent the prototypical and existing natural substrates of MDR pumps. The simplest MDRs of the SMR family are unique in that no specific translocases are found in this family, and they might have been the first dedicated MDRs to evolve (Lewis, 1994). These are also the only group of MDRs to have amphipathic cations as their exclusive substrates.

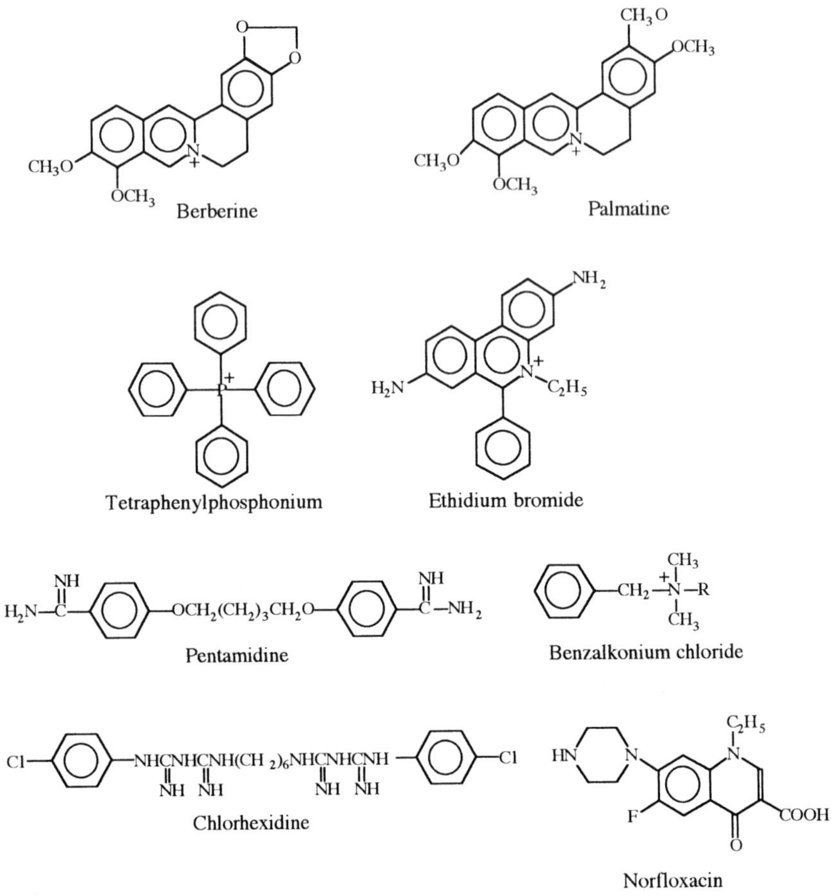

Fig. 5. Structure of cationic MDR substrates. Amphipathic cations are indicated by their positive charge. Berberine and palmatine are plant isoquinoline alkaloids. Tetraphenylphosphonium has been used as a probe to measure the membrane potential. Benzalkonium chloride is an antiseptic and disinfectant. Weak bases: pentamidine is a systemic antiprotozoan and chlorhexidine is an antiseptic.

There are many different MDR pumps in the MF family, and most of them exclusively extrude amphipathic cations. For example, the QacA pump only extrudes cations; the NorA pump of *S. aureus* extrudes cations and to a lesser extent quinolones; the BMR pump of *B. subtilis* extrudes primarily cations and neutral chloramphenicol (review, Paulsen *et al.*, 1996a). The BmrR regulator that activates BMR transcription has a design suggesting it specifically evolved to detect a wide range of amphipathic cations.

Like EmrAB, the MDRs of the RND family (resistance–nodulation–division) are *trans*-envelope translocases of Gram-negative bacteria. They export toxins across the outer membrane, which is a permeability barrier for hydrophobic compounds (Fig. 2). The substrate spectrum of RND pumps is extremely broad and in the case of the *E. coli* AcrAB includes cationic acridine, anionic detergent SDS, and neutral β-lactam antibiotics, for example (Nikaido, 1996). All tested RND pumps extrude amphipathic cations.

The preferred substrates for ABC MDRs LmrA and P-glycoprotein are amphipathic cations, but some neutral compounds can be extruded as well. There are many ABC MDRs in yeast, and at least nine functional ABC MDRs are present in *Saccharomyces cerevisiae* alone. The substrates of these pumps are amphipathic cations, and neutral substances such as anti-yeast azoles (Kolaczkowski & Goffeau, 1997; Paulsen *et al.*, 1998).

The following picture emerges from this analysis. Simple MDRs like SMRs only export amphipathic cations; MF MDRs export mainly cations; RND and ABC are the most complex of the MDRs and have broad spectra of specificity that include amphipathic cations as preferred substrates. MDRs clearly prefer amphipathic cations to other substances, even though they belong to four unrelated protein families. Not only are MDRs unrelated, but even the general mechanisms of drug transport are different for different MDRs, as discussed above. It appears that a similar need to protect the cell from amphipathic cations evolved in different groups of MDRs (and in different organisms) in spite of a lack of overall homology or similarity in the mechanism of action.

Quite surprisingly, one does not find amphipathic cations in a general list of natural antimicrobials. (The known cationic antibiotics of the amino-glycoside group such as streptomycin and kanamycin are hydrophilic substances that get smuggled into the cell via specific translocases and are not substrates of MDR pumps.) At the same time, amphipathic cations should be among the most potent antimicrobials. A positive charge will lead to a considerable accumulation of a substance in the cell. According to the Nernst equation, there is a 10-fold accumulation of a cation (cat) for every 60 mV of the membrane potential:

$$\Delta\psi \text{ (mV)} = (RT/nF)\ln\{[cat]in/[cat]out\} = 60lg\{[cat]in/[cat]out\}$$

The $\Delta\psi$ in bacteria is around 140 mV and around 180 mV in yeast plasma membrane (see Skulachev, 1988, for a review), which would result in a 100–1000-fold accumulation of an antibiotic. Note that weak amphipathic bases (such as chlorhexidine) are also MDR substrates, but one would not expect these compounds to be among natural antibiotics, since they are extruded from the cell by the pH gradient and are therefore intrinsically less potent than neutral compounds or strong cations. The fact that strong amphipathic cations are conspicuously absent from known natural antibiotics is especially puzzling given that these substances are the preferred substrates for most MDRs. We have argued that it is precisely the existence of MDR pumps that is responsible for this apparent paradox (Hsieh *et al.*, 1998). If MDRs evolved in response to natural antimicrobial amphipathic cations, then these substances would be difficult to discover in standard screens that employ cells carrying MDR pumps. In the process of drug discovery, the concentration of antimicrobials is prone to be low, and MDR substrates will be overlooked. We therefore proposed to use MDR mutants as sensitive tools for drug discovery, and in collaboration with scientists from Phytera, Inc. used a number of MDR⁻ strains to screen compound libraries for antimicrobials. This approach yielded a number of novel compounds that were missed in a screen with wild-type strains (not shown). MDR mutants were also used by scientists from Microcide Pharmaceuticals to successfully screen for new antimicrobials (Lomovskaya, 1998).

It seemed possible to further increase sensitivity to amphipathic cations by increasing the pH of the growth medium, and model experiments yielded interesting information on the nature of MDR substrates. A Gram-positive organism, *S. aureus*, was used as a test organism, for this is an important pathogen and its NorA pump (of the MF family) is responsible for resistance to cationic antiseptics. *norA* was disrupted by homologous recombination with a DNA fragment, containing *norA* interrupted with a *cam* cassette. Energy-dependent extrusion of a model substrate, ethidium bromide, was observed in the wild-type, but not in the mutant strain (Hsieh *et al.*, 1998) (Fig. 6).

The mutant strain had a higher sensitivity to a number of quinolones and amphipathic bases and cations such as the antimicrobial agent pentamidine and antiseptics chlorhexidine and benzalkonium chloride (Tables 1 and 2). The absolute increase in sensitivity ranged between 4- and 30-fold. It appeared that weak bases and amphipathic cations are the preferred substrates for NorA. It seemed possible to further improve the sensitivity of screening for these compounds by simply increasing the pH of the medium. At high pH (9.0), the pH gradient inverts, leading to an active accumulation of weak bases (Padan & Schuldiner, 1994). At the same time, there is an

Table 1. *NorA dependence of antimicrobial resistance (MIC) in S. aureus*[a]

Drug ($\mu g\ ml^{-1}$)	RN 4222 (WT)	KLE 820 (*norA*)	Sensitivity (*norA* vs WT)
Pentamidine	10	0.3	33
EtBr	5	0.33	16
TPP	15.6	1.9	8
Acriflavine	20	2.5	8
Norfloxacin	1.25	0.33	4
Ciprofloxacin	0.67	0.16	4
Kanamycin	0.33	0.33	1

[a] The MIC was determined by turbidity measurements in microtitre plates after an overnight incubation of cells in growth medium with a test substance.

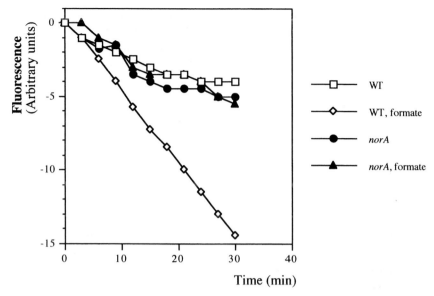

Fig. 6. Efflux of ethidium bromide (EtBr) from wild-type and *norA* strains of *S. aureus*. Cells were de-energized with uncoupler and loaded with EtBr. The graph depicts the loss of EtBr from cells diluted into the medium. Cells were energized by the addition of formate. Fluorescence of EtBr is expressed in arbitrary units.

increase in the membrane potential, which leads to an increased accumulation of amphipathic cations. The sensitivity to all amphipathic cations increased significantly at pH 9 (Hsieh *et al.*, 1998) (Table 2). For example, sensitivity to pentamidine increased approximately 30-fold at pH 9 in the wild-type. The NorA pump is electrogenic (L. Grinius, personal communication) and continues to protect the cell at alkaline pH. When combined, pH 9

Table 2. *Synergistic effect of high pH and a norA mutation on S. aureus cells*[a]

| Drug (μg ml⁻¹) | RN 4222 (WT) | | KLE 820 (*norA*) | | Sensitivity (*norA* at pH 9 vs WT at pH 7.2) |
	pH 7.2	pH 9	pH 7.2	pH 9	
Pentamidine	10	0.3	0.3	0.01	1030
EtBr	5	0.078	0.3	0.01	500
Berberine	60	3.75	7.5	0.23	260
Palmatine	200	12.5	50	0.78	256
TPP	15.6	0.5	1.98	0.12	128
Benzalkonium chloride	0.67	0.08	0.16	0.01	64
Chlorhexidine	0.5	0.06	0.125	0.016	32
Puromycin	20	2.5	20	2.5	8
Norfloxacin	1.25	1.25	0.33	0.33	4

[a] Resistance is expressed in MIC.

and the *norA* mutation provided a strong synergistic effect, increasing sensitivity by an impressive 1000-fold. This decreased the MIC for pentamidine to 10 ng ml^{-1}. These results suggest that a combination of high pH and an MDR mutation can be used for very sensitive screening for cationic antimicrobials. It is also evident that an MDR inhibitor and high pH will produce a very effective synergistic combination for increasing the potency of cationic antiseptics. One might also expect that in nature, a high pH environment will make cationic antimicrobials especially dangerous, and alkaliphilic micro-organisms will probably express MDR pumps at high levels.

While using MDR mutants is a reasonable (if somewhat unpredictable) way to discover possible cationic antimicrobials, another approach is to search for possible MDR substrates among known compounds. Many natural substances have been described as a result of systematic chemical analysis of organisms, rather than in particular bioassay-driven purifications. One would then look for substances that are strong amphipathic cations of natural origin that have little or no antimicrobial activity. Using these criteria, we identified a group of plant alkaloids whose members have little or no antimicrobial activity. These are the isoquinoline alkaloids (Fig. 5), which are widely spread among the plant world and are found among many *Ranunculales* species, for example (Colombo & Bosisio, 1996). These substances bear a resemblance to artificial MDR substrates such as ethidium bromide or benzalkonium chloride (Fig. 5). They are amphipathic and have a positive charge which is delocalized by the conjugated ring structure, an essential feature of a good permeant cation (Skulachev, 1988).

We have chosen two representative substances of this group, berberine and palmatine, to test whether they are MDR substrates. Palmatine had very low (>200 μg ml^{-1} MIC) and berberine poor (120–60 μg ml^{-1}) activity against wild-type *S. aureus* (Hsieh *et al.*, 1998). The antimicrobial activity of the alkaloids increased sharply in a *norA* mutant, with an MIC of 50 μg ml^{-1} for palmatine and 7.5 for berberine (Table 2). Sensitivity to alkaloids increased further in the presence of an MDR inhibitor, INF271 (kindly provided by Dr P. N. Markham, Influx, Inc.; K. Lewis, unpublished). It appears that the inhibitor disables both NorA and possible additional unknown MDR(s). Thus in the presence of the MDR inhibitor berberine becomes an extremely potent antibiotic (MIC 0.5 μg ml^{-1}), about 10 times stronger than streptomycin. Our preliminary experiments show that berberine is also the substrate of the plasmid-borne QacA pump of *S. aureus* (MIC 500 μg ml^{-1} vs 1 μg ml^{-1} in the presence of INF271). In the presence of the MDR inhibitor, yeast becomes very sensitive to berberine as well (MIC 120 μg ml^{-1} vs 1 μg ml^{-1}).

Why should plants keep on making isoquinoline alkaloids if micro-organisms have MDR pumps that can render these substances essentially ineffective? One possibility is that isoquinoline alkaloids are not antimicrobial compounds *in vivo* and have a different function, such as antiherbivoral. A more interesting possibility is that in response to bacterial resistance mechanisms, plants have developed MDR inhibitors that act synergistically with isoquinoline alkaloids. We have tested this hypothesis using a berberine-producing *Mahonia fremontii*. An extract of the plant has at least two different MDR inhibitors that act synergistically with berberine. The active concentration of the more hydrophobic purified substance (F. R. Stermitz and others, unpublished) was 1 μg ml^{-1}, suggesting it is a potent MDR inhibitor.

CONCLUDING REMARKS

Basic research

The mechanism of action remains the central question regarding MDRs. The immediate future of this direction of research does not look particularly promising. We are unlikely to learn the mechanism of drug discrimination by an MDR pump unless a crystal structure becomes available. So far, attempts to crystallize any solute translocase (as opposed to a channel) have been unsuccessful. All membrane proteins that have been crystallized, such as cytochrome oxidase or bacteriorhodopsin, have fixed pathways for electron or proton, or a fixed channel, which do not require the protein to undergo major conformational changes. Limited flexibility helps pack proteins in a crystal domain. A breakthrough in the methods for membrane protein crystallization which will allow the crystallization of translocases including MDRs will be required for us to get a glimpse at the drug discrimination mechanism. Meanwhile, our best hope for structural information on drug binding/discrimination will come from the study of multidrug sensors.

The study of MDR function is becoming an exciting area of research. The use of MDR mutants in pathogens will show whether virulence depends upon MDRs, and will allow for the identification of host defence molecules that are MDR substrates. Screening natural compound libraries, which is being performed by drug discovery companies in search of new antibiotics, will produce a catalogue of natural MDR substrates. This survey will show whether hydrophobic cations represent the main class of natural MDR substrates and a previously overlooked group of antibiotics.

Clinical role

The very significant level of multidrug resistance observed in *P. aeruginosa* is an example of the potential threat of resistance based on MDR pumps. Azole-resistant pathogenic yeast *C. albicans* overexpressing MDRs is also essentially invulnerable to currently available antimicrobials and presents a serious health threat. Most pathogenic micro-organisms form biofilms which are resistant to all antimicrobials and are a major cause of therapy failure. It will be interesting to learn whether MDRs have a role in this puzzling resistance. These examples underline the need to detect and survey pathogens for MDR efflux-based drug resistance. The fact that yeasts are resistant to the newest azoles and *P. aeruginosa* is resistant to the newest quinolones is alarming but not surprising. An MDR pump is perhaps the microbe's 'perfect defence' – any amphipathic molecule with an intracellular target will be subject to efflux. One may conclude that micro-organisms already possess a potential resistance mechanism to all future drugs. Yet pathogens are not invincible in nature, and as our studies of the berberine-producing *M. fremontii* show, the host learned to make an MDR inhibitor that acts synergistically with the antibiotic. Our best hope is indeed to identify efficient MDR inhibitors that can be used together with conventional antimicrobials as an intelligent response to microbes' perfect defence.

REFERENCES

Ahmed, M., Borsch, C. M., Neyfakh, A. A. & Schuldiner, S. (1993). Mutants of the *Bacillus subtilis* multidrug transporter Bmr with altered sensitivity to the antihypertensive alkaloid reserpine. *Journal of Biological Chemistry* **268**, 11086–11089.

Ahmed, M., Borsch, C. M., Taylor, S. S., Vazquez-Laslop, N. & Neyfakh, A. A. (1994). A protein that activates expression of a multidrug efflux transporter upon binding the transporter substrates. *Journal of Biological Chemistry* **269**, 28506–28513.

Blattner, F. R., Plunkett, G., III, Bloch, C. A. & 14 other authors (1997). The complete genome sequence of *Escherichia coli* K-12 [comment]. *Science* **277**, 1453–1474.

Bolhuis, H., van Veen, H. W., Brands, J. R., Putman, M., Poolman, B., Driessen, A. J. M. & Konings, W. N. (1996a). Energetics and mechanism of drug transport mediated by the lactococcal multidrug transporter LmrP. *Journal of Biological Chemistry* **271**, 24123–24128.

Bolhuis, H., van Veen, H. W., Molenaar, D., Poolman, B., Driessen, A. J. & Konings, W. N. (1996b). Multidrug resistance in *Lactococcus lactis*: evidence for ATP-dependent drug extrusion from the inner leaflet of the cytoplasmic membrane. *EMBO Journal* **15**, 4239–4245.

Brooun, A., Tomashek, J. J. & Lewis, K. (1999). Purification and ligand binding of EmrR, a regulator of a multidrug transporter. *Journal of Bacteriology* (in press).

Caballero, J. L., Martinez, E., Malpartida, F. & Hopwood, D. A. (1991). Organisation and functions of the *actVA* region of the actinorhodin biosynthetic gene cluster of *Streptomyces coelicolor*. *Molecular & General Genetics* **230**, 401–412.

Colmer, J. A., Fralick, J. A. & Hamood, A. N. (1998). Isolation and characterization of a putative multidrug resistance pump from *Vibrio cholerae*. *Molecular Microbiology* **27**, 63–72.

Colombo, M. L. & Bosisio, E. (1996). Pharmacological activities of *Chelidonium majus* L. (*Papaveraceae*). *Pharmacological Research* **33**, 127–134.

Fralick, J. A. (1996). Evidence that TolC is required for functioning of the Mar/AcrAB efflux pump of *Escherichia coli*. *Journal of Bacteriology* **178**, 5803–5805.

Furukawa, H., Tsay, J. T., Jackowski, S., Takamura, Y. & Rock, C. O. (1993).

Thiolactomycin resistance in *Escherichia coli* is associated with the multidrug resistance efflux pump encoded by *emrAB*. *Journal of Bacteriology* **175**, 3723–3729.

Gomez-Diaz, C., Rodriguez-Aguilera, J. C., Barroso, M. P., Villalba, J. M., Navarro, F., Crane, F. L. & Navas, P. (1997). Antioxidant ascorbate is stabilized by NADH-coenzyme Q10 reductase in the plasma membrane. *Journal of Bioenergetics and Biomembranes* **29**, 251–257.

Gramajo, H. C., White, J., Hutchinson, C. R. & Bibb, M. J. (1991). Overproduction and localization of components of the polyketide synthase of *Streptomyces glaucescens* involved in the production of the antibiotic tetracenomycin C. *Journal of Bacteriology* **173**, 6475–6483.

Greener, T., Govezensky, D. & Zamir, A. (1993). A novel multicopy suppressor of a *groEL* mutation includes two nested open reading frames transcribed from different promoters. *EMBO Journal* **12**, 889–896.

Grinius, L. L. & Goldberg, E. B. (1994). Bacterial multidrug resistance is due to a single membrane protein which functions as a drug pump. *Journal of Biological Chemistry* **269**, 29998–30004.

Grkovic, S., Brown, M. H., Roberts, N. J., Paulsen, I. T. & Skurray, R. A. (1998). QacR is a repressor protein that regulates expression of the *Staphylococcus aureus* multidrug efflux pump QacA. *Journal of Biological Chemistry* **273**, 18665–18673.

Guilfoile, P. G. & Hutchinson, C. R. (1992). Sequence and transcriptional analysis of the *Streptomyces glaucescens tcmAR* tetracenomycin C resistance and repressor gene loci. *Journal of Bacteriology* **174**, 3651–3658.

Hagman, K. E., Pan, W., Spratt, B. G., Balthazar, J. T., Judd, R. C. & Shafer, W. M. (1995). Resistance of *Neisseria gonorrhoeae* to antimicrobial hydrophobic agents is modulated by the *mtrRCDE* efflux system. *Microbiology* **141**, 611–622.

van Helvoort, A., Smith, A. J., Sprong, H., Fritzsche, I., Schinkel, A. H., Borst, P. & van Meer, G. (1996). MDR1 P-glycoprotein is a lipid translocase of broad specificity, while MDR3 P-glycoprotein specifically translocates phosphatidylcholine. *Cell* **87**, 507–517.

Higgins, C. F. & Gottesman, M. M. (1992). Is the multidrug transporter a flippase? *Trends in Biochemical Sciences* **17**, 8–21.

Higgins, C. F., Callaghan, R., Linton, K. J., Rosenberg, M. F. & Ford, R. C. (1997). Structure of the multidrug resistance P-glycoprotein. *Seminars in Cancer Biology* **8**, 135–142.

Hsieh, P. C., Siegel, S. A., Rogers, B., Davis, D. & Lewis, K. (1998). Bacteria lacking a multidrug pump: a sensitive tool for drug discovery. *Proceedings of the National Academy of Sciences, USA* **95**, 6602–6606.

Kaatz, G. W., Seo, S. M. & Ruble, C. A. (1993). Efflux-mediated fluoroquinolone resistance in *Staphylococcus aureus*. *Antimicrobial Agents and Chemotherapy* **37**, 1086–1094.

Klyachko, K. A. & Neyfakh, A. A. (1998). Paradoxical enhancement of the activity of a bacterial multidrug transporter caused by substitutions of a conserved residue. *Journal of Bacteriology* **180**, 2817–2821.

Klyachko, K. A., Schuldiner, S. & Neyfakh, A. A. (1997). Mutations affecting substrate specificity of the *Bacillus subtilis* multidrug transporter Bmr. *Journal of Bacteriology* **179**, 2189–2193.

Kolaczkowski, M. & Goffeau, A. (1997). Active efflux by multidrug transporters as one of the strategies to evade chemotherapy and novel practical implications of yeast pleiotropic drug resistance. *Pharmacology & Therapeutics* **76**, 219–242.

Lewis, K. (1994). Multidrug resistance pumps in bacteria: variations on a theme. *Trends in Biochemical Sciences* **19**, 119–123.

Li, X. Z., Ma, D., Livermore, D. M. & Nikaido, H. (1994). Role of efflux pump(s) in intrinsic resistance of *Pseudomonas aeruginosa*: active efflux as a contributing factor to beta-lactam resistance. *Antimicrobial Agents and Chemotherapy* **38**, 1742–1752.

Li, X. Z., Nikaido, H. & Poole, K. (1995). Role of *mexA-mexB-oprM* in antibiotic efflux in *Pseudomonas aeruginosa*. *Antimicrobial Agents and Chemotherapy* **39**, 1948–1953.

Liu, R. & Sharom, F. J. (1996). Site-directed fluorescence labeling of P-glycoprotein on cysteine residues in the nucleotide binding domains. *Biochemistry* **35**, 11865–11873.

Lomovskaya, O. (1998). Efflux pumps as targets and tools for drug discovery. In *Society for Industrial Microbiology Annual Meeting*, Abstr. S47, p. 64.

Lomovskaya, O. & Lewis, K. (1992). Emr, an *Escherichia coli* locus for multidrug resistance. *Proceedings of the National Academy of Sciences, USA* **89**, 8938–8942.

Lomovskaya, O., Lewis, K. & Matin, A. (1995). EmrR is a negative upstream regulator of the *E. coli* multidrug resistance pump EmrA. *Journal of Bacteriology* **177**, 2328–2334.

Ma, D., Cook, D. N., Alberti, M., Pon, N. G., Nikaido, H. & Hearst, J. E. (1993). Molecular cloning and characterization of *acrA* and *acrE* genes of *Escherichia coli*. *Journal of Bacteriology* **175**, 6299–6313.

Ma, D., Cook, D. N., Hearst, J. E. & Nikaido, H. (1994). Efflux pumps and drug resistance in gram-negative bacteria. *Trends in Microbiology* **2**, 489–493.

Ma, D., Cook, D. N., Alberti, M., Pon, N. G., Nikaido, H. & Hearst, J. E. (1995). Genes *acrA* and *acrB* encode a stress-induced efflux system of *Escherichia coli*. *Molecular Microbiology* **16**, 45–55.

McMurry, L. M. & Levy, S. B. (1995). The NH2-terminal half of the Tn10-specified tetracycline efflux protein TetA contains a dimerization domain. *Journal of Biological Chemistry* **270**, 22752–22757.

Marger, M. D. & Saier, M. H., Jr (1993). A major superfamily of transmembrane facilitators that catalyze uniport, symport and antiport. *Trends in Biochemical Sciences* **18**, 13–20.

Markham, P. N. & Neyfakh, A. A. (1996). Inhibition of the multidrug transporter NorA prevents emergence of norfloxacin resistance in *Staphylococcus aureus* [letter]. *Antimicrobial Agents and Chemotherapy* **40**, 2673–2674.

Miller, P. F. & Sulavik, M. C. (1996). Overlaps and parallels in the regulation of intrinsic multiple-antibiotic resistance in *Escherichia coli*. *Molecular Microbiology* **21**, 441–448.

Neyfakh, A. A., Bidnenko, V. E. & Chen, L. B. (1991). Efflux-mediated multidrug resistance in *Bacillus subtilis*: similarities and dissimilarities with the mammalian system. *Proceedings of the National Academy of Sciences, USA* **88**, 4781–4785.

Neyfakh, A. A., Borsch, C. M. & Kaatz, G. W. (1993). Fluoroquinolone resistance protein NorA of *Staphylococcus aureus* is a multidrug efflux transporter. *Antimicrobial Agents and Chemotherapy* **37**, 128–129.

Ng, E. Y., Trucksis, M. & Hooper, D. C. (1994). Quinolone resistance mediated by *norA*: physiologic characterization and relationship to *flqB*, a quinolone resistance locus on the *Staphylococcus aureus* chromosome. *Antimicrobial Agents and Chemotherapy* **38**, 1345–1355.

Nikaido, H. (1996). Multidrug efflux pumps of Gram-negative bacteria. *Journal of Bacteriology* **178**, 5853–5859.

Nikaido, H., Basina, M., Nguyen, V. & Rosenberg, E. Y. (1998). Multidrug efflux pump AcrAB of *Salmonella typhimurium* excretes only those beta-lactam antibiotics containing lipophilic side chains. *Journal of Bacteriology* **180**, 4686–4692.

Padan, E. & Schuldiner, S. (1994). Molecular physiology of Na^+/H^+ antiporters, key transporters in circulation of Na^+ and H^+ in cells. *Biochimica et Biophysica Acta* **1185**, 129–151.

Palumbo, J. D., Kado, C. I. & Phillips, D. A. (1998). An isoflavonoid-inducible efflux pump in *Agrobacterium tumefaciens* is involved in competitive colonization of roots. *Journal of Bacteriology* **180**, 3107–3113.

Paulsen, I. T., Brown, M. H., Dunstan, S. J. & Skurray, R. A. (1995). Molecular characterization of the staphylococcal multidrug resistance export protein QacC. *Journal of Bacteriology* **177**, 2827–2833.

Paulsen, I. T., Brown, M. H. & Skurray, R. A. (1996a). Proton-dependent multidrug efflux systems. *Microbiological Reviews* **60**, 575–608.

Paulsen, I. T., Skurray, R. A., Tam, R., Saier, M. H., Jr, Turner, R. J., Weiner, J. H., Goldberg, E. B. & Grinius, L. L. (1996b). The SMR family: a novel family of multidrug efflux proteins involved with the efflux of lipophilic drugs. *Molecular Microbiology* **19**, 1167–1175.

Paulsen, I. T., Sliwinski, M. K., Nelissen, B., Goffeau, A. & Saier, M. H., Jr (1998). Unified inventory of established and putative transporters encoded within the complete genome of *Saccharomyces cerevisiae*. *FEBS Letters* **430**, 116–125.

Poole, K. (1994). Bacterial multidrug resistance – emphasis on efflux mechanisms and *Pseudomonas aeruginosa*. *Journal of Antimicrobial Chemotherapy* **34**, 453–456.

Rosenberg, M. F., Callaghan, R., Ford, R. C. & Higgins, C. F. (1997). Structure of the multidrug resistance P-glycoprotein to 2.5 nm resolution determined by electron microscopy and image analysis. *Journal of Biological Chemistry* **272**, 10685–10694.

Rouch, D. A., Cram, D. S., DiBerardino, D., Littlejohn, T. G. & Skurray, R. A. (1990). Efflux-mediated antiseptic resistance gene *qacA* from *Staphylococcus aureus*: common ancestry with tetracycline- and sugar-transport proteins. *Molecular Microbiology* **4**, 2051–2062.

Rubin, R. A., Levy, S. B., Heinrikson, R. L. & Kezdy, F. J. (1990). Gene duplication in the evolution of the two complementing domains of gram-negative bacterial tetracycline efflux proteins. *Gene* **87**, 7–13.

Ruetz, S. & Gros, P. (1994). Phosphatidylcholine translocase: a physiological role for the *mdr*2 gene. *Cell* **77**, 1071–1081.

Schwaiger, M., Lebendiker, M., Yerushalmi, H., Coles, M., Groger, A., Schwarz, C., Schuldiner, S. & Kessler, H. (1998). NMR investigation of the multidrug transporter EmrE, an integral membrane protein. *European Journal of Biochemistry* **254**, 610–619.

Shafer, W. M., Qu, X., Waring, A. J. & Lehrer, R. I. (1998). Modulation of *Neisseria gonorrhoeae* susceptibility to vertebrate antibacterial peptides due to a member of the resistance/nodulation/division efflux pump family. *Proceedings of the National Academy of Sciences, USA* **95**, 1829–1833.

Sharom, F. J. (1997). The P-glycoprotein efflux pump: how does it transport drugs? *Journal of Membrane Biology* **160**, 161–175.

Skulachev, V. P. (1988). *Membrane Bioenergetics*. Berlin & New York: Springer.

Srikumar, R., Li, X. Z. & Poole, K. (1997). Inner membrane efflux components are responsible for beta-lactam specificity of multidrug efflux pumps in *Pseudomonas aeruginosa*. *Journal of Bacteriology* **179**, 7875–7881.

Thanassi, D. G., Cheng, L. W. & Nikaido, H. (1997). Active efflux of bile salts by *Escherichia coli*. *Journal of Bacteriology* **179**, 2512–2518.

Ueda, K., Taguchi, Y. & Morishima, M. (1997). How does P-glycoprotein recognize its substrates? *Seminars in Cancer Biology* **8**, 151–159.

Varela, M. F., Sansom, C. E. & Griffith, J. K. (1995). Mutational analysis and molecular modelling of an amino acid sequence motif conserved in antiporters but not symporters in a transporter superfamily [published erratum appears in *Molecular Membrane Biology* 1996 Jan–Mar; 13(1), 66]. *Molecular Membrane Biology* **12**, 313–319.

van Veen, H. W., Venema, K., Bolhuis, H., Oussenko, I., Kok, J., Poolman, B.,

Driessen, A. J. & Konings, W. N. (1996). Multidrug resistance mediated by a bacterial homolog of the human multidrug transporter MDR1. *Proceedings of the National Academy of Sciences, USA* **93**, 10668–10672.

van Veen, H. W., Callaghan, R., Soceneantu, L., Sardini, A., Konings, W. N. & Higgins, C. F. (1998). A bacterial antibiotic-resistance gene that complements the human multidrug-resistance P-glycoprotein gene. *Nature* **391**, 291–295.

Woolridge, D. P., Vazquez-Laslop, N., Markham, P. N., Chevalier, M. S., Gerner, E. W. & Neyfakh, A. A. (1997). Efflux of the natural polyamine spermidine facilitated by the *Bacillus subtilis* multidrug transporter Blt. *Journal of Biological Chemistry* **272**, 8864–8866.

Zheleznova, E. E., Markham, P. N., Neyfakh, A. A. & Brennan, R. G. (1997). Preliminary structural studies on the multi-ligand-binding domain of the transcription activator, BmrR, from *Bacillus subtilis*. *Protein Science* **6**, 2465–2468.

Zheleznova, E. E., Markham, P. N., Neyfakh, A. A. & Brennan, R. G. (1999). Structural basis of multidrug recognition by BmrR, a transcription activator of a multidrug transporter. *Cell* **96**, 353–362.

REGULATION OF SOLUTE ACCUMULATION IN BACTERIA AND ITS PHYSIOLOGICAL SIGNIFICANCE

BERT POOLMAN

Department of Microbiology, Groningen Biomolecular Sciences and Biotechnology Institute, University of Groningen, Kerklaan 30, 9751 NN Haren, The Netherlands

INTRODUCTION

Bacterial solute transport systems can be subdivided in three categories on the basis of their energetics: (i) *primary transport systems* use the free energy that is released upon the hydrolysis of ATP; (ii) *secondary transport systems* use the free energy that is stored in the electrochemical gradients of protons, sodium ions or other solutes across the membrane; (iii) *group translocation systems* chemically modify the substrate concomitant with translocation. The latter group constitutes the bacterial phospho*enol*pyruvate sugar:phosphotransferase systems (PEP-PTS).

Both the primary and secondary transport systems may catalyse either uptake or efflux of solutes (Poolman & Konings, 1993). The uptake systems serve important functions in the nutrition of the cell by facilitating the transport of sugars, amino acids, peptides, vitamins, inorganic ions or others. The efflux systems function in the excretion of metabolic end products but also in the protection of the cell against unwanted toxic compounds present in the environment. The primary ATP-dependent solute *uptake* systems in bacteria all belong to the *A*TP-*b*inding *c*assette (ABC) family and use a binding protein to capture the substrate and deliver it to the translocator complex (Fig. 1; Higgins, 1992). For a long time it was thought that the use of binding proteins was restricted to these systems, but recently it has been shown that some secondary transport systems also employ such accessory proteins (Fig. 1; Jacobs *et al.*, 1996). The primary ATP-dependent *efflux* systems fall into at least four classes, i.e. those that belong to the ABC family, the F/V-type and P-type ATPases, and the family represented by the oxyanion ATPase Ars (see Konings *et al.*, 1996).

Energetics of primary solute transport systems

The driving force for an ATP-dependent transport system is given by the Gibbs free energy made available during ATP hydrolysis; ΔG_p in kJ mol^{-1}

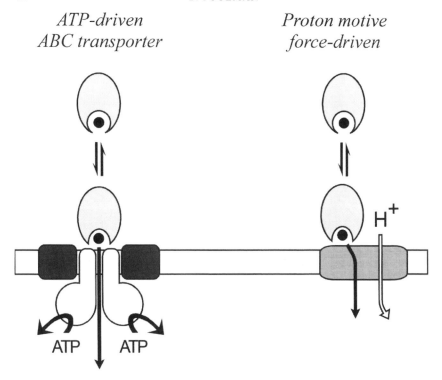

ATP-driven
ABC transporter

Proton motive
force-driven

Fig. 1. Binding-protein-dependent transport systems in bacteria.

or $\Delta G_p/F$ when expressed in mV:

$$\frac{\Delta G_p}{F} = \frac{\Delta G^\circ}{F} + \frac{RT}{F}\ln\left(\frac{[\text{ADP}] \times [\text{P}_i]}{[\text{ATP}]}\right)$$

in which $\Delta G^\circ/F$ is the standard Gibbs free energy (~ -320 mV), F is the Faraday constant (9.6485×10^4 J V^{-1} mol^{-1}), R is the gas constant (8.3144 J mol^{-1} K^{-1}) and T is the absolute temperature (K). In the growing cell, the concentrations of ATP, ADP and inorganic phosphate (P$_i$) are such that $\Delta G_p/F \sim -540$ mV. The solute concentration gradient ($[\text{S}]_{in}/[\text{S}]_{out}$), that is when the system reaches thermodynamic equilibrium, can now be calculated from the equation

$$\frac{\Delta G_p}{F} = -n \times \Delta pS = -\frac{nRT}{F}\ln\left(\frac{[\text{S}]_{in}}{[\text{S}]_{out}}\right)$$

in which n represents the reaction stoichiometry of the ATP-driven transport system. Although the value for n has not been determined unambiguously for any ABC-type ATP-dependent solute transport system, it is most likely 1 or 2. If one assumes that 1 ATP is hydrolysed per solute taken up ($n = 1$) then $[\text{S}]_{in}/[\text{S}]_{out}$ becomes 1×10^9; for $n = 2$ $[\text{S}]_{in}/[\text{S}]_{out}$ is $\sim 3.2 \times 10^4$. At a medium concentration of 1 mM and at thermodynamic equilibrium, $[S]_{in}$ would become 10^6 ($n = 1$) or 32 ($n = 2$) M, both of which are unrealistically high. The underlying mechanisms that prevent the solutes accumulating to such high values can be manifold, for example passive leakage across the membrane of accumulated solute (external leak), uncoupled backflux via the transporter (internal leak), product inhibition or tailored regulation. External leak pathways only contribute significantly to a lowered accumulation when the solute is relatively hydrophobic. The evidence for internal leak pathways is scarce, and in our opinion, the most prominent mechanism to regulate the internal solute concentration involves product inhibition; tailored regulation is known for some ATP-driven transport systems and these examples will be discussed below. For transport processes, product inhibition is usually referred to as *trans*-inhibition as the solute is compartmentalized (moved to the *trans* compartment) and not chemically modified. *Trans*-inhibition manifests itself as a decrease in the rate of uptake with increasing internal solute concentration. Several examples of *trans*-inhibition of ATP-dependent transport are known in the literature, but only a special one will be treated in this review (see 'Uptake of compatible solutes' section).

Energetics of secondary solute transport systems

The secondary transport systems can be subdivided into *symport*, *antiport* and *uniport* systems (Fig. 2). Most often *symport* systems couple the uptake of a solute to that of a proton or sodium ion, in which case the systems are driven by the proton and sodium motive force, respectively. The translocation reaction catalysed by secondary transport systems is essentially bidirectional. The net direction of transport is dependent on the direction of the driving force. Since under normal physiological conditions the membrane potential ($\Delta\psi$) is inside negative relative to outside, and the pH gradient (ΔpH) is inside alkaline relative to outside, a solute–H^+ symport system will catalyse the accumulation of the solute at the expense of the proton motive force (Δp).

$$\Delta p = \Delta\psi - \frac{2.3RT}{F}\Delta pH = \Delta\psi - Z\Delta pH$$

The same holds true for sodium-motive-force (Δs)-driven systems as $[Na]_{out} > [Na]_{in}$.

$$\Delta s = \Delta\psi + \frac{2.3RT}{F}\Delta pNa = \Delta\psi + Z\Delta pNa$$

As for the primary transport systems, external or internal leaks, *trans*-inhibition and/or tailored regulation may lead to steady state levels of accumulation that are lower than predicted from the driving forces of transport reactions. The consequences of external and internal leaks have

Symport:

Antiport:

Uniport:

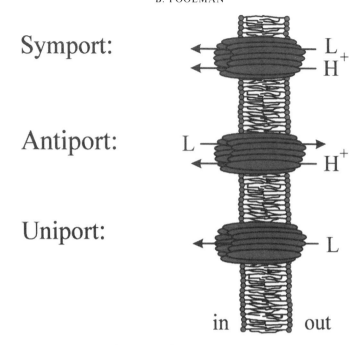

Fig. 2. Secondary transport proteins.

been described in previous papers (Poolman & Konings, 1993; Lolkema & Poolman, 1995); a particular case of *trans*-inhibition concerns the inactivation of proton-linked transport systems by a low internal pH (see below). Unlike the situation in yeast, inhibition of secondary transport by accumulated solute is not frequently observed in bacteria.

Under special conditions, the solute concentration gradient (ΔpS; $[S]_{in} > [S]_{out}$) may exceed the proton (or sodium) motive force, which results in the generation of a Δp (or Δs) when the solute leaves the cell in symport with a proton (or sodium ion). How much metabolic energy in the form of a proton (or sodium) motive force is actually generated depends on the stoichiometry of the symport reaction.

In fermentative bacteria, the carbon and/or energy sources are often only partially degraded and the pathway product(s) are frequently structurally related to the substrate or precursor of the pathway (Poolman, 1990; Lolkema *et al.*, 1996). When the secondary transport system that facilitates the uptake of the precursor also recognizes the pathway product, both may be exchanged which ensures a tight coupling of the uptake and excretion processes to the metabolism. This type of transport has been referred to as precursor/product exchange (Poolman, 1990). A system is referred to as antiporter when the coupling of the transport of the precursor and product is obligatory. Examples are the hexose-phosphate/phosphate and arginine/

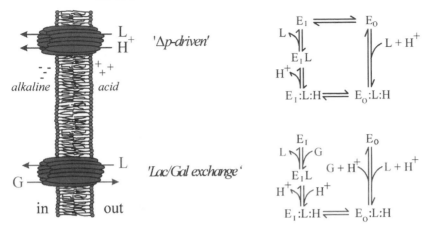

Fig. 3. Schematic representation of the lactose–H^+ symport (Δp-driven) and lactose/galactose exchange reactions. E, L and H^+ refer to enzyme (transporter), ligand and proton, respectively; $_O$ and $_I$ indicate the outer and inner surface of the membrane, respectively.

ornithine antiporters found in various bacteria (Maloney *et al.*, 1990; Driessen *et al.*, 1987). In other cases, precursor/product exchange occurs by an ion-linked secondary transport system, but for kinetic reasons the exchange is favoured over the symport mode of transport. This is readily seen in the kinetic scheme of the lactose transport protein LacS of *Streptococcus thermophilus* (Fig. 3). The LacS protein catalyses the uptake of lactose in symport with a proton ('Δp-driven') or catalyses the uptake of lactose in exchange for galactose. The latter reaction is one to two orders of magnitude faster than the lactose–H^+ symport and relevant for the bacterium, because galactose is a metabolic end product. The symport mode of transport is slow because the isomerization from E_I to E_O is highly rate-determining, i.e. the reorientation of unloaded binding site is slow (Poolman *et al.*, 1995b). This step is avoided in the exchange mode of transport and the two half reactions (lactose from 'out to in' and galactose from 'in to out') involve the same steps in opposite directions, and isomerization of the binding site proceeds via the ternary E:L:H complex. We stress that the main features of this kinetic scheme also apply for many other secondary transport proteins, of which the LacY protein of *Escherichia coli* represents the best-studied example (Kaback, 1997).

In special cases, precursor/product exchange leads to the generation of metabolic energy in the form of a proton motive force as exemplified by the transporters in the pathways for oxalate fermentation in *Oxalobacter formigenes* and malolactic fermentation in lactic acid bacteria (Anantharam *et al.*, 1989; Poolman *et al.*, 1991); many other related pathways have been discovered in recent years (Lolkema *et al.*, 1996). An important feature of these pathways is that the translocation reaction, *precursor/product exchange*, generates a membrane potential, whereas a pH gradient is formed as a result of the conversion of precursor into product. These systems represent 'ordinary' secondary transport systems for which the exchange mode of transport is much faster than the symport or uniport mode.

In general, the energetics and kinetics of the different types of transport mechanisms are known quite precisely from *in vitro* studies in membrane vesicles or proteoliposomes. Within the context of the cell there can be regulatory mechanisms superimposed on the systems, which then determine the solute accumulation levels as well as the actual rates of transport. This will lead to solute concentration gradients that are far from thermodynamic equilibrium. The focus of this monograph is on these regulatory mechanisms and their specific physiological functions in bacteria. This review is by no means comprehensive, but it serves to illustrate some of the main principles of regulation of membrane transport.

REGULATION BY THE INTERNAL pH

The net direction of transport by an ion-coupled secondary transport system is determined by the magnitude of the ion motive force and the solute concentration gradient. The net flux is zero under conditions of thermodynamic equilibrium, but this situation is rarely met in growing cells as (part of) the solute is usually needed for biosynthetic purposes. This constant drain of solute from the intracellular pool results in a kinetic rather than a thermodynamic steady state for the transport reaction. Upon energy starvation of the cell, the ATP levels and ion motive force drop in time and accumulated solutes are expected to leak out in response to this lowering of the energy status of the cell, i.e. at least when secondary transport systems for these solutes are present. For many solutes (nutrients), however, it has been observed that they are retained for prolonged periods of time even in the absence of the ion motive force. It was found that these transport systems are highly sensitive to the actual value of the internal pH (for a review see Poolman *et al.*, 1987a). The activity of these systems drops sharply as the internal pH decreases below the physiological one. Thus, despite the fact that there is a large outwardly directed solute concentration gradient, the solute does not leak out because the carrier is in the inactive state. This has not only been shown in *in vitro* studies in membrane vesicles, but also *in vivo*, i.e. upon energy starvation of *Lactococcus lactis* a number of amino acids were retained by the cell even after the proton motive force had dropped to zero (Poolman *et al.*, 1987b). A strong dependence of the activity on the internal pH has not only been observed for secondary, but also for the ATP-dependent primary transport systems. Since the ATP-dependent transport systems catalyse an essentially unidirectional transport, the pH control mechanism has no apparent role in the retention of solutes inside the cell under conditions of metabolic energy deprivation.

REGULATION BY ACCESSORY PROTEIN(S)

Gram-negative bacteria: role of IIA protein

Transport of sugars in micro-organisms is often regulated at the level of gene expression and protein activity. In bacteria, the latter is referred to as catabolite inhibition or inducer exclusion (Saier, 1989; Postma et al., 1993). The dual regulation allows a fast response of the organism to the presence or absence of a specific sugar (inducer exclusion) and a slow response which involves switching on/off the expression of certain genes (catabolite repression). The phospho*enol*pyruvate : sugar transferase system (PTS) is involved in both processes (Fig. 4), and the regulation is mediated by the phosphorylation state of the phosphoryl transfer protein (IIA, previously enzyme III or III^{Glc}) (Saier, 1989; Postma et al., 1988; Peterkofsky et al., 1989). The inducer exclusion mechanism controls allosterically secondary transporters such as LacY and MelB (Osumi & Saier, 1982; Nelson et al., 1983; Kuroda et al., 1992), the ATP-driven transport system for maltose (Dean et al., 1990) and the cytoplasmic enzyme glycerol kinase (Hurley et al., 1993). The result of this regulation is that when *E. coli* or *Salmonella typhimurium* grows in the presence of either lactose, melibiose, maltose or glycerol **plus** glucose, diauxic growth is observed with glucose being utilized first in each case.

What is the mechanism of inducer exclusion? It has been shown that IIA, but not IIA ~ P, binds stoichiometrically to the LacY protein (Osumi & Saier, 1982; Nelson et al., 1983); this type of regulation is depicted schematically in Fig. 4 by the inhibition of the sugar–cation symport system by

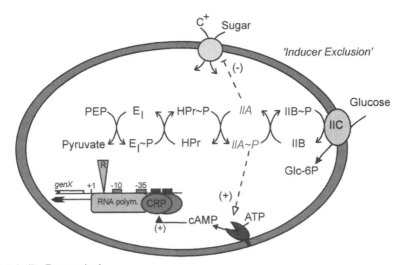

'Catabolite Repression'

Fig. 4. Catabolite repression and inducer exclusion in Gram-negative bacteria. CRP, catabolite regulator protein; R, regulator (e.g. LacI).

IIA. The binding of IIA to the lactose carrier is enhanced by the presence of galactosides, and, accordingly, binding of IIA causes a decrease in the K_D^{App} for galactosides (αNPG). Most importantly, however, binding of IIA to the lactose carrier causes inhibition of galactoside transport, and this inhibition has been interpreted as the molecular basis for inducer exclusion. In a similar manner IIA affects glycerol kinase, i.e. it binds to and inhibits purified glycerol kinase only in the presence of glycerol, and most likely also the melibiose and maltose transport proteins.

How does inducer exclusion occur *in vivo*? The PTS catalyses phosphoryl transfer from phospho*enol*pyruvate (PEP) to sugars (e.g. glucose) via a number of energy coupling proteins. Under conditions that the cell is metabolizing at a high rate a PTS sugar, IIA~P (phosphorylated by HPr~P) is rapidly dephosphorylated as a consequence of phosphoryl transfer to the sugar. Under these conditions IIA is largely in the non-phosphorylated form (Nelson *et al.*, 1986), which inhibits LacY and other transporters as indicated above. In the absence of PTS sugars, IIA remains phosphorylated by HPr~P since the phosphoryl group cannot be transferred to a sugar, which relieves inhibition of LacY, MelB, MalK (peripheral inner membrane protein of the maltose transport system) or glycerol kinase. Since interaction of IIA with LacY or one of the other systems requires the presence of substrate, IIA does not have to bind to each of these systems at the same time, which prevents escape from inducer exclusion by depletion of IIA. Furthermore, in the absence of appropriate substrate (inducer) the level of expression of the transport systems remains low.

Gram-positive bacteria: role of HPr(Ser-P) protein

In contrast to the Gram-negative enteric bacteria, where the glucose-specific phosphocarrier IIA^{Glc} has a central role in the regulation of carbohydrate uptake, in Gram-positive bacteria the serine-phosphorylated form of HPr [HPr(Ser-P)] seems to control carbohydrate metabolism both at the protein and the gene level. As explained more fully below HPr(Ser-P) regulates transcription and also modulates transport activities by causing a switch in the transport mechanism that prevents further uptake (*inducer exclusion*) or efflux of sugar (*inducer expulsion*) (Fig. 5). In Gram-positive bacteria, HPr cannot only be phosphorylated by PEP/EI on a histidine (His15~P), but, in addition, a metabolite-activated ATP-dependent protein kinase can phosphorylate a serine residue at position 46 in HPr (Deutscher & Saier, 1983). The activity of this HPr(Ser) kinase is dependent on divalent cations as well as on one of several intermediary metabolites, e.g. fructose 1,6-bisphosphate or gluconate 6-phosphate (Reizer *et al.*, 1984; Deutscher & Engelmann, 1984). Thus, the activity of HPr(Ser) kinase is directly linked to the glycolysis and/or the pentose phosphate pathway, thereby forming an intricate network to control sugar metabolism.

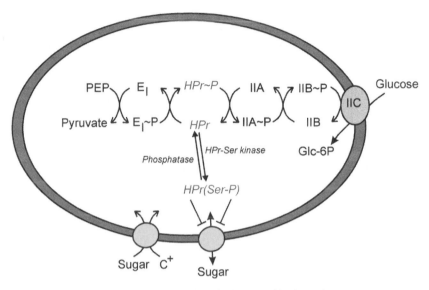

Fig. 5. Inducer expulsion in Gram-positive bacteria.

The *inducer exclusion* mechanism in Gram-negative bacteria prevents the sugar ('inducer of the operon') from entering the cell. In *inducer expulsion*, accumulated sugar is released from the cell following a switch in the coupling mechanism of the transport protein, e.g. from symport to uniport (Fig. 5). As the symport mode of transport allows the sugar to be accumulated ($[S]_{in} > [S]_{out}$), a subsequent switch to the uniport mode will lead to exit of sugar down the concentration gradient. *Lactobacillus brevis* and other heterofermentative lactobacilli exhibit this metabolite-activated sugar expulsion. Transport experiments suggest that the H^+–glucose and H^+–lactose symporters in the heterofermentative lactobacilli are mechanistically converted into uniport systems upon binding of HPr(Ser-P) to the cytoplasmic surface of the carrier protein (Romano *et al.*, 1987; Ye *et al.*, 1994). Binding of HPr(Ser-P) to these symporters only occurs in the presence of substrate (Ye & Saier, 1995a, b). As HPr(Ser-P) is high in cells that rapidly metabolize a PTS sugar, it will lead to the preferential utilization of this sugar over lactose or glucose.

In summary, for both Gram-positive and Gram-negative bacteria, it is clear that components of the PTS are major factors (signal transducers) in the regulation of carbohydrate metabolism. In Gram-negative bacteria external sugars are sensed, whereas in Gram-positive bacteria intracellular metabolites are sensed as a primary signal. A second signal is then transmitted to effect allosteric regulation of the target proteins. This signal is transmitted via the proteins IIB/HPr by (de)phosphorylation of IIA^{Glc} in Gram-negative bacteria, whereas in Gram-positive bacteria a phosphorylation signal is transmitted to HPr via the HPr kinase.

REGULATION BY PHOSPHORYLATION OF THE TRANSPORTER MOLECULE

Although the 'eukaryotic-type' of protein kinases are known in bacteria (Kennelly & Potts, 1996), there is no evidence for direct regulation of bacterial transport by serine, threonine or tyrosine phosphorylation. This type of kinase-mediated regulation is perhaps more rule than exception in the eukaryotic transport systems. Some prokaryotic secondary sugar transporters, however, are regulated by histidine phosphorylation in a domain that is homologous to IIA^{Glc} of *E. coli*. To this family of transporters belong the lactose transport protein LacS of *Streptococcus thermophilus*, *Lactobacillus bulgaricus* and *Leuconostoc lactis* and the raffinose transport protein RafP of *Pediococcus pentosaceus*. Interestingly, these proteins are, amongst others, homologous to the melibiose transport proteins of *Salmonella typhimurium* and *E. coli*, which lack a IIA-like domain, but are regulated by IIA^{Glc} as discussed above (Poolman et al., 1996).

The IIA domain of LacS (and other members of the '*LacS subfamily*') has several structural features in common with the corresponding PTS proteins and can be phosphorylated in the presence of PEP, enzyme I and HPr (Poolman et al., 1992). Phosphorylation of the *Streptococcus thermophilus* LacS IIA domain results in an inhibition of lactose–H^+ symport (Poolman et al., 1995a). Since in vivo HPr $\sim$ P is formed not only from the metabolism of a PTS sugar but also from lactose, it has been proposed that regulation of LacS activity by HPr $\sim$ P serves to prevent the unbridled uptake of lactose (control of glycolysis). This negative feedback control, resulting from the phosphorylation of the IIA domain of LacS, contrasts the regulation by IIA^{Glc} in enteric bacteria, which determines the hierarchy of sugar utilization (Poolman et al., 1995a).

OSMOTIC REGULATION OF TRANSPORT

Uptake of compatible solutes

To survive osmotic stresses, micro-organisms need to adapt by accumulating specific solutes under hyperosmotic conditions and releasing them under hypoosmotic conditions. Such solutes include K^+, amino acids (e.g. glutamate, proline), amino acid derivatives (peptides, *N*-acetylated amino acids), quaternary amines (e.g. glycine betaine, carnitine), sugars (e.g. sucrose, trehalose) and tetrahydropyrimidines (ectoines) (Csonka, 1989; Galinski & Trüper, 1994). These solutes are often referred to as *compatible solutes* because they can be accumulated to high levels by *de novo* synthesis or transport without interference with vital cellular processes. In fact, many compatible solutes proved to be effective stabilizers of enzymes, providing protection not only against high salt, but also against high temperature, freeze–thawing and drying (Yancey et al., 1982). The cells may synthesize (some of the) compatible solutes following an osmotic upshock and degrade them following an osmotic downshift, but the initial response is much more rapid if compatible solutes can be taken up from the medium and/or released into the medium via semi-constitutive transport systems that are regulated by the medium osmolality.

Upon a change in the medium osmolality, a membrane transport system may be (in)activated by one or more of the following physico-chemical parameters: (i) *external osmolality* or *water activity*; (ii) *turgor pressure*, i.e. the difference between the extra- and intracellular potential of all osmotically active solutes, which affects the compression of the membrane against the cell wall; (iii) *membrane strain*, which occurs in response to a change in turgor pressure and affects the compression/expansion of the bilayer in the plane of the membrane; (iv) *internal hydrostatic pressure*; (v) *internal osmolality* or *water activity*; (vi) *cytoplasmic volume*; and (vii) *concentration of specific cytoplasmic signal molecule(s)*. The complexity of osmoregulated transport is illustrated by describing the properties of the major systems for uptake and efflux of compatible solutes in *Lactobacillus plantarum* and *Listeria mono-cytogenes*.

In *Lactobacillus plantarum* glycine betaine, carnitine and proline are taken up via one and the same system (QacT) that is activated upon osmotic upshock (Glaasker et al., 1996a, b). The increase in uptake rate from 'basal' to 'activated' upon an increase in medium osmolality reflects an increase in maximal activity (V_{max} increases five- to tenfold), but it involves more than a single effect (Glaasker et al., 1998b). It appears that the increase in V_{max} upon osmotic upshock is due to a diminished *trans*-inhibition (by proline and/or glycine betaine) (Fig. 6; Glaasker et al., 1998b). The inhibition by intracellular glycine betaine, carnitine and proline forms a level of control against excessive accumulation of these solutes. This regulatory mechanism is probably used by all the ATP-dependent solute uptake systems. The linkage of the *trans*-inhibitory effect to the osmotic strength of the environment is rather unique. It allows the cell to tune the intracellular osmolality and maintain the cell turgor within certain limits.

A similar mechanism of regulation is found for the uptake of glycine betaine and carnitine in *Listeria monocytogenes*. In this organism, however, glycine betaine and carnitine are taken up via separate systems, i.e. an ATP-driven carnitine and an ion-motive-force-driven betaine uptake system (Verheul et al., 1997). These systems, although highly specific for their substrates at the outer surface of the membrane, are inhibited by both glycine betaine and carnitine at the cytoplasmic face (*trans* site) of the membrane. Without intracellular glycine betaine and/or carnitine the activity

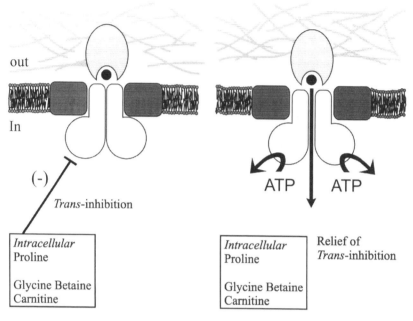

Fig. 6. Osmoregulated *trans*-inhibition of transporter for compatible solutes.

is maximal and not affected by medium osmolality. The inhibition by intracellular glycine betaine and carnitine is relieved upon osmotic upshock, which allows the cells to accumulate these compatible solutes further and restore turgor more rapidly. In kinetic terms, the activation of the glycine betaine and carnitine uptake systems upon osmotic upshock is thought to reflect an increase in K_I^{App} for the compatible solutes at the inner surface of the membrane. Apparently, a decrease in turgor alters the internal binding site for glycine betaine and carnitine.

In summary, the driving force for the glycine betaine or carnitine transport reaction is not a determining factor for the final level of accumulation of these compatible solutes. Rather the osmotic state of the cells controls the extent of accumulation through *trans*-inhibition of the transporters, a phenomenon that can be observed for both ATP-driven and ion-linked uptake systems.

Excretion of compatible solutes

It has been shown for a number of micro-organisms that compatible solutes are rapidly released from the cells upon a hypoosmotic shock. For instance, in *E. coli* a rapid release of K^+, glutamate and trehalose is observed upon an osmotic downshock, whereas solutes such as alanine, lysine, arginine and sucrose are fully retained by the cells (Schleyer *et al.*, 1993). When *Lactobacillus plantarum* is subjected to an osmotic downshock, a rapid efflux of glycine betaine, proline and some glutamate occurs, whereas the pools of other amino acids remain unaffected (Glaasker *et al.*, 1996a). Osmoregulated efflux activity with specificity for compatible solutes has also been described in *Corynebacterium glutamicum* (Lambert *et al.*, 1995; Ruffert *et al.*, 1997). Although the molecular nature of these efflux activities is unknown, the systems exhibit properties that mimic mechanosensitive channels (Sukharev *et al.*, 1997; see below). Some features that discriminate the systems from 'ordinary' secondary carrier proteins (Poolman & Konings, 1993) are the following: (i) efflux is extremely fast and effected by an osmotic downshock as well as amphipaths that insert into the membrane; (ii) efflux is independent of metabolic energy; (iii) efflux is unaffected by substrate at the *trans* site of the membrane; (iv) in many cases efflux is inhibited by gadolinium ions (Gd^{3+}), an unspecific channel blocker (Berrier *et al.*, 1992; Glaasker *et al.*, 1996b; Lambert *et al.*, 1995; Ruffert *et al.*, 1997). In the case of the glutamate excretion system from *C. glutamicum*, there is direct evidence that the effects of the osmotic gradients and the amphipath tetracaine are mutually compensative, i.e. the higher the medium osmolality the more tetracaine is needed to elicit efflux (Lambert *et al.*, 1995). These and other experiments suggest that osmotic changes and amphipaths exert a similar type of mechanical stress on the membrane (Martinac *et al.*, 1990; Sheetz *et al.*, 1976; Sheetz & Singer, 1974).

The efflux of glycine betaine and proline by *Lactobacillus plantarum* upon osmotic downshock is characterized by two kinetic components, i.e. one with a $t_{0.5} < 1$ s and a slow one with a $t_{0.5}$ of 4–5 min. Similar observations have been made for the efflux of glycine betaine and carnitine in *Listeria monocytogenes*. The component with the slow kinetics is affected by the metabolic state of the cell and may represent a specific efflux system or, alternatively, exit of compatible solutes via the same system that effects the uptake. The glycine betaine uptake system of *Listeria monocytogenes* is a secondary transport protein that under the appropriate conditions (e.g. hypoosmotic shock) may have a role in the release of compatible solutes. Thus, a lowering of the ion motive force as a result of an osmotic downshock would lead to net efflux. The QacT system of *Lactobacillus plantarum* and the carnitine uptake system of *Listeria monocytogenes* are ATP-driven, and these systems are generally thought to operate unidirectionally and not to mediate efflux. However, it is worth mentioning that in *Rhizobium leguminosarum* an ABC-type binding-protein-dependent amino acid uptake system (Aap) has been described that also affects efflux (Walshaw & Poole, 1996). At present it cannot be excluded that Aap-mediated efflux is indirect, i.e. through regulation of another channel or transport system, but the experimental data favour a direct role as the system mediated uptake and efflux in the heterologous host *E. coli*.

In summary, osmoregulation in bacteria involves the tailored regulation of uptake and efflux systems, which allows the organism to survive and grow over a broad range of medium osmolarities. It seems that some osmosensing mechanism forms an inherent property of the osmoregulated transport systems and channel proteins. It allows these systems to sense the osmotic value of the environment (or related parameter) and transduce the signal into an activity change of the transporter.

ACKNOWLEDGEMENTS

I would like to thank the members of my laboratory for valuable discussions. I thank Dr R. H. E. Friesen for critical reading of the manuscript. The work of the author was supported by grants from the Human Frontier Science Program Organization (HFSPO), the European Union (BIO-4-CT-960129/960439/960380) and the Dutch Science Foundation (NWO) as represented by Scheikundig Onderzoek Nederland and Algemene Levenswetenschappen.

REFERENCES

Anantharam, V., Allison, M. J. & Maloney, P. C. (1989). Oxalate:formate exchange: the basis for energy coupling in *Oxalobacter*. *Journal of Biological Chemistry* **264**, 7244–7250.

Berrier, C., Coulombe, A., Szabo, I., Zoratti, M. & Ghazi, A. (1992). Gadolinium ion inhibits loss of metabolites induced by osmotic shock and large stretch-activated channels in bacteria. *European Journal of Biochemistry* **206**, 559–565.

Csonka, L. N. (1989). Physiological and genetic responses of bacteria to osmotic stress. *Microbiological Reviews* **53**, 121–147.

Dean, D. A., Reizer, J., Nikaido, H. & Saier, M. H., Jr (1990). Regulation of the maltose transport system of *Escherichia coli* by the glucose-specific enzyme III of the phospho*enol*pyruvate:sugar phosphotransferase system. Characterization of inducer exclusion-resistant mutants and reconstitution of inducer exclusion in proteoliposomes. *Journal of Biological Chemistry* **265**, 21005–21010.

Deutscher, J. & Engelmann, R. (1984). Purification and characterization of an ATP-dependent protein kinase from *Streptococcus faecalis*. *FEMS Microbiology Letters* **23**, 157–162.

Deutscher, J. & Saier, M. H., Jr (1983). ATP-dependent protein kinase-catalyzed phosphorylation of a seryl residue in HPr, a phosphate carrier protein of the phosphotransferase system in *Streptococcus pyogenes*. *Proceedings of the National Academy of Sciences, USA* **80**, 6790–6794.

Driessen, A. J. M., Poolman, B., Kiewit, R. & Konings, W. N. (1987). Arginine transport in *Streptococcus lactis* is catalyzed by an arginine:ornithine antiporter. *Proceedings of the National Academy of Sciences, USA* **84**, 6093–6097.

Galinski, E. A. & Trüper, H. G. (1994). Microbial behaviour in salt-stressed ecosystems. *FEMS Microbiology Reviews* **15**, 95–108.

Glaasker, E., Konings, W. N. & Poolman, B. (1996a). Osmotic regulation of intracellular solute pools in *Lactobacillus plantarum*. *Journal of Bacteriology* **178**, 575–582.

Glaasker, E., Konings, W. N. & Poolman, B. (1996b). Glycine-betaine fluxes in *Lactobacillus plantarum* during osmostasis and hyper- and hypoosmotic shock. *Journal of Biological Chemistry* **271**, 10060–10065.

Glaasker, E., Tjan, F. S. B., Ter Steeg, P. F., Konings, W. N. & Poolman, B. (1998a). The physiological response of *Lactobacillus plantarum* towards salt and non-electrolyte stress. *Journal of Bacteriology* **180**, 4718–4723.

Glaasker, E., Heuberger, E. H. M. L., Konings, W. N. & Poolman, B (1998b). The mechanism of osmotic activation of the quaternary ammonium compound transporter (QacT) of *Lactobacillus plantarum*. *Journal of Bacteriology* **180**, 5540–5546.

Higgins, C. F. (1992). ABC transporters: from microorganisms to man. *Annual Review of Cell Biology* **8**, 67–113.

Hurley, J. H., Faber, H. R., Worthylake, D., Meadow, N. D., Roseman, S., Pettegrew, D. W. & Remington, S. J. (1993). Structure of the regulatory complex of *Escherichia coli* III^{Glc} with glycerol kinase. *Science* **259**, 673–677.

Jacobs, M. H. J., van der Heide, T., Driessen, A. J. M. & Konings, W. N. (1996). Glutamate transport in *Rhodobacter sphaeroides* is mediated by a novel binding protein-dependent secondary transport system. *Proceedings of the National Academy of Sciences, USA* **93**, 12786–12790.

Kaback, H. R. (1997). A molecular mechanism for energy coupling in a membrane transport protein, the lactose permease of *Escherichia coli*. *Proceedings of the National Academy of Sciences, USA* **94**, 5539–5543.

Kennelly, P. J. & Potts, M. (1996). Fancy meeting you here! A fresh look at 'prokaryotic' protein phosphorylation. *Journal of Bacteriology* **178**, 4759–4764.

Konings, W. N., Kaback, H. R. & Lolkema, J. S. (1996). *Transport Processes in Eukaryotic and Prokaryotic Organisms*, vol. 2. Amsterdam: Elsevier.

Kuroda, M., de Waard, S., Mizushima, K., Tsuda, M., Postma, P. & Tsuchiya, T. (1992). Resistance of the melibiose carrier to inhibition by the phosphotransferase system due to substitutions of amino acids in the carrier of *Salmonella typhimurium*. *Journal of Biological Chemistry* **267**, 18336–18341.

Lambert, C., Erdmann, A., Eikmanns, M. & Krämer, R. (1995). Triggering glutamate excretion in *Corynebacterium glutamicum* by modulating the membrane state with local anesthetics and osmotic gradients. *Applied and Environmental Microbiology* **61**, 4334–4342.

Lolkema, J. S. & Poolman, B. (1995). Uncoupling in secondary transport proteins: a mechanistic explanation for mutants of *lac* permease with an uncoupled phenotype. *Journal of Biological Chemistry* **270**, 12670–12676.

Lolkema, J. S., Poolman, B. & Konings, W. N. (1996). Secondary transporters and metabolic energy generation in bacteria. In *Transport Processes in Eukaryotic and Prokaryotic Membranes*, pp. 229–260. Edited by W. N. Konings, H. R. Kaback & J. S. Lolkema. Amsterdam: Elsevier.

Maloney, P. C., Ambudkar, S. V., Anantharam, V., Sonna, L. A. & Varadhachary, A. (1990). Anion exchange mechanisms in bacteria. *Microbiological Reviews* **54**, 1–17.

Martinac, B., Adler, J. & Kung, C. (1990). Mechanosensitive ion channels of *E. coli* activated by amphipaths. *Nature* **348**, 261–263.

Nelson, S. O., Wright, J. K. & Postma, P. W. (1983). The mechanism of inducer exclusion. Direct interaction between purified III^{Glc} of the bacterial phosphoenol-pyruvate:sugar phosphotransferase system. *EMBO Journal* **2**, 715–720.

Nelson, S. O., Schuitema, A. R. J. & Postma, P. W. (1986). The phosphoenolpyruvate:sugar phosphotransferase system of *Salmonella typhimurium*: the phosphorylated form of III^{Glc}. *European Journal of Biochemistry* **154**, 337–341.

Osumi, T. & Saier, M. H., Jr (1982). Regulation of lactose permease activity by the phospho*enol*pyruvate:sugar phosphotransferase system: evidence for direct binding of the glucose-specific enzyme III to the lactose permease. *Proceedings of the National Academy of Sciences, USA* **79**, 1457–1461.

Peterkofsky, A., Svenson, I. & Amin, N. (1989). Regulation of *Escherichia coli* adenylate cyclase activity by the phospho*enol*pyruvate:sugar phosphotransferase system. *FEMS Microbiology Reviews* **63**, 103–108.

Poolman, B. (1990). Precursor/product antiport in bacteria. *Molecular Microbiology* **4**, 1629–1636.

Poolman, B. & Konings, W. N. (1993). Secondary solute transport in bacteria. *Biochimica et Biophysica Acta* **1183**, 5–39.

Poolman, B., Driessen, A. J. M. & Konings, W. N. (1987a). Regulation of solute transport in *Streptococci* by external and internal pH values. *Microbiological Reviews* **51**, 498–508.

Poolman, B., Smid, E. J., Veldkamp, H. & Konings, W. N. (1987b). Bioenergetic consequences of lactose starvation for continuously cultured *Streptococcus cremoris*. *Journal of Bacteriology* **169**, 1460–1468.

Poolman, B., Molenaar, D., Smid, E. J., Ubbink, T., Abee, T., Renault, P. P. & Konings, W. N. (1991). Malolactic fermentation: electrogenic malate uptake and malate/lactate antiport generate metabolic energy. *Journal of Bacteriology* **173**, 6030–6037.

Poolman, B., Modderman, R. & Reizer, J. (1992). Lactose transport system of *Streptococcus thermophilus*: the role of histidine residues. *Journal of Biological Chemistry* **267**, 9150–9157.

Poolman, B., Knol, J., Mollet, B., Nieuwenhuis, B. & Sulter, G. (1995a). Regulation of bacterial sugar-H^+ symport by phosphoenolpyruvate-dependent enzyme I/HPr-mediated phosphorylation. *Proceedings of the National Academy of Sciences, USA* **92**, 778–782.

Poolman, B., Knol, J. & Lolkema, J. S. (1995b). Kinetic analysis of lactose and proton coupling in Glu-379 mutants of the lactose transport system of *Streptococcus thermophilus*. *Journal of Biological Chemistry* **270**, 12995–13003.

Poolman, B., Knol, J., van der Does, C., Henderson, P. J. F., Liang, W.-J., Leblanc, G., Pourcher, T. & Mus-Veteau, I. (1996). Cation and sugar selectivity determinants in a novel family of transport proteins. *Molecular Microbiology* **19**, 911–922.

Postma, P. W., Broekhuizen, C. P., Schuitema, A. R. J., Vogler, A. P. & Lengeler, J. W. (1988). Carbohydrate transport and metabolism in *Escherichia coli* and *Salmonella typhimurium*: regulation by the phospho*enol*pyruvate:sugar phosphotransferase system. In *Molecular Basis of Biomembrane Transport*, pp. 43–52. Edited by F. Palmieri & E. Quagliariello. Amsterdam: Elsevier.

Postma, P. W., Lengeler, J. W. & Jacobson, G. J. (1993). Phospho*enol*pyruvate:carbohydrate phosphotransferase systems of bacteria. *Microbiological Reviews* **57**, 543–594.

Reizer, J., Novotny, M. J., Hengstenberg, W. & Saier, M. H., Jr (1984). Properties of ATP-dependent protein kinase from *Streptococcus pyogenes* that phosphorylates a seryl residue in HPr, a phosphocarrier protein of the phosphotransferase system. *Journal of Bacteriology* **160**, 333–340.

Romano, A. H., Brino, G., Peterkofsky, A. & Reizer, J. (1987). Regulation of β-galactoside transport and accumulation in heterofermentative lactic acid bacteria. *Journal of Bacteriology* **169**, 5589–5596.

Ruffert, S., Lambert, C., Peter, H., Wendisch, V. F. & Kramer, R. (1997). Efflux of compatible solutes in *Corynebacterium glutamicum* mediated by osmoregulated channel activity. *European Journal of Biochemistry* **247**, 572–580.

Saier, M. H., Jr (1989). Protein phosphorylation and allosteric control of inducer exclusion and catabolite repression by the bacterial phospho*enol*pyruvate:sugar phosphotransferase system. *Microbiological Reviews* **53**, 109–120.

Schleyer, M., Schmid, R. & Bakker, E. P. (1993). Transient, specific and extremely rapid release of osmolytes from growing cells of *Escherichia coli* K-12 exposed to hypoosmotic shock. *Archives of Microbiology* **160**, 424–431.

Sheetz, M. P. & Singer, S. J. (1974). Biological membranes as bilayer couples. A molecular mechanism of drug-erythrocyte interactions. *Proceedings of the National Academy of Sciences, USA* **71**, 4457–4461.

Sheetz, M. P., Painter, R. G. & Singer, S. J. (1976). Biological membranes as bilayer couples. III. Compensatory shape changes induced in membranes. *Journal of Cell Biology* **70**, 193–203.

Sukharev, S. I., Blount, P., Martinac, B. & Kung, C. (1997). Mechanosensitive channels of *Escherichia coli*: the MscL gene, protein, and activities. *Annual Review of Physiology* **59**, 633–657.

Verheul, A., Glaasker, E., Poolman, B. & Abee, T. (1997). Betaine and L-carnitine transport in response to osmotic signals in *Listeria monocytogenes* Scott A. *Journal of Bacteriology* **179**, 6979–6985.

Walshaw, D. L. & Poole, P. S. (1996). The general L-amino acid permease of *Rhizobium leguminosarum* is an ABC uptake system that also influences efflux of solutes. *Molecular Microbiology* **21**, 1239–1252.

Yancey, P. H., Clark, M. E., Hand, S. C., Bowlus, R. D. & Somero, G. N. (1982). Living with water stress: evolution of osmolyte systems. *Science* **217**, 1214–1222.

Ye, J.-J. & Saier, M. H., Jr (1995a). Cooperative binding of lactose and the phosphorylated phosphocarrier protein HPr(Ser-P) to the lactose/H^+ symport permease of *Lactobacillus brevis*. *Proceedings of the National Academy of Sciences, USA* **92**, 417–421.

Ye, J.-J. & Saier, M. H., Jr (1995b). Allosteric regulation of the glucose:H^+ symporter of *Lactobacillus brevis*: cooperative binding of glucose and HPr(Ser-P). *Journal of Bacteriology* **177**, 1900–1902.

Ye, J.-J., Reizer, J., Cui, X. & Saier, M. H., Jr (1994). ATP-dependent phosphorylation of serine-46 in the phosphocarrier protein HPr regulates lactose/H^+ symport in *Lactobacillus brevis*. *Proceedings of the National Academy of Sciences, USA* **91**, 3102–3106.

ARSENIC TRANSPORT SYSTEMS FROM ESCHERICHIA COLI TO HUMANS

HIRANMOY BHATTACHARJEE, MALLIKA GHOSH, RITA MUKHOPADHYAY AND BARRY P. ROSEN

Department of Biochemistry and Molecular Biology, Wayne State University, School of Medicine, Detroit, MI 48201, USA

INTRODUCTION

As we enter the post-antibiotic era, where treatment of infectious diseases is limited by our ability to develop new antimicrobial agents, we should reflect on the fact that metal resistances were more common in the pre-antibiotic era than were antibiotic resistances (Hughes & Datta, 1983). This reflects the prevalence of arsenic and other metals in the environment, from both natural and man-made sources. Arsenic is common in geological formations and frequently contaminates water supplies. It is an active ingredient in a variety of commonly used insecticides, rodenticides and herbicides; for example, calcium methylarsenate is the sole active ingredient of Ortho Crabgrass Killer, Formula II. Arsenate is commonly used in ripening sprays for citrus crops, which contributes towards arsenic contamination of fruits and vegetables. Arsenicals are also present in wallpaper, paint, ceramics, glass and certain metal alloys. Finally, organic arsenicals were among the original antimicrobial agents synthesized by Paul Ehrlich, who won the Nobel Prize in 1908 for development of his Silver Bullet, the arsenical Salvarsan, the first effective treatment for syphilis.

Thus it should not be surprising that resistance to arsenicals has been found in bacteria, fungi, parasites and animals (Rosen, 1999). From genomic analysis of an ever-growing number of organisms, it now appears that arsenic resistance genes are found in nearly every member of all three kingdoms. Here we describe some of the recent developments in the area of resistance to the salts of the metalloids arsenic and antimony. It is likely that the *primordial soup* was rich in dissolved metals, so evolution of resistance to toxic metals was an imperative in the early development of life. One of the primary mechanisms for resistance to metals, xenobiotics and drugs is active extrusion of the toxic compounds from the cytosol (Dey & Rosen, 1995b). In bacteria this usually means extrusion from the cell, but in eukaryotes it can also be due to sequestration in an intracellular organelle.

Both chromosomal- and plasmid-encoded *ars* operons confer resistance to arsenite and antimonite in both Gram-negative and Gram-positive bacteria

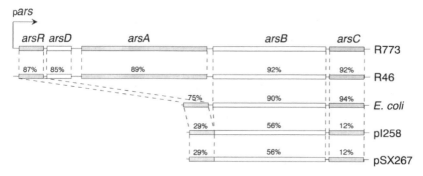

Fig. 1. Bacterial *ars* operons. Five representative *ars* operons are shown, two five-gene operons and three three-gene operons. In the top line the five genes of the *ars* operon of R773 are shown with the direction of transcription indicated by the arrow, starting with the promoter, p_{ars}. Genes are indicated by boxes, with the intergenic spaces as single lines. The genes of homologous *ars* operons of R46, the *E. coli* chromosomal operon, and staphylococcal pI258 and pSX267 are aligned below. The similarities of the gene products to the R773 proteins are given as % identity.

(Xu *et al.*, 1998). Resistance results from the active arsenical extrusion (Mobley & Rosen, 1982; Silver & Keach, 1982). The *ars* operons of *Escherichia coli* resistance factors R773 and R46 confer high-level arsenical resistance. Each has five genes, *arsR*, *arsD*, *arsA*, *arsB* and *arsC* (Bruhn *et al.*, 1996; Chen *et al.*, 1986) (Fig. 1). In contrast, others, including the chromosomal *ars* operon of *E. coli*, which confers a low level of arsenical resistance (Carlin *et al.*, 1995), and the *ars* operons from staphylococcal plasmids pI258 (Ji & Silver, 1992) and pSX267 (Rosenstein *et al.*, 1992), have only three genes, *arsR*, *arsB* and *arsC*. It appears that the acquisition of the two additional genes increased the effectiveness of the resistance determinants. The mechanisms of the products of the three- and five-gene operons differ in their ability to couple energy to the extrusion of arsenite and antimonite. The primary resistance protein produced by both operons is the 45 kDa ArsB arsenite carrier that, in the absence of ArsA, catalyses extrusion of the toxic metalloid oxyanion from cells coupled to the membrane potential (Fig. 2) (Dey & Rosen, 1995a; Kuroda *et al.*, 1997). In cells with five-gene operons the 63 kDa ArsA ATPase is produced and forms a complex with ArsB (Dey *et al.*, 1994b). This complex is an ATP-coupled arsenite extrusion pump (Dey *et al.*, 1994a). The fact that pumps can form much larger concentration gradients than secondary carriers most likely accounts for the higher resistance conferred by the five-gene operons. ArsR and ArsD are transcription repressors that respond to environmental arsenicals and antimonials (Chen & Rosen, 1997; San Francisco *et al.*, 1990; Shi *et al.*, 1994; Wu & Rosen, 1991, 1993). Finally, ArsC is a reductase that transforms arsenate [As(V)] to arsenite [As(III)], the substrate of the extrusion pump (Oden *et al.*, 1994). Thus ArsC expands the range of resistance to include more oxidized arsenic compounds.

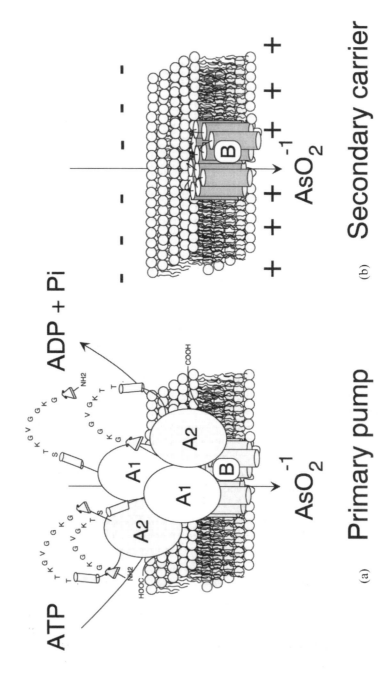

ATP

ADP + Pi

(a) Primary pump

(b) Secondary carrier

AsO_2^{-1}

AsO_2^{-1}

Fig. 2. Dual modes of energy coupling of the Ars arsenite transporter. (a) The complex of the ArsA and ArsB proteins forms an arsenite efflux ATPase. ArsA has two homologous halves, A1 (N-terminal) and A2 (C-terminal). Although two ArsA subunits are shown, the stoichiometry of the subunits of the pump has not been determined. ArsA is the catalytic subunit, exhibiting oxyanion-stimulated ATPase activity. The primary sequence of the two phosphate loops of the nucleotide-binding domains in the A1 and A2 halves is shown. ArsB, an inner membrane protein in *E. coli*, is the anion-conducting subunit of the ATP-coupled pump. (b) In the absence of ArsA, ArsB is a $\Delta\psi$-driven carrier protein.

ARSA STRUCTURE, FUNCTION AND EVOLUTION

ArsA encoded by R773 has been studied in detail. Under physiological conditions, ArsA is a part of a complex with ArsB in the inner membrane of *E. coli*. When expressed at high levels, ArsA is found predominantly as a soluble protein in the cytosol. This property has allowed for large-scale purification of ArsA protein, facilitating its characterization (Hsu & Rosen, 1989). The purified protein exhibits an ATPase activity that is allosterically activated specifically by arsenite or antimonite. Among the nucleotides, only ATP is a substrate, and only ATP, ADP and ATP analogues bind to the enzyme.

Analysis of the predicted amino acid sequence (583 residues) of ArsA indicates that the N-terminal A1 half of the protein exhibits similarity to the C-terminal A2 half (Fig. 3). This suggests that ArsA protein evolved by duplication and fusion of a primordial gene which was half the size of the current *arsA* gene (Chen *et al.*, 1986). Each half of ArsA contains a consensus sequence for the phosphate-binding loop (P-loop or Walker A motif) of a nucleotide-binding site (Walker *et al.*, 1982). Both A1 and A2 Walker A consensus sites in ArsA bind ATP, and both are required for resistance and enzyme activity, but it is not certain whether both sites are catalytic (Karkaria *et al.*, 1990; Karkaria & Rosen, 1991; Kaur & Rosen, 1992, 1994).

From a second site suppressor analysis, it appears that the two sites must interact to produce an active enzyme (J. Li *et al.*, 1996). The codon for Gly15 in the N-terminal consensus nucleotide-binding sequence was mutated to cysteine codon. Cells expressing an $arsA_{G15C}$ mutation resulted in substantial reductions in arsenite resistance, transport and ATPase activity. Selection for suppression of the G15C substitution that restored arsenite resistance yielded an A344V substitution. Ala344 is located adjacent to the C-terminal nucleo-tide-binding sequence. Cells expressing the G15C/A344V double mutant regained arsenite extrusion. These results again suggest a spatial proximity of Gly15 and Ala344 and support a model for interaction of the nucleotide-binding sites in ArsA. In further experiments, Gly15 and Ala344 were replaced with a variety of residues representing negatively charged, positively charged, neutral, polar or aromatic amino acids (Li & Rosen, 1998). Steric limitations on the interaction of ArsA ATP-binding sites were observed. The larger the residue volume at 15, the lower the resistance, with a G15R substitution producing the least resistance. On the other hand, the larger the residue at 344, the greater the suppression effect to G15C or G15R. These results are consistent with a physical interaction of the two nucleotide-binding domains and indicate that the geometry at the interface between the N- and C-terminal nucleotide-binding sites places spatial constraints on allowable residues at that interface.

Although the above experiments are suggestive of interactions of subunits, biochemical data clearly demonstrate that the ArsA protein functions as an

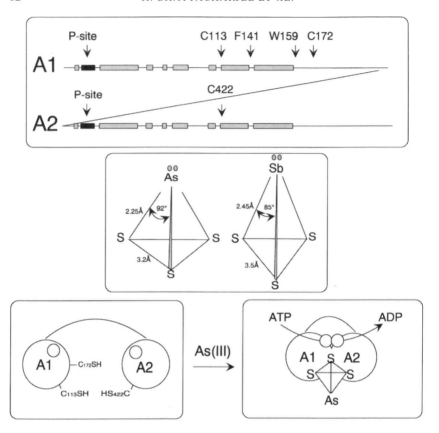

Fig. 3. Allosteric regulation of the ArsA ATPase. Top: ArsA has two homologous halves, N-terminal A1 and C-terminal A2, with the regions of greatest similarity boxed. Each half has a consensus sequence for the binding site of phosphoryl groups of ATP (P-site). The location of the three cysteine residues in the allosteric binding site is indicated. Also shown are Phe141 and Trp159, which enclose the DTAP domain. Middle: The structure of the soft metal–thiol complex in ArsA is postulated from the bond angles and distances found by crystallographic analysis of small molecules containing As–S or Sb–S bonds and from X-ray absorption spectroscopic results. This trigonal pyramidal structure contains three sulfur thiolates liganded to As(III) or Sb(III), with the metal at the apex. In ArsA the three sulfurs are the thiolates of Cys113, Cys172 and Cys422. Bottom: ArsA ATPase activity is allosterically regulated by As(III) or Sb(III). When metalloid is bound at the allosteric site, the two domains of the protein are brought together such that an A1 domain from one subunit forms an interface with an A2 domain from the other subunit. This brings the two ATP-binding sites into close proximity, promoting catalysis.

oligomer. Data from cross-linking experiments with *N*-ethoxycarbonyl-2-ethoxy-1,2-dihydroquinoline and light-scattering measurements suggested that ArsA is functionally a homodimer (Hsu *et al.*, 1991). Both intergenic and intragenic complementation experiments support the concept of interacting A1 and A2 ATP-binding sites (Kaur & Rosen, 1993; J. Li *et al.*, 1996), as do biochemical reconstitution experiments with the A1 and A2 domains expressed as separate polypeptides (Kaur & Rosen, 1994). We would therefore speculate that there are two active catalytic sites in an ArsA dimer, each formed by a pair of interacting ATP-binding sites. This hypothesis is similar to the mechanism of the H^+-translocating F_1-ATPase, in which residues from the α and β subunits form a catalytic site (Abrahams et al., 1994). It is not known whether the A1 and A2 ATP-binding sites are from intra- or intersubunit interactions. The possibility of intersubunit interaction was tested by mixing active and inactive ArsA proteins. Dimer formation between active and inactive subunits would decrease the overall activity of the ArsA mixture. However, no inhibition was observed (J. Li & B. P. Rosen, unpublished results), suggesting intrasubunit interaction between A1 and A2 sites within each monomer.

As discussed above, ArsA is allosterically activated by either arsenite or antimonite. The primary sequence of ArsA did not reveal the presence of any conserved metal-binding motifs. An indication that cysteines might be important for the activity of ArsA came from the extreme sensitivity to thiol reagents such as N-ethylmaleimide and methyl methanethiosulfonate (Bhattacharjee et al., 1995). Inhibition of enzymes by arsenite usually requires binding to the thiol groups of spatially proximate cysteine residues. This implies that for arsenite or antimonite to inhibit (or activate) an enzyme, the protein would be expected to have two or more cysteine residues in close proximity in the tertiary structure, if not in the primary sequence. In ArsA there are four cysteines, Cys26, Cys113, Cys172 and Cys422, all located in different regions of the primary sequence (Fig. 3). For arsenite or antimonite to act as an allosteric modulator, two or more of the cysteines should be close together in the folded protein to form As(III)– or Sb(III)–S bonds.

To investigate the role of cysteine residues in the allosteric activation of the ArsA protein, each of the four cysteines was altered to a serine residue by site-directed mutagenesis of the arsA gene (Bhattacharjee et al., 1995). The C26S protein had essentially the same properties as wild-type, indicating that Cys26 is not required for function. Cells expressing the other three mutations lost resistance to arsenite and antimonite. The C113S, C172S and C422S enzymes each had relatively normal K_m values for ATP. However, the concentration of oxyanion required for activation was substantially increased, most likely reflecting a decrease in affinity for oxyanion. Alteration of Cys422 appears to have more of an effect than alteration of Cys113 or Cys172. The C113S and C172S proteins exhibited a 20-fold increase in the concentration of antimonite required for half maximal activation, while the C422S required 200-fold more. These results suggest that Cys113, Cys172 and Cys422 are each involved in metalloactivation of ArsA. A reasonable model is that As(III) or Sb(III) preferentially bond with the sulfur ligands of Cys113, Cys172 and Cys422 to lock the enzyme into the active conformation (Fig. 3).

To be able to form this novel structure, Cys113, Cys172 and Cys422 must be within 3–4 Å of each other in the tertiary structure of ArsA. To determine the distance between cysteine residues, wild-type ArsA and ArsA proteins with cysteine to serine substitutions were treated with the bifunctional alkylating agent dibromobimane (Bhattacharjee & Rosen, 1996). Dibromobimane reacts with thiol pairs within 3–6 Å of each other to form a fluorescent adduct. ArsA proteins in which only one of the three essential cysteines was altered by site-directed mutagenesis still formed fluorescent adducts; thus in each altered enzyme there were still two cysteines that could react with dibromobimane. Proteins in which two of the three essential cysteine residues were substituted did not form fluorescent adducts, showing that there had to be at least two of the three to produce fluorescence. These results demonstrate that Cys113, Cys172 and Cys422 are within 6 Å of each other in the native enzyme, consistent with their forming a novel trigonal pyramidal three coordinate As(III)– or Sb(III)–S structure.

Wild-type ArsA was titrated with a stoichiometric amount of arsenite (or antimonite) and then analysed by arsenic (or antimony) X-ray absorption spectroscopy (R. A. Scott, H. Bhattacharjee & B. P. Rosen, unpublished results). This technique provides direct local structural information about the coordination environment of a selected element. The X-ray absorption spectroscopy studies clearly indicate that As(III) is bound to three sulfur ligands in ArsA, each with an As(III)–S distance of 2.25 Å. There are no oxygen ligands; thus neither serine nor threonine residues can be arsenic ligands. Similar results were obtained with Sb(III); the Sb(III)–S distance was found to be 2.45 Å. These experiments suggest that a novel As(III) [or Sb(III)]–thiol structure is involved in metalloregulation of ArsA. ArsA is the first and only example of a protein that requires As(III) or Sb(III) for maximal activity.

How does binding of As(III) or Sb(III) at the allosteric site stimulate ATP hydrolysis? Obviously, there must be communication between the metal-binding domain and ATP-binding sites. A 12-residue consensus sequence (DTAPTGHTIRLL) has been identified in ArsA homologues from eubacteria, archaea, fungi, plants and animals (Zhou & Rosen, 1997). The high degree of sequence conservation implies that this DTAP motif has a conserved function. In ArsA there are two DTAP motifs, one in A1 and the other in the A2 half of the protein. Intrinsic tryptophan fluorescence was used as a spectroscopic probe to monitor conformational changes in ArsA during catalysis (Zhou *et al.*, 1995; Zhou & Rosen, 1997). ArsA has four tryptophans, Trp159, Trp253, Trp522 and Trp524. Of the four, only Trp159, located at the C-terminal side of the DTAP motif in A1, responds to addition of ATP or effectors. To determine the conformational changes of the DTAP domain during ATP hydrolysis, two single-tryptophan-containing ArsA proteins were constructed and expressed. One of the altered ArsA enzymes had a single tryptophan at position 141, the N-terminal side of the

DTAP domain, while the second enzyme had a single tryptophan at position 159, the C-terminal side of the domain (Fig. 3). Fluorescence experiments show that in the absence of ligands, Trp141 is in a relatively nonpolar environment, and Trp159 is in a relatively hydrophilic environment. During ATP hydrolysis the C-terminal end of the conserved domain moves into a less polar environment, whereas the N-terminal end moves into a more hydrophilic environment as product is formed. These results suggest that the conserved domain experiences a rotational movement during the catalytic cycle.

A model for allosteric activation is that the two halves of the protein must be in contact with each other to be active. In the absence of effector, arsenite or antimonite, the A1 and A2 domains of ArsA protein are free to move independently of each other. At this point ArsA has a basal level of ATPase activity. Arsenite or antimonite pulls the two domains together by interacting with Cys113 and Cys172 in the A1 half and Cys422 in the A2 half. This in turn brings the two nucleotide-binding sites into close contact with each other, thereby accelerating catalysis. The release of energy from ATP hydrolysis in ArsA is transduced into the ArsB subunit of the pump, thereby driving transport of oxyanions.

Sequence analyses have now revealed the ubiquitous presence of ArsA homologues in all organisms – prokaryotes, archaea and eukaryotes. However, the physiological functions of these proteins are largely unknown. It appears that three types of proteins with common motifs arose from a common ancestor. One branch led to proteins involved in cell division in bacteria, such as MinD (de Boer *et al.*, 1991). The second branch includes proteins involved in nitrogen fixation, such as the *nifH* gene product, dinitrogen reductase (Murphy *et al.*, 1993). The third branch consists of proteins closely related to bacterial ArsA (Fig. 4). The ArsA family can be subdivided into representatives from the three kingdoms. The bacterial ArsAs arose from gene duplication and fusion of the common ancestor A*, producing a gene product with A1 and A2 domains. Examples include the ArsAs from R46 (Bruhn *et al.*, 1996) and R773 (Chen *et al.*, 1986) and the chromosomally encoded ArsA from *Acidiphilium multivorum* (Suzuki *et al.*, 1998), which share approximately 70% sequence similarity.

The eukaryotic ArsAs are a subclass within the ArsA superfamily. All known eukaryotic sequences are single A* domain proteins. Mammalian ArsA homologues have been identified in humans (Kurdi-Haidar *et al.*, 1996, 1998) and mice (H. Bhattacharjee & B. P. Rosen, unpublished results). From genome projects other eukaryotic homologues have been identified, including ones from the soil nematode *Caenorhabditis elegans*, the malaria parasite *Plasmodium falciparum* and the yeast *Saccharomyces cerevisiae*.

A common feature of the ArsA homologues is the presence of the nucleotide-binding domain (GKGGVGKT). As discussed earlier, *E. coli* R773 ArsA has two such motifs, one in the N-terminal and the other in the

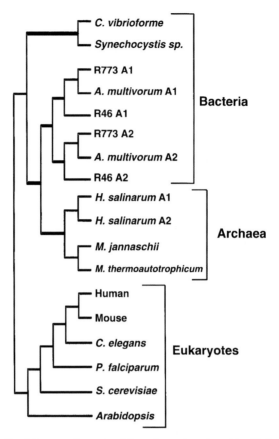

Fig. 4. ArsA family of ATPases. Many ArsA homologues have been identified, and representative ones are shown. Accession numbers are given in parentheses below. The dendrogram was made using MegAlign (DNAstar). The branch lengths in the phylogenetic tree are proportional to the number of amino acid substitutions separating each pair. Arsenic resistance ATPases include the closely related proteins encoded by R773 (J02591), R46 (U38947) and *Acidiphilium multivorum* (AB004659), which are the result of gene duplication and fusion, producing A1 and A2 halves, each of which has a consensus nucleotide-binding sequence. The protein from the archaeon *Halobacterium salinarum* (AF016485) has the same duplicated domain structure. All of the other ArsA homologues have a single A* and are approximately half the size of the R773 enzyme. Additional bacterial enzymes are from *Chlorobium vibrioforme* (U09867) and *Synechocystis* sp. (D64004). Additional archaeal enzymes are from *Methanococcus jannaschii* (2128876) and *Methanobacterium thermoautotrophicum* (AE000911). Eukaryotic ArsA homologues include those from *Homo sapiens* (U60276), *Mus musculus* (AF039405), *Caenorhabditis elegans* (P30632), *Plasmodium falciparum* (AL010250), *Saccharomyces cerevisiae* (S67642) and *Arabidopsis thaliana* (AB005246).

C-terminal half of the protein, both necessary for ArsA activity, while all eukaryotic ArsA homologues have just a single ATP-binding domain. In addition to the nucleotide-binding domains, there is a highly conserved region called the DTAP motif. As discussed above, in the R773 ArsA, this domain undergoes significant conformational changes during ATP hydrolysis and may be involved in coupling the allosteric site to the catalytic site. There are other conserved sequences within these ArsA homologues. For example, the consensus sequence (S/T)(S/T)DPA is found in equivalent positions in the bacterial arsenite resistance ArsAs and in all eukaryotic homologues. However, their functions are unknown. Although the eukaryotic homologues have the conserved nucleotide-binding domain and the DTAP domain, the location of a putative metalloregulatory site in this class of proteins is not clear. Cys113, Cys172 and Cys422, which form the site of interaction with As(III) or Sb(III) in the R773 ArsA, are not conserved in the eukaryotic counterparts. In contrast, eukaryotic homologues have conserved CXC and CXXC motifs that are absent in the prokaryotic enzymes and could serve as high-affinity metalloid-binding sites.

Very little is known about the biological function or the biochemical properties of eukaryotic ArsA homologues. Biochemical analysis of the human protein (hArsA) shows that it has a low level of ATPase activity that is activated 1.5-fold by arsenite (Kurdi-Haidar et al., 1996, 1998). This protein has been suggested to be involved in resistance to arsenite and cisplatin, but the mechanism is not clear. A mouse homologue (mArsA) has been cloned and expressed in our laboratory. The purified protein exhibits a low level of ATPase activity. The mouse enzyme exhibits 98% homology with hArsA and has an additional 16 residues at its N-terminus that are absent in the human enzyme. A second version of the human enzyme (accession number AF047469) has been recently reported that has an N-terminus identical to the mouse protein. A gene encoding an ArsA homologue has also been located in chromosome IV of the yeast Sacch. cerevisiae (Boskovic et al., 1996). The putative yArsA has 354 amino acid residues and shows 29% identity and 49% similarity with either the A1 or A2 domains of the E. coli R773-encoded ArsA. This gene was disrupted in our laboratory, but no metal-related phenotype has been observed (J. Shen & B. P. Rosen, unpublished results). Thus the physiological role of the yeast ArsA homologue remains unknown.

Archaeal ArsA homologues identified from sequencing projects show interesting sequence similarities to both the prokaryotic and eukaryotic proteins. All have P-loops and (S/T)(S/T)DPA and DTAP consensus sequences. The Halobacterium salinarum ArsA has an A1–A2 organization and also has a cysteine triad corresponding to the R773 ArsA. In contrast, both methanogenic archaea Methanococcus jannaschii and Methanobacterium thermoautotrophicum have the single A* domain structure of the eukaryotic ArsA and also lack the cysteine triad of the bacterial enzyme. Instead they have the CXXC sequence found in the eukaryotic homologues but do not have the conserved CXC sequence. In conclusion, the ubiquity of the arsA gene suggests a function of the gene product that is required by most organisms. However, until the physiological function of eukaryotic ArsA homologues is determined, a role of these proteins in metal resistance remains speculative.

From the energetics of arsenite extrusion, measured both *in vivo* and *in vitro*, the ArsA–ArsB complex was described as a primary efflux pump coupled to ATP (Dey *et al.*, 1994a; Dey & Rosen, 1995a; Mobley & Rosen, 1982) (Fig. 2). The existence of *ars* operons lacking an *arsA* gene suggested that ArsB might be capable of functioning in the absence of ArsA. From its hydropathic profile, ArsB was predicted to be an integral membrane protein that presumably forms the anion-conducting pathway for arsenite transport. The topological arrangement of ArsB was determined by using a series of gene fusions to *phoA*, *blaM* and *lacZ* genes (Wu *et al.*, 1992), with 12 membrane-spanning α-helices (TMs) and the N- and C-termini located in the cytosol. Interestingly, this secondary structure is more similar to that of secondary transporters than primary pumps (Maloney & Wilson, 1993). The data from studies of the energetics of ArsB-mediated arsenite transport support the suggestion that ArsB (in the absence of ArsA) catalyses arsenite transport coupled to the membrane potential, perhaps as an electrophoretic uniporter (Dey & Rosen, 1995a; Kuroda *et al.*, 1997). Since the membrane potential is the driving force, the transported species is most likely an anion such as arsenite rather than the soft metal form As(III).

Resistance conferred by the staphylococcal plasmids pI258 and pSX267 also results from active extrusion of arsenite, and the Gram-positive homologues of ArsB most likely function as secondary anion carriers (Broer *et al.*, 1993). Although the R773 and pI258 ArsB proteins are only 58% similar in sequence, secondary structural predictions demonstrate that the two proteins are topologically much more similar, and functional chimeras of the two have been constructed, suggesting that the ArsB proteins function similarly at the biochemical level (Dou *et al.*, 1994). Thus it is clear that the ArsBs of Gram-positive and Gram-negative bacteria are both evolutionarily and functionally related, forming a single family of arsenite transporters (Fig. 5).

Recently another family of arsenite resistance membrane proteins has been identified (Fig. 5). An *ars* operon from *Bacillus subtilis* encodes a 38 kDa membrane protein that exhibits no significant sequence similarity to ArsB (Sato & Kobayashi, 1998; Takemaru *et al.*, 1995). This 346-residue protein is homologous (30% identity) to the 404-residue *ACR3* gene product of *Sacch. cerevisiae*. *ACR3* is part of a cluster of three arsenic resistance genes (*ACR1*, *ACR2* and *ACR3*) on chromosome XVI (Bobrowicz *et al.*, 1997). *ACR1* encodes a putative transcription factor. As discussed below, *ACR2* is required for arsenate resistance and encodes a 130-residue arsenate reductase (Mukhopadhyay & Rosen, 1998). None of the three gene products shows any significant sequence similarity to the R773 or pI258 *ars* gene products.

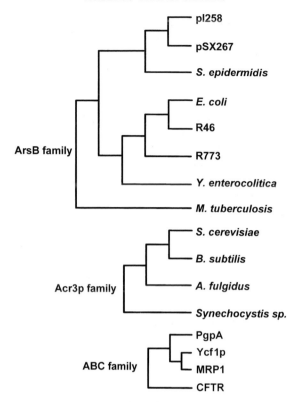

Fig. 5. Families of arsenite carrier proteins. Shown are representative members of the three independently evolved families of arsenite transporters, with accession numbers in parentheses below. The dendrogram was made using MegAlign (DNAstar). The branch lengths in the phylogenetic tree are proportional to the number of amino acid substitutions separating each pair. Top: The ArsB family. pI258, *Staphylococcus aureus* pI258 (M86824); pSX267, *Staphylococcus xylosus* pSX267 (M80565); *Staphylococcus epidermidis* (S57521); *E. coli* (U00039); *E. coli* R46 (U38947); *E. coli* R773 (J02591); *Yersinia enterocolitica* (U58366); *Mycobacterium tuberculosis* (Z96072). Middle: The Acr3p family. *Saccharomyces cerevisiae* Acr3p (Q06598); *Bacillus subtilis* skin element (Z99117); *Archaeoglobus fulgidus* (AE001071); *Synechocystis* sp. (D90914). Bottom: ABC ATPases. Three members of this large superfamily have been shown to confer arsenite resistance, including PgpA of *Leishmania tarentolae* (P21441); the *Sacch. cerevisiae* Ycf1p cadmium resistance protein (S51863); and the human multidrug-resistance-associated protein 1 (MRP1) (2828206). A homologue with a completely different function is the human cystic fibrosis transmembrane conductance regulator (CFTR) (L49339).

Acr3p is most likely an arsenite exporter with a function similar to the bacterial ArsB proteins (Wysocki *et al.*, 1997). In addition to its homology with the *B. subtilis ars* gene product, Acr3p also shares 37% identity with a putative membrane protein of unknown function from *Mycobacterium tuberculosis* and 26% identity with a hypothetical protein from the cyanobacterium *Synechocystis* (Wysocki *et al.*, 1997). In contrast to the ArsB proteins, which have 12 TMs, Acr3p and its homologues are predicted to have only 10 TMs. It appears that the specificity of the yeast resistance determinant is different from the bacterial *ars* operons: a strain with an *ACR3* disruption was sensitive to arsenite but not antimonite, in contrast to the antimonial resistance conferred by the bacterial *ars* operons.

We have recently demonstrated that wild-type yeast cells accumulate only small amounts of $^{73}AsO_2^{-1}$, reflecting active extrusion of the metalloid (Ghosh *et al.*, 1999). In contrast, *ACR3*-disrupted cells accumulate high amounts of radioactive arsenite, demonstrating that they have lost the ability to extrude the oxyanion. This is consistent with the function of Acr3p being a plasma membrane arsenite efflux protein. To examine the Acr3p activity *in vitro*, we prepared crude membrane vesicles from yeast. We anticipated that these vesicles would accumulate $^{73}AsO_2^{-1}$, as we have shown for ArsB-catalysed transport in everted plasma membrane vesicles of *E. coli*. Indeed a crude membrane vesicle preparation from yeast exhibited transport of $^{73}AsO_2^{-1}$. Transport required both ATP and GSH, suggesting ATP-driven uptake of As(GS)₃. Contrary to our expectations, however, disruption of *ACR3* had no effect on transport in this crude membrane preparation.

Obviously a transport system other than Acr3p had to be responsible for the transport in membrane vesicles. We have recently demonstrated that the protein that catalyses this transport reaction is Ycf1p, the yeast cadmium resistance factor (Szczypka *et al.*, 1994). Ycf1p is a member of the ABC superfamily of transport ATPases and has been shown to pump $Cd(GS)_2$ conjugates into the yeast vacuole, producing cadmium resistance (Z. S. Li *et al.*, 1996, 1997). We have found that *YCF1*-disrupted yeast cells are as sensitive to arsenate and arsenite as *ACR3*-disrupted cells, and a double *ACR3/YCF1* disruption is hypersensitive to arsenicals. Interestingly, the *YCF1* disruption was antimonite sensitive, while the *ACR3* disruption was not. This suggests that Ycf1p pumps antimonite, as does the ArsAB complex. Cells with a *YCF1* disruption were found to extrude $^{73}AsO_2^{-1}$ as efficiently as its parent, as measured by the lack of uptake of arsenite. This is in contrast to an *ACR3*-disrupted strain, which accumulated high amounts of arsenite, indicating lack of ability to extrude arsenite. Crude membranes prepared from the *YCF1*-disrupted strain exhibited loss of accumulation of $^{73}AsO_2^{-1}$ in vesicles. The membranes were fractionated into plasma and vacuolar membranes. All of the transport activity was localized to the vacuolar membranes, and the vacuoles from a *YCF1*-disrupted strain had no transport. Thus, in both bacteria and yeast, there is a plasma membrane exporter for arsenite, either ArsB or Acr3p. However, in yeast, arsenite is also pumped into intracellular organelles by Ycf1p (Fig. 6). We propose that this model will be applicable to eukaryotes in general. The closest homologue of Ycf1p is the human MRP (multidrug-resistance-associated protein) (63 % identity), which also confers arsenite resistance (Cole *et al.*, 1992). In normal human physiology MRP is responsible for export of GS-conjugates such as leukotriene C4 (LTC4) (Deeley & Cole, 1997). MRP-catalysed export of

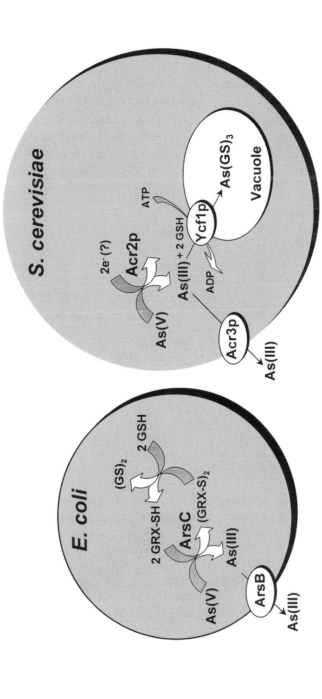

Fig. 6. Arsenical detoxification mechanisms in prokaryotes and eukaryotes. In both *E. coli* and yeast arsenite is extruded from cells by a carrier protein, ArsB and Acr3p, respectively. In both organisms arsenate is reduced to arsenite by a reductase, ArsC and Acr2p, respectively. Although the pathways of resistance are similar in overall design, neither of the two prokaryotic proteins is related evolutionarily to its eukaryotic analogue. In addition, a second arsenite resistance mechanism exists in *Sacch. cerevisiae*, which sequesters As(GS)$_3$ in the vacuole catalysed by the ABC ATPase Ycf1p.

glutathione from cells was increased by arsenite, suggesting that MRP functions as an As(GS)$_3$ carrier (Zaman *et al.*, 1995). MRP homologues in eukaryotic microbes also confer arsenite resistance. In *Leishmania* increased expression of *pgpA*, which encodes another homologue of Ycf1p (33% identical) and MRP (35% identical), also produces arsenite resistance (Ouellette & Borst, 1991; Ouellette *et al.*, 1990), suggesting that, like Ycf1p, it pumps arsenite into the parasite vacuole. Thus this sub-group of the ABC superfamily appears to form a third family of arsenical resistance transporters (Fig. 5). Considering the ubiquity of arsenic in the environment, nearly all organisms are continually exposed to this metalloid, making resistance mechanisms a necessity. We predict that eukaryotes – from *Leishmania* to humans – have both an Acr3p-like plasma membrane arsenite exporter and an ABC transporter that pumps arsenite into an intracellular compartment.

ARSC STRUCTURE, FUNCTION AND EVOLUTION

While both arsenate and arsenite are salts of arsenic, they contain the $+V$ and $+III$ oxidation states of arsenic, respectively, and are chemically dissimilar. Yet, in both prokaryotes and eukaryotes, organisms resistant to one oxidation state of the metalloid are cross-resistant to the other (Cole *et al.*, 1994; Dey *et al.*, 1994c; Wang & Lee, 1993; Wang *et al.*, 1996; Wang & Rossman, 1993). As discussed above, arsenite resistance is the result of active extrusion, while arsenate must first be reduced to arsenite to produce resistance. This suggests that many organisms have arsenate reductases.

Arsenate reductases appear to have evolved independently three times. The *Staphylococcus aureus* pI258 ArsC exhibits no significant sequence similarity to the R773 enzyme. The two enzymes appear to be the result of independent evolution and are members of different families of proteins (Fig. 7). Interestingly, many of the members of the Gram-positive group are low molecular weight protein tyrosine phosphatases (lmw PTPases) (Wo *et al.*, 1992a, b). As described below, a third has recently been identified in yeast (Fig. 7). This family includes a protein, Acr2p, that we have demonstrated to be an arsenate reductase, as described below. Acr2p is related to the Cdc25 superfamily of protein phosphotyrosyl phosphatases involved in cell cycle control (Fauman *et al.*, 1998; Hofmann *et al.*, 1998). Although both proteins are phosphotyrosyl phosphatases, it is not clear whether the lmw PTPases and Cdc25 phosphatase families are related to each other. The fact that two different arsenate reductases are related to phosphatases suggests common features in the enzymic mechanisms of reductases and phosphatases.

The ArsC enzyme encoded by *E. coli* R773 has been extensively characterized. It has been shown to require glutathione *in vivo* (Oden *et al.*, 1994). The enzyme has been purified and shown to require both GSH and glutaredoxin 1 (Grx1) as the source of reductant (Gladysheva *et al.*, 1994; Liu & Rosen,

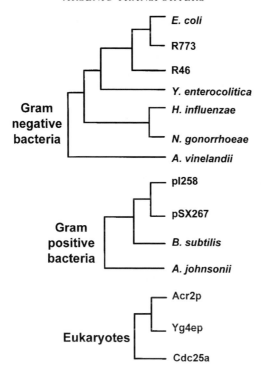

Fig. 7. Families of arsenate reductases. Three families of arsenate reductases have been identified. Accession numbers are given in parentheses below. The dendrogram was made using MegAlign (DNAstar). The branch lengths in the phylogenetic tree are proportional to the number of amino acid substitutions separating each pair. Top: In Gram-negative bacteria such as *E. coli* the ArsC enzyme (U00039) is a 15.8 kDa glutaredoxin-linked reductase. Other representative Gram-negative arsenate reductases include those from *E. coli* R773 (J02591); *E. coli* R46 (U38947); *Yersinia enterocolitica* (U58366); *Haemophilus influenzae* (U32710); *Neisseria gonorrhoeae* (U76418); and *Azotobacter vinelandii* (J03411). Middle: In Gram-positive bacteria ArsC is a 14.8 kDa thioredoxin-linked enzyme such as those encoded by pI258 (P30330) from *Staphylococcus aureus*, pSX267 (M80565) from *Staphylococcus xylosus*; and the *Bacillus subtilis* skin element (P45947). These are homologous to a family of low molecular weight phosphotyrosyl phosphatases that includes *Acinetobacter johnsonii* ptp (Y15162). Bottom: In *Sacch. cerevisiae* Acr2p (Q06597) is a 14.8 kDa arsenate reductase homologous to another yeast protein of unknown function, Yg4ep (S53926), and to the human Cdc25a (3660166), a protein phosphotyrosyl phosphatase.

1997). No *in vivo* requirement for Grx1 was demonstrated: a *grx1* disruption had no effect on *ars*-mediated arsenate resistance. However, *E. coli* has three glutaredoxin genes (Åslund *et al.*, 1994). We have recently shown that both Grx2 and Grx3 will substitute for Grx1 in enzymic assays of ArsC activity (J. Shin, T. Gladysheva & B. P. Rosen, unpublished results). Thus, all three glutaredoxins can participate in arsenate reduction. The glutaredoxins are rereduced by GSH. *In vitro* the ArsC enzyme of *Staph. aureus* pI258 enzyme uses thioredoxin as a reductant (Ji *et al.*, 1994). Thus the source of reductant differs between Gram-positive and Gram-negative bacteria: the R773 ArsC cannot use thioredoxin, and the pI258 ArsC cannot use glutaredoxin. The source of reducing potential in different organisms may relate more to the reductants available in the host strains than to the mechanism of arsenate reduction. Like all Gram-positive bacteria, *Staph. aureus* contains little or no glutathione, but glutathione is the major intracellular reductant in *E. coli*, present in millimolar concentrations (Fahey & Sundquist, 1991).

From the substrate specificity of the reduction reaction, ArsC apparently discriminates between arsenate and other oxyanions, including antimonate, nitrate, phosphate and sulfate. Although the other oxyanions tested were not reduced, phosphate and sulfate inhibited reduction. Inhibition appeared competitive, indicating that the oxyanion-binding site of ArsC may be rather nonspecific (Gladysheva *et al.*, 1994). The pathway of electron transfer (Fig. 6) and the inhibition by *N*-ethylmaleimide suggests that a redox-active thiol may participate in ArsC-catalysed reduction, and Cys12 was shown to be required for reductase activity (Liu *et al.*, 1995).

Resistance to arsenate in eukaryotes is less well characterized. As discussed above, a gene cluster on chromosome XVI of *Sacch. cerevisiae* termed *ACR1*, *ACR2* and *ACR3* confers resistance to both arsenite and arsenate (Bobrowicz *et al.*, 1997). Expression of *ACR2* from a plasmid in wild-type yeast resulted in a slight increase in arsenate resistance, leading to the suggestion that *ACR2* might be an arsenate reductase (Wysocki *et al.*, 1997). We have disrupted *ACR2* and showed that it results in sensitivity to arsenate but not to arsenite (Mukhopadhyay & Rosen, 1998). The *ACR2* gene was cloned and introduced into wild-type and *ACR2*-disrupted cells on a plasmid. The arsenate-sensitive phenotype of *ACR2*-disrupted cells was complemented by expression of *ACR2* on a plasmid. To examine the properties of Acr2p, the *ACR2* gene was fused to the bacterial *malE* gene at the 5' end and to six histidine codons at the 3' end. The gene fusion was expressed in *E. coli*, and the chimeric protein was purified. When supplemented with cytosol from *Sacch. cerevisiae*, the purified protein reduced arsenate to arsenite. The source of reducing potential for the Acr2p-catalysed reaction has yet to be determined. These results suggest that Acr2p is a eukaryotic equivalent of the bacterial ArsC arsenate reductases (Fig. 6).

This raises the possibility that arsenate reductases are required for arsenate resistance in other eukaryotes. It has been shown that glutathione will nonenzymically reduce arsenate to arsenite and then spontaneously form $As(GS)_3$ (Delnomdedieu *et al.*, 1993, 1994; Scott *et al.*, 1993). Since both prokaryotes and eukaryotes all have high levels of cytosolic reduced thiols, usually in the form of GSH, perhaps nonenzymic reduction is sufficient for resistance. This is clearly not true in bacteria, where ArsC is obligatory for arsenate resistance. Both *Leishmania* (Dey *et al.*, 1994c) and hamster cells (Wang *et al.*, 1996; Wang & Rossman, 1993) selected for arsenite resistance are cross-resistant to arsenate, as are human cells overexpressing MRP (Cole *et al.*, 1994). Is a reductase involved in eukaryotic arsenate resistance? In yeast *ACR2* is clearly required for resistance to arsenate, and, as described above, Acr2p has arsenate reductase activity. It is reasonable to conclude first that nonenzymic reduction is too slow to be biologically relevant, and second that arsenate resistance in eukaryotes must be enzymic. Furthermore, we predict that this enzyme is also required for drug activation in treatment of parasitic disease. Pentostam, the drug of choice for treatment of leishmaniasis, is Sb(V) chelated to gluconate. It is likely that Sb(V) must be reduced to form the active form of the drug. Since Pentostam-resistant *Leishmania* cell lines are cross-resistant to As(III), Sb(III), As(V) and Sb(V) (Ouellette *et al.*, 1991), Pentostam reduction may require an enzyme similar to Acr2p.

CONCLUDING REMARKS

In conclusion, pathways for resistance to arsenicals and antimonials appear to have evolved at least three times. The ArsB arsenite carrier proteins of prokaryotes confer resistance to salts of both metalloids. Subsequent evolution of ArsA subunits converted ArsB into a high-level resistance pump. Acr3p and its homologues are independently evolved carriers that extrude arsenite but not antimonite. ABC ATPases such as Ycf1p transport As(GS)$_3$ [and probably Sb(GS)$_3$] into intracellular compartments such as the yeast vacuole. Similarly, arsenate reductases appear to have evolved independently at least three times: the Gram-negative glutaredoxin-coupled ArsC, the Gram-positive thioredoxin-coupled ArsC, and the yeast Acr2p reductase.

REFERENCES

Abrahams, J. P., Leslie, A. G., Lutter, R. & Walker, J. E. (1994). Structure at 2.8 Å resolution of F$_1$–ATPase from bovine heart mitochondria. *Nature* **370**, 621–628.

Åslund, F., Ehn, B., Miranda-Vizuete, A., Pueyo, C. & Holmgren, A. (1994). Two additional glutaredoxins exist in *Escherichia coli*: glutaredoxin 3 is a hydrogen donor for ribonucleotide reductase in a thioredoxin/glutaredoxin 1 double mutant. *Proceedings of the National Academy of Sciences, USA* **91**, 9813–9817.

Bhattacharjee, H. & Rosen, B. P. (1996). Spatial proximity of Cys113, Cys172, and Cys422 in the metalloactivation domain of the ArsA ATPase. *Journal of Biological Chemistry* **271**, 24465–24470.

Bhattacharjee, H., Li, J., Ksenzenko, M. Y. & Rosen, B. P. (1995). Role of cysteinyl residues in metalloactivation of the oxyanion-translocating ArsA ATPase. *Journal of Biological Chemistry* **270**, 11245–11250.

Bobrowicz, P., Wysocki, R., Owsianik, G., Goffeau, A. & Ulaszewski, S. (1997). Isolation of three contiguous genes, *ACR1*, *ACR2* and *ACR3*, involved in resistance to arsenic compounds in the yeast *Saccharomyces cerevisiae*. *Yeast* **13**, 819–828.

de Boer, P. A., Crossley, R. E., Hand, A. R. & Rothfield, L. I. (1991). The MinD protein is a membrane ATPase required for the correct placement of the *Escherichia coli* division site. *EMBO Journal* **10**, 4371–4380.

Boskovic, J., Soler-Mira, A., Garcia-Cantalejo, J. M., Ballesta, J. P., Jimenez, A. & Remacha, M. (1996). The sequence of a 16,691 bp segment of *Saccharomyces cerevisiae* chromosome IV identifies the DUN1, PMT1, PMT5, SRP14 and DPR1 genes, and five new open reading frames. *Yeast* **12**, 1377–1384.

Broer, S., Ji, G., Broer, A. & Silver, S. (1993). Arsenic efflux governed by the arsenic resistance determinant of *Staphylococcus aureus* plasmid pI258. *Journal of Bacteriology* **175**, 3480–3485.

Bruhn, D. F., Li, J., Silver, S., Roberto, F. & Rosen, B. P. (1996). The arsenical resistance operon of IncN plasmid R46. *FEMS Microbiology Letters* **139**, 149–153.

Carlin, A., Shi, W., Dey, S. & Rosen, B. P. (1995). The *ars* operon of *Escherichia coli* confers arsenical and antimonial resistance. *Journal of Bacteriology* **177**, 981–986.

Chen, Y. & Rosen, B. P. (1997). Metalloregulatory properties of the ArsD repressor. *Journal of Biological Chemistry* **272**, 14257–14262.

Chen, C. M., Misra, T. K., Silver, S. & Rosen, B. P. (1986). Nucleotide sequence of the structural genes for an anion pump. The plasmid-encoded arsenical resistance operon. *Journal of Biological Chemistry* **261**, 15030–15038.

Cole, S. P., Bhardwaj, G., Gerlach, J. H. & 7 other authors (1992). Overexpression of a transporter gene in a multidrug-resistant human lung cancer cell line. *Science* **258**, 1650–1654.

Cole, S. P., Sparks, K. E., Fraser, K., Loe, D. W., Grant, C. E., Wilson, G. M. & Deeley, R. G. (1994). Pharmacological characterization of multidrug resistant MRP-transfected human tumour cells. *Cancer Research* **54**, 5902–5910.

Deeley, R. G. & Cole, S. P. (1997). Function, evolution and structure of multidrug resistance protein (MRP). *Seminars in Cancer Biology* **8**, 193–204.

Delnomdedieu, M., Basti, M. M., Otvos, J. D. & Thomas, D. J. (1993). Transfer of arsenite from glutathione to dithiols: a model of interaction. *Chemical Research in Toxicology* **6**, 598–602.

Delnomdedieu, M., Basti, M. M., Otvos, J. D. & Thomas, D. J. (1994). Reduction and binding of arsenate and dimethylarsinate by glutathione: a magnetic resonance study. *Chemico-Biological Interactions* **90**, 139–155.

Dey, S. & Rosen, B. P. (1995a). Dual mode of energy coupling by the oxyanion-translocating ArsB protein. *Journal of Bacteriology* **177**, 385–389.

Dey, S. & Rosen, B. P. (1995b). Mechanisms of drug transport in prokaryotes and eukaryotes. In *Drug Transport in Antimicrobial and Anticancer Chemotherapy*, pp. 103–132. Edited by N. H. Georgopapadakou. New York: Dekker.

Dey, S., Dou, D. & Rosen, B. P. (1994a). ATP-dependent arsenite transport in everted membrane vesicles of *Escherichia coli*. *Journal of Biological Chemistry* **269**, 25442–25446.

Dey, S., Dou, D., Tisa, L. S. & Rosen, B. P. (1994b). Interaction of the catalytic and the membrane subunits of an oxyanion-translocating ATPase. *Archives of Biochemistry and Biophysics* **311**, 418–424.

Dey, S., Papadopoulou, B., Haimeur, A., Roy, G., Grondin, K., Dou, D., Rosen, B. P. & Ouellette, M. (1994c). High level arsenite resistance in *Leishmania tarentolae* is mediated by an active extrusion system. *Molecular and Biochemical Parasitology* **67**, 49–57.

Dou, D., Dey, S. & Rosen, B. P. (1994). A functional chimeric membrane subunit of an ion-translocating ATPase. *Antonie van Leeuwenhoek* **65**, 359–368.

Fahey, R. C. & Sundquist, A. R. (1991). Evolution of glutathione metabolism. In *Advances in Enzymology and Related Areas of Molecular Biology*, pp. 1–53. Edited by A. Meister. New York: Wiley.

Fauman, E. B., Cogswell, J. P., Lovejoy, B., Rocque, W. J., Holmes, W., Montana, V. G., Piwnica-Worms, H., Rink, M. J. & Saper, M. A. (1998). Crystal structure of the catalytic domain of the human cell cycle control phosphatase, Cdc25A. *Cell* **93**, 617–625.

Ghosh, M., Shen, J. & Rosen, B. P. (1999). Pathways of As(III) detoxification in

Saccharomyces cerevisiae. Proceedings of the National Academy of Sciences, USA **96**, (in press).

Gladysheva, T. B., Oden, K. L. & Rosen, B. P. (1994). Properties of the arsenate reductase of plasmid R773. *Biochemistry* **33**, 7288–7293.

Hofmann, K., Bucher, P. & Kajava, A. V. (1998). A model of Cdc25 phosphatase catalytic domain and Cdk-interaction surface based on the presence of a rhodanese homology domain. *Journal of Molecular Biology* **282**, 195–208.

Hsu, C. M. & Rosen, B. P. (1989). Characterization of the catalytic subunit of an anion pump. *Journal of Biological Chemistry* **264**, 17349–17354.

Hsu, C., Kaur, P., Karkaria, C. E., Steiner, R. F. & Rosen, B. P. (1991). Substrate-induced dimerization of the ArsA protein, the catalytic component of an anion-translocating ATPase. *Journal of Biological Chemistry* **266**, 2327–2332.

Hughes, V. M. & Datta, N. (1983). Conjugative plasmids in bacteria of the 'pre-antibiotic' era. *Nature* **302**, 725–726.

Ji, G. & Silver, S. (1992). Regulation and expression of the arsenic resistance operon from *Staphylococcus aureus* plasmid pI258. *Journal of Bacteriology* **174**, 3684–3694.

Ji, G., Garber, E. A., Armes, L. G., Chen, C. M., Fuchs, J. A. & Silver, S. (1994). Arsenate reductase of *Staphylococcus aureus* plasmid pI258. *Biochemistry* **33**, 7294–7299.

Karkaria, C. E. & Rosen, B. P. (1991). Trinitrophenyl-ATP binding to the ArsA protein: the catalytic subunit of an anion pump. *Archives of Biochemistry and Biophysics* **288**, 107–111.

Karkaria, C. E., Chen, C. M. & Rosen, B. P. (1990). Mutagenesis of a nucleotide-binding site of an anion-translocating ATPase. *Journal of Biological Chemistry* **265**, 7832–7836.

Kaur, P. & Rosen, B. P. (1992). Mutagenesis of the C-terminal nucleotide-binding site of an anion-translocating ATPase. *Journal of Biological Chemistry* **267**, 19272–19277.

Kaur, P. & Rosen, B. P. (1993). Complementation between nucleotide binding domains in an anion-translocating ATPase. *Journal of Bacteriology* **175**, 351–357.

Kaur, P. & Rosen, B. P. (1994). In vitro assembly of an anion-stimulated ATPase from peptide fragments. *Journal of Biological Chemistry* **269**, 9698–9704.

Kurdi-Haidar, B., Aebi, S., Heath, D., Enns, R. E., Naredi, P., Hom, D. K. & Howell, S. B. (1996). Isolation of the ATP-binding human homolog of the *arsA* component of the bacterial arsenite transporter. *Genomics* **36**, 486–491.

Kurdi-Haidar, B., Heath, D., Aebi, S. & Howell, S. B. (1998). Biochemical characterization of the human arsenite-stimulated ATPase (hASNA-I). *Journal of Biological Chemistry* **273**, 22173–22176.

Kuroda, M., Dey, S., Sanders, O. I. & Rosen, B. P. (1997). Alternate energy coupling of ArsB, the membrane subunit of the Ars anion-translocating ATPase. *Journal of Biological Chemistry* **272**, 326–331.

Li, J. & Rosen, B. P. (1998). Steric limitations in the interaction of the ATP binding domains of the ArsA ATPase. *Journal of Biological Chemistry* **273**, 6796–6800.

Li, J., Liu, S. & Rosen, B. P. (1996). Interaction of ATP binding sites in the ArsA ATPase, the catalytic subunit of the Ars pump. *Journal of Biological Chemistry* **271**, 25247–25252.

Li, Z. S., Szczypka, M., Lu, Y. P., Thiele, D. J. & Rea, P. A. (1996). The yeast cadmium factor protein (YCF1) is a vacuolar glutathione S-conjugate pump. *Journal of Biological Chemistry* **271**, 6509–6517.

Li, Z. S., Lu, Y. P., Zhen, R. G., Szczypka, M., Thiele, D. J. & Rea, P. A. (1997). A new pathway for vacuolar cadmium sequestration in *Saccharomyces cerevisiae:*

YCF1-catalyzed transport of bis(glutathionato)cadmium. *Proceedings of the National Academy of Sciences, USA* **94**, 42–47.

Liu, J. & Rosen, B. P. (1997). Ligand interactions of the ArsC arsenate reductase. *Journal of Biological Chemistry* **272**, 21084–21089.

Liu, J., Gladysheva, T. B., Lee, L. & Rosen, B. P. (1995). Identification of an essential cysteinyl residue in the ArsC arsenate reductase of plasmid R773. *Biochemistry* **34**, 13472–13476.

Maloney, P. C. & Wilson, T. H. (1993). The evolution of membrane carriers. *Society of General Physiology Serials* **48**, 147–160.

Mobley, H. L. & Rosen, B. P. (1982). Energetics of plasmid-mediated arsenate resistance in *Escherichia coli. Proceedings of the National Academy of Sciences, USA* **79**, 6119–6122.

Mukhopadhyay, R. & Rosen, B. P. (1998). The *Saccharomyces cerevisiae ACR2* gene encodes an arsenate reductase. *FEMS Microbiology Letters* **68**, 127–136.

Murphy, S. T., Jackman, D. M. & Mulligan, M. E. (1993). Cloning and nucleotide sequence of the gene for dinitrogenase reductase (*nifH*) from the heterocyst-forming cyanobacterium *Anabaena* sp. L31. *Biochimica et Biophysica Acta* **1171**, 337–340.

Oden, K. L., Gladysheva, T. B. & Rosen, B. P. (1994). Arsenate reduction mediated by the plasmid-encoded ArsC protein is coupled to glutathione. *Molecular Microbiology* **12**, 301–306.

Ouellette, M. & Borst, P. (1991). Drug resistance and P-glycoprotein gene amplification in the protozoan parasite *Leishmania. Research in Microbiology* **142**, 737–746.

Ouellette, M., Fase-Fowler, F. & Borst, P. (1990). The amplified H circle of methotrexate-resistant *Leishmania tarentolae* contains a novel P-glycoprotein gene. *EMBO Journal* **9**, 1027–1033.

Ouellette, M., Hettema, E., Wust, D., Fase-Fowler, F. & Borst, P. (1991). Direct and inverted DNA repeats associated with P-glycoprotein gene amplification in drug resistant *Leishmania. EMBO Journal* **10**, 1009–1016.

Rosen, B. P. (1999). Families of arsenic transporters. *Trends in Microbiology* **7**, 207–212.

Rosenstein, R., Peschel, A., Wieland, B. & Götz, F. (1992). Expression and regulation of the antimonite, arsenite, and arsenate resistance operon of *Staphylococcus xylosus* plasmid pSX267. *Journal of Bacteriology* **174**, 3676–3683.

San Francisco, M. J., Hope, C. L., Owolabi, J. B., Tisa, L. S. & Rosen, B. P. (1990). Identification of the metalloregulatory element of the plasmid-encoded arsenical resistance operon. *Nucleic Acids Research* **18**, 619–624.

Sato, T. & Kobayashi, Y. (1998). The *ars* operon in the skin element of *Bacillus subtilis* confers resistance to arsenate and arsenite. *Journal of Bacteriology* **180**, 1655–1661.

Scott, N., Hatlelid, K. M., MacKenzie, N. E. & Carter, D. E. (1993). Reactions of arsenic(III) and arsenic(V) species with glutathione. *Chemical Research in Toxicology* **6**, 102–106.

Shi, W., Wu, J. & Rosen, B. P. (1994). Identification of a putative metal binding site in a new family of metalloregulatory proteins. *Journal of Biological Chemistry* **269**, 19826–19829.

Silver, S. & Keach, D. (1982). Energy-dependent arsenate efflux: the mechanism of plasmid-mediated resistance. *Proceedings of the National Academy of Sciences, USA* **79**, 6114–6118.

Suzuki, K., Wakao, N., Kimura, T., Sakka, K. & Ohmiya, K. (1998). Expression and regulation of the arsenic resistance operon of *Acidiphilium multivorum* AIU301 plasmid pKW301 in *Escherichia coli. Applied and Environmental Microbiology* **64**, 411–418.

Szczypka, M. S., Wemmie, J. A., Moye-Rowley, W. S. & Thiele, D. J. (1994). A yeast metal resistance protein similar to human cystic fibrosis transmembrane conductance regulator (CFTR) and multidrug resistance-associated protein. *Journal of Biological Chemistry* **269**, 22853–22857.

Takemaru, K., Mizuno, M., Sato, T., Takeuchi, M. & Kobayashi, Y. (1995). Complete nucleotide sequence of a *skin* element excised by DNA rearrangement during sporulation in *Bacillus subtilis*. *Microbiology* **141**, 323–327.

Walker, J. E., Saraste, M., Runswick, M. J. & Gay, N. J. (1982). Distantly related sequences in the α- and β-subunits of ATP synthase, myosin, kinases and other ATP-requiring enzymes and a common nucleotide binding fold. *EMBO Journal* **1**, 945–951.

Wang, H. F. & Lee, T. C. (1993). Glutathione S-transferase pi facilitates the excretion of arsenic from arsenic-resistant Chinese hamster ovary cells. *Biochemical and Biophysical Research Communications* **192**, 1093–1099.

Wang, Z. & Rossman, T. G. (1993). Stable and inducible arsenite resistance in Chinese hamster cells. *Toxicology and Applied Pharmacology* **118**, 80–86.

Wang, Z., Dey, S., Rosen, B. P. & Rossman, T. G. (1996). Efflux-mediated resistance to arsenicals in arsenic-resistant and -hypersensitive Chinese hamster cells. *Toxicology and Applied Pharmacology* **137**, 112–119.

Wo, Y. Y., McCormack, A. L., Shabanowitz, J., Hunt, D. F., Davis, J. P., Mitchell, G. L. & Van Etten, R. L. (1992a). Sequencing, cloning, and expression of human red cell-type acid phosphatase, a cytoplasmic phosphotyrosyl protein phosphatase. *Journal of Biological Chemistry* **267**, 10856–10865.

Wo, Y. Y., Zhou, M. M., Stevis, P., Davis, J. P., Zhang, Z. Y. & Van Etten, R. L. (1992b). Cloning, expression, and catalytic mechanism of the low molecular weight phosphotyrosyl protein phosphatase from bovine heart. *Biochemistry* **31**, 1712–1721.

Wu, J. & Rosen, B. P. (1991). The ArsR protein is a trans-acting regulatory protein. *Molecular Microbiology* **5**, 1331–1336.

Wu, J. & Rosen, B. P. (1993). The *arsD* gene encodes a second trans-acting regulatory protein of the plasmid-encoded arsenical resistance operon. *Molecular Microbiology* **8**, 615–623.

Wu, J., Tisa, L. S. & Rosen, B. P. (1992). Membrane topology of the ArsB protein, the membrane subunit of an anion-translocating ATPase. *Journal of Biological Chemistry* **267**, 12570–12576.

Wysocki, R., Bobrowicz, P. & Ulaszewski, S. (1997). The *Saccharomyces cerevisiae* *ACR3* gene encodes a putative membrane protein involved in arsenite transport. *Journal of Biological Chemistry* **272**, 30061–30066.

Xu, C., Zhou, T., Kuroda, M. & Rosen, B. P. (1998). Metalloid resistance mechanisms in prokaryotes. *Journal of Biochemistry (Tokyo)* **123**, 16–23.

Zaman, G. J., Lankelma, J., van Tellingen, O., Beijnen, J., Dekker, H., Paulusma, C., Oude Elferink, R. P., Baas, F. & Borst, P. (1995). Role of glutathione in the export of compounds from cells by the multidrug-resistance-associated protein. *Proceedings of the National Academy of Sciences, USA* **92**, 7690–7694.

Zhou, T. & Rosen, B. P. (1997). Tryptophan fluorescence reports nucleotide-induced conformational changes in a domain of the ArsA ATPase. *Journal of Biological Chemistry* **272**, 19731–19737.

Zhou, T., Liu, S. & Rosen, B. P. (1995). Interaction of substrate and effector binding sites in the ArsA ATPase. *Biochemistry* **34**, 13622–13626.

TYPE II PROTEIN SECRETION: THE MAIN TERMINAL BRANCH OF THE GENERAL SECRETORY PATHWAY

ALAIN FILLOUX

Laboratoire d'Ingéniérie des Systèmes Macromoléculaires, CNRS-IBSM-UPR9027, 31 Chemin Joseph Aiguier, 13402 Marseille Cedex 20, France

INTRODUCTION

Bacterial membranes are essential barriers, preserving the integrity of the organism. However, the cell envelope must be sufficiently permeable to allow the intense traffic of molecules into and out of the cells. This is essential for the acquisition of nutrients from the environment, the uptake of DNA, which favours gene transfer, the delivery of degradative enzymes or toxins to their targets, and the assembly of organelles such as pili and flagella on the cell surface. The cell envelope of Gram-negative bacteria consists of two membranes, delimiting an intermediate compartment, the periplasm. Nutrient molecules are small enough to diffuse through porins located within the outer membrane and are subsequently actively transported across the inner membrane into the cytoplasmic compartment. Extracellular enzymes and toxins are much larger molecules. The presence of large holes in the membranes to allow the transport of such molecules would threaten the survival of the cell. Consequently, three major types of specialized pathway have been developed for protein secretion in Gram-negative bacteria. The type I, or ABC transporter (ABC = $\underline{A}$TP-$\underline{b}$inding $\underline{c}$assette), is typified by the *Escherichia coli* α-haemolysin (Broome-Smith & Mitsopoulos, this volume), and the type III, or contact site, by Yop (*Yersinia* $\underline{o}$uter $\underline{p}$rotein) secretion from the human pathogen *Yersinia enterocolitica* (Anderson and others, this volume). In these two cases, it is believed that exoproteins cross the bacterial cell envelope in a single step, directly from the cytoplasm to the outside of the cell. The general secretory pathway (GSP) has been suggested as an alternative mode of translocation, in which the exoproteins cross successively, in two steps, the inner and outer membranes. The first step, involving translocation across the cytoplasmic membrane, is reminiscent of the mechanisms of protein translocation into the lumen of the endoplasmic reticulum (Schatz & Dobberstein, 1996; Young and others, this volume). The transported protein is synthesized as a precursor containing an N-terminal cleavable signal peptide. This precursor is targeted and transported through the membrane via a proteinaceous complex constituting the translocator.

This complex, also called preprotein translocase, consists, in *E. coli*, of a dimer of SecA peripherally bound to a membrane-embedded subcomplex of the SecY, SecE, SecG, SecD and YajC proteins (Duong *et al.*, 1997). Most proteins are kept in an unfolded conformation by the SecB chaperone and are brought to the translocase by interaction with SecA. SecA then binds to the core of the translocase integral membrane domain, SecYE. SecA has a central role in preprotein translocation because it also hydrolyses ATP and provides energy for the translocation process. SecA undergoes profound conformational changes, which lead to its insertion into the integral membrane subcomplex, and a series of insertion–deinsertion cycles of SecA introduce the preprotein into the membrane complex and move it to the other side of the membrane. The signal peptide is then cleaved by the leader peptidase (LepB) and the mature protein is released into the subsequent compartment, the periplasm. This first series of events is called the general export pathway. At this stage, the exoprotein requires a second set of machinery specifically to assist its translocation across the outer membrane and its release into the extracellular environment. This second series of events, the secretion step, is called the terminal branch of the GSP. Several branches have been identified. The simplest, also called 'autotransporter' (Henderson *et al.*, 1998), was first described for the IgA1 protease of *Neisseria gonorrhoeae* (Fig. 1). The protein consists of two domains and the C-terminal part is the transporter. This region consists of amphipathic β-sheets forming a barrel in the outer membrane, through which the N-terminal domain is translocated. The N-terminus, which is the functional domain of the enzyme, also known as the passenger domain, is finally released into the medium by specific cleavage. The transporter domain can also be encoded by a separate gene. In the case of the ShlA haemolysin from *Serratia marcescens*, the outer membrane protein ShlB is the only accessory factor required to achieve the final release of the protein into the extracellular medium (Braun *et al.*, 1992). Some terminal branches of the GSP are more complex, and involve a multiprotein complex spanning the bacterial cell envelope. A mechanism, tentatively called type IV, has been demonstrated in the case of pertussis toxin from *Bordetella pertussis*. The pertussis toxin consists of five types of subunit, S1–S5, all of which carry a signal peptide. This pathway may therefore be considered to be one terminal branch of the GSP. The secretory apparatus consists of at least nine proteins homologous to those involved in DNA transfer between *Agrobacterium tumefaciens* and plants (Zupan *et al.*, 1998), and those involved in the transfer of conjugative bacterial plasmids (Winans *et al.*, 1996). This chapter will focus on a particular transport process, the type II secretion mechanism or main terminal branch of the GSP, which involves 12–14 different proteins. The components of this pathway were first discovered in *Klebsiella oxytoca* (d'Enfert *et al.*, 1987), and their subsequent identification in *Pseudomonas aeruginosa* demonstrated that they are conserved among most Gram-

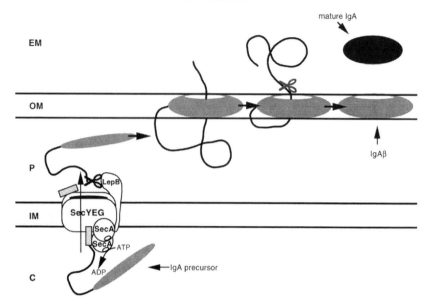

Fig. 1. IgA protease secretion from *N. gonorrhoeae*, the autotransporter process. The IgA precursor is shown with three domains. The signal peptide corresponds to the dotted box, the N-terminal domain is shown in black, and the C-terminal domain is shown in grey. The precursor is translocated from the cytoplasm (C) across the inner membrane (IM) via the Sec machinery, and is processed by removal of the signal peptide by leader peptidase (LepB, black scissors). Once in the periplasm (P), the C-terminal domain folds and inserts into the outer membrane (OM) as a β-barrel (IgAβ) through which the N-terminal domain can reach the extracellular medium (EM). Following autoproteolytic cleavage (grey scissors), the mature fully folded IgA protease (shown as a black ellipse) is released into the extracellular medium.

negative bacteria (Table 1) (Filloux *et al.*, 1990). Gsp is the general term used to describe those proteins of the type II secretory apparatus that will now be described in further detail.

THE GSP PROTEINS

The genetic determinant of all known type II systems is a set of genes clustered together and organized into one large operon (Fig. 2). In the cases of *P. aeruginosa* and *Pseudomonas alcaligenes*, these genes are clustered but are organized into two divergently transcribed operons (Filloux *et al.*, 1998; Gerritse *et al.*, 1998). Mutations in any of these genes abort the secretion process, resulting in the accumulation of exoproteins within the periplasm. Thus all *gsp* genes are essential for the outer membrane translocation step. Surprisingly, the deduced amino acid sequences and initial characterization of the proteins indicated that only two of the 14 are effectively located in the outer membrane (Fig. 3). All but one of the other proteins seem to be anchored in the cytoplasmic membrane, with the remaining protein peripherally associated

Table 1. *Characteristics of Gsp components in the type II secretory pathway and their presence in various Gram-negative bacteria*[a]

Characteristic	P. ae Xcp	P. alc Xcp	P. pu Xcp	K. ox Pul	E. ch Out	E. ca Out	A. hy Exe	V. ch Eps	X. ca Xps	E. co Gsp
ATP binding motif/interacts with B; Inner membrane							A			A
TonB homologue?[b]/interacts with A; Inner membrane/energy transducer				B	B	B	B			
Inner membrane/bitopic	P	P	P	C	C	C	C	C	C	C
Secretin channel; Outer membrane/homomultimers	Q	Q	Q	D	D	D	D	D	D	D
ATP-binding motif/interacts with L (Y); Inner-membrane-associated	R	R	R	E	E	E	E	E	E	E
Inner membrane/polytopic; Three transmembrane segments	S	S	S	F	F	F	F	F	F	F
Pseudopilin	T	T		G	G	G	G	G	G	G
Pseudopilin	U	U		H	H	H	H	H	H	H
Pseudopilin	V	V		I	I	I	I	I	I	I
Pseudopilin	W	W	W	J	J	J	J	J	J	J
Atypical pseudopilin/lacks E + 5	X	X		K	K	K	K	K	K	K
Inner membrane/bitopic; Docks E (R) to the inner membrane; Interacts with M (Z)	Y	Y	Y	L	L	L	L	L	L	L
Inner membrane/bitopic/interacts with L (Y)	Z	Z	Z	M	M	M	M	M	M	M
Inner membrane/bitopic		N	N	N		N	N	N		
Prepilin peptidase/inner membrane; Eight transmembrane segments	A (PilD)	A	A	O	O	O	TapD			O
Chaperone/outer membrane/lipoprotein; Secretin pilot				S	S	S				

[a] Each line represents a family of homologous proteins, with their characteristics summarized in the left column. Proteins are identified by the fourth letter of their designation, the first three letters being indicated in the top line, except for the prepilin peptidase TapD of A. hydrophila. In V. cholerae, the prepilin peptidase TcpJ has been identified, but was not shown to be involved in secretion (only in piliation), and is therefore not indicated. XcpA from P. aeruginosa is also called PilD. PulB is not required for secretion of K. oxytoca pullulanase. Components from P. aeruginosa (P. ae), P. alcaligenes (P. alc), P. putida (P. pu), K. oxytoca (K. ox), Erw. chrysanthemi (E. ch), Erw. carotovora (E. ca), A. hydrophila (A. hy), V. cholerae (V. ch), X. campestris (X. ca) and E. coli (E. co) are listed.

[b] Homology with TonB has been detected only for ExeB, not for the other B proteins.

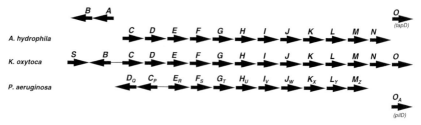

Fig. 2. Main types of genetic organization in *gsp* gene clusters for *A. hydrophila*, *K. oxytoca* and *P. aeruginosa*. Continuous thin lines between genes indicate short intervening stretches, whereas genes shown on different levels are thought to be separated by long undetermined stretches of DNA. All the clusters described in Table 1 are organized as for *K. oxytoca*, except for *P. alcaligenes*, which is organized similarly to *P. aeruginosa*. In *P. aeruginosa* only a short DNA region, about 200 bp long, including both promoters, is present between the *gspE* and *gspC* genes, whereas in *P. alcaligenes* two *orfs* have been identified between these same two genes. In *P. putida* the *gspK* gene is missing and thus *gspL* immediately follows *gspJ* within the operon. In *P. aeruginosa* the letters in subscript represent the nomenclature normally used for *xcp* genes (see Table 1). *pilD* and *tapD* is the nomenclature used for these genes with respect to type IV piliation.

with the cytoplasmic side of the inner membrane (Fig. 3). This particular cell envelope distribution of the Gsp proteins makes it difficult to investigate their roles in the molecular mechanism of outer membrane translocation.

GspO, a specialized peptidase

P. aeruginosa and *Aeromonas hydrophila gsp* gene clusters have an unusual feature in that *gspO*, usually the last gene in the operon, is not clustered with the other *gsp* genes (Fig. 2). It is found at another chromosomal location, clustered with genes required for the assembly of type IV pili, and called *pilD* and *tapD*, respectively (Nunn *et al.*, 1990; Pepe *et al.*, 1996). Type IV pili were originally defined as long cell surface appendages located at the poles and involved in a particular type of motion called twitching motility (Henrichsen, 1983). PilD and TapD are the prepilin peptidases required for the processing of the N-terminal leader peptide of the pilin subunit before its assembly into pilus (Nunn & Lory, 1991; Pepe *et al.*, 1996) (Fig. 4a). The leader peptide of these subunits structurally resembles the classically defined signal peptide of Sec-dependent transported proteins (von Heijne, 1985). It consists of a short stretch of 6–7 residues with an overall positive charge, preceding a hydrophobic domain (Strom & Lory, 1993) (Fig. 4b). However, this hydrophobic domain is not followed by the typical cleavage site of the leader peptidase (A-X-A). Instead the processing site is located immediately before the hydrophobic region, after a highly conserved glycine residue (G − 1) (Fig. 4b). Mature pilin subunits are then helicoidally packed via interactions between their hydrophobic N-terminal domains and are assembled into pilus. A conserved glutamate residue, at position + 5 within the hydrophobic domain (E + 5) (Fig. 4b), may be a key element in the registration mechan-

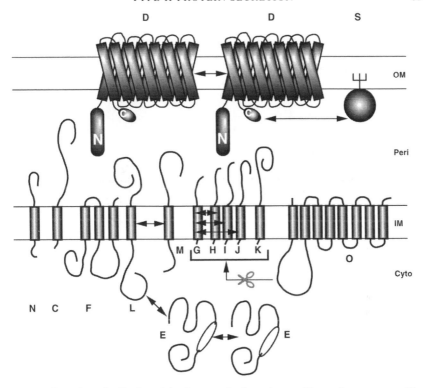

Fig. 3. Cell envelope distribution of the Gsp proteins from the type II secretion apparatus. The GspD protein is shown in the outer membrane (OM), with a large C-terminal domain forming an inserted β-barrel (series of tilted rectangles connected by small loops). The N-terminus and the extreme C-terminus of the GspD protein are also indicated with an N and a C, respectively. The GspS lipoprotein is present at the outer membrane, to which it is anchored via its fatty-acyl chain, represented by a trident. The inner membrane (IM) proteins are represented with their transmembrane segments as rectangles and their hydrophilic domains as black lines. Bitopic inner membrane proteins (GspN, -C, -L, -M, -G, -H, -I, -J and -K) have an N_{in}-C_{out} topology, i.e. the N-terminus in the cytoplasm (Cyto) and the C-terminus in the periplasm (Peri). The polytopic inner membrane proteins GspF and GspO are shown with their N-termini at the left end of the molecules. The GspE protein is shown as being present in the cytoplasm but it is peripherally associated with the cytoplasmic side of the inner membrane (see text). The internal bulky domain in the GspE protein indicates the presence of an ATP-binding motif. Double-headed arrows indicate proteins known to interact with each other. The pseudopilins GspG, -H, -I, -J and -K are N-terminally processed by the prepilin peptidase, GspO, as shown with the grey scissors. Processing may cause a change in the location of pseudopilin (see text).

ism associated with the assembly of the pilin subunits (Parge *et al.*, 1995). The new N-terminal residue of pilin, a conserved phenylalanine (F + 1) (Fig. 4b) or a hydrophobic residue such as the methionine of the TcpA pilin of *Vibrio cholerae*, is methylated during processing. This second post-translational modification is also catalysed by the prepilin peptidase, which is thus a bifunctional enzyme (Strom & Lory, 1993). The G − 1 residue is essential for cleavage of the pilin subunits and assembly into pilus (Strom & Lory, 1991).

(a)

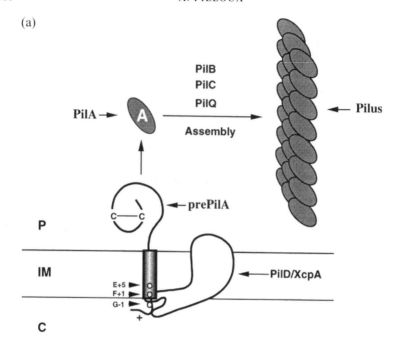

(b)

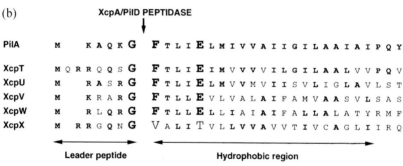

Fig. 4. (a) Successive series of events during assembly of the pilin subunit, PilA, of *P. aeruginosa* into pilus. The prepilin peptidase, XcpA/PilD, cleaves the positively charged (+) N-terminal leader peptide of prePilA between a glycine (G − 1) and a phenylalanine (F + 1) residue. The mature subunit (PilA) is then assembled into a pilus with the help of several accessory proteins, including PilB, -C and -Q. This representation of the bacterial inner membrane (IM) shows the topology of prePilA and the location of XcpA/PilD, but is not intended to depict the location of the mature PilA subunit and the pilus. The white circles mark the approximate positions of the conserved G − 1, F + 1 and E + 5 residues within the PilA protein. The Cys-bond is also shown (C–C). P, Periplasm; C, cytoplasm. (b) Sequence alignment of the N-terminal domains of the PilA pilin subunit and pseudopilins of the Xcp-type II secretory pathway from *P. aeruginosa*. XcpT, -U, -V, -W and -X are also more generally named GspG, -H, -I, -J and -K, respectively. The residues in the pseudopilins that are identical to those in the PilA sequence are shown in bold. (Here R and K, and also I and L, are treated as 'identical' residues.) The conserved G − 1, F + 1 and E + 5 residues are shown in larger upper-case letters. The position of the leader sequence cleavage site is indicated by an arrow. The leader peptide and the hydrophobic region are shown.

There is no strict requirement for phenylalanine at position $+1$ or for glutamate at position $+5$ for prepilin processing. However, although an $E + 5V$ substitution did not affect leader peptide cleavage, it did abolish methylation and piliation. In the case of an $F + 1S$ substitution, pilin was assembled into pili despite the complete absence of N-methylation. This observation raises questions as to the function of methylation during pilus biogenesis.

Mutations in the prepilin peptidase *pilD* and *tapD* genes not only affect type IV piliation but also abolish the protein secretion process. These proteins, despite the location of their genes within the genome, are thus real Gsp components, namely GspO. The GspO proteins are polytopic inner membrane proteins with eight transmembrane segments, as demonstrated for *Erwinia chrysanthemi* OutO (Reeves *et al.*, 1994). The first cytoplasmic loop of these proteins is large and contains a tetracysteine consensus motif, C-X-X-C....X_{21}....C-X-X-C. The enzymic activity of the peptidase, including its methyltransferase activity, involves these conserved cysteine residues, as shown with PilD (Strom & Lory, 1993). This implies that the enzyme interacts with its substrate from the cytoplasmic side of the membrane (Fig. 4a).

GspG, H, I, J and K, the so-called pseudopilins

Four of the Gsp proteins, GspG–J, have the particular N-terminus associated with type IV pilin subunits (Fig. 4b). The sequence similarity of this subset of Gsp proteins has led to their being called the pseudopilin family, despite the C-terminal domain of pseudopilins and pilins being rather different. One relevant difference is that the two conserved cysteines at the extreme C-terminus of the pilin subunit, which form a disulphide bridge, are not present in pseudopilins. The Cys-bond formed in pilin is important for bacterial adhesion involving type IV pili (Hahn, 1997).

The GspO-dependent cleavage of N-terminal pseudopilin leader peptides has been demonstrated in several organisms (Bally *et al.*, 1992; Nunn & Lory, 1992; Pugsley & Dupuy, 1992; Howard *et al.*, 1993). As with the pilins, the new N-terminal residue is methylated (Nunn & Lory, 1993). Topology studies of the pseudopilin indicate that they are bitopic inner membrane proteins with a single transmembrane segment (Reeves *et al.*, 1994). The *in vivo* location of these proteins is more ambiguous, because if they are overproduced they may partly cofractionate with the outer membrane, particularly in the case of the mature form (Bally *et al.*, 1992). This suggests that there may be a redistribution of the pilin subunit after its processing. This may involve the effective relocation of the molecule to the outer membrane, or the assembly of pseudopilins into a macromolecular complex, thereby changing fractionation behaviour. The similarity between pilins and pseudopilins suggests that pseudopilins assemble into a similar macromolecular complex tentatively called the 'pseudopilus' (Fig. 5) (Hobbs & Mattick, 1993). It is unknown whether such a structure truly exists. Pseudopilins, like pilins, have been reported to form homomultimers stabilized by chemical cross-linking (Pugsley, 1996). However, multimers of pseudopilins such as GspG from *K. oxytoca* have been obtained, even in conditions in which pilins are not assembled, including the replacement of the conserved hydrophobic N-terminal domain with the MalE signal peptide (Pugsley, 1996). However, the crystal structure of a gonococcal type IV pilin dimer shows that the monomers are held together by interactions involving the hydrophobic N-terminal region (Parge *et al.*, 1995). Therefore, it is reasonable to conclude that cross-linked pseudopilin multimers are unrelated to type IV pili. These multimers probably form a rudimentary *trans*-periplasmic structure connecting the inner and outer membranes. *In vivo* cross-linking experiments with *P. aeruginosa* cells expressing the *gsp* gene cluster from the chromosome have detected mainly pseudopilin dimers (Lu *et al.*, 1997). These dimers are GspG homodimers and heterodimers containing the other pseudopilins, GspH, I and J. It has been suggested that the dimeric forms are intermediates between the monomeric membrane pools of monomers and the fully assembled secretory apparatus. In addition, the composition of the cross-linked species depends on the size of the individual pools, which may reflect an active process involving the assembly and disassembly of individual subunits. The relative amounts of *P. aeruginosa* GspG, GspH, GspI and GspJ are approximately 16:1:1:4, respectively (Nunn & Lory, 1993). Depolymerization of the assembled type IV pilus is thought to be involved in twitching motility and retraction during the infection of *P. aeruginosa* by pili-specific phages (Whitchurch & Mattick, 1994). If the pseudopilus structure does exist, then an analogous process may drive the

---

Fig. 5. Model for Gsp machinery assembly (left), and comparison with a model for type IV pilus assembly in *P. aeruginosa* (right). On the left, the Gsp-dependent exoprotein, shown as a black line with an N-terminal signal peptide (dotted box), is exported across the inner membrane (IM), from the cytoplasm (Cyto) to the periplasm (Peri), via the Sec machinery. The exoprotein is kept unfolded in the cytoplasm, by binding to SecB, and folds within the periplasm after peptide signal cleavage by the leader peptidase LepB (scissors). Folding may involve the function of a chaperone protein (Ch). The exoprotein is subsequently recognized by the Gsp machinery and translocated across the outer membrane (OM) via the secretin GspD and into the extracellular medium (EM). The secretin GspD is shown as a homomultimeric ring forming a channel with a large central opening. Most Gsp proteins are shown according to their membrane topology, and for those for which an interaction has been proposed, and supported by experimental evidence, the corresponding parts are shown in contact. The GspG pseudopilin is processed (removal of the leader peptide, shown as a small black circle) by the leader peptidase GspO. The pseudopilus is arbitrarily represented as a succession of GspG homo- and heterodimers. The GspA and GspB proteins have not been represented. For more details about the Gsp machinery, see the text. The homologous components of the type IV piliation system in *P. aeruginosa* (Pil) are shown on the right for comparison. Only the PilA subunit is represented within the pilus structure, because it is not clear whether the newly identified pseudopilins of this system, described in the text, are part of the final structure. Similarity in function between PilP and GspS has been shown for *N. gonorrhoeae*.

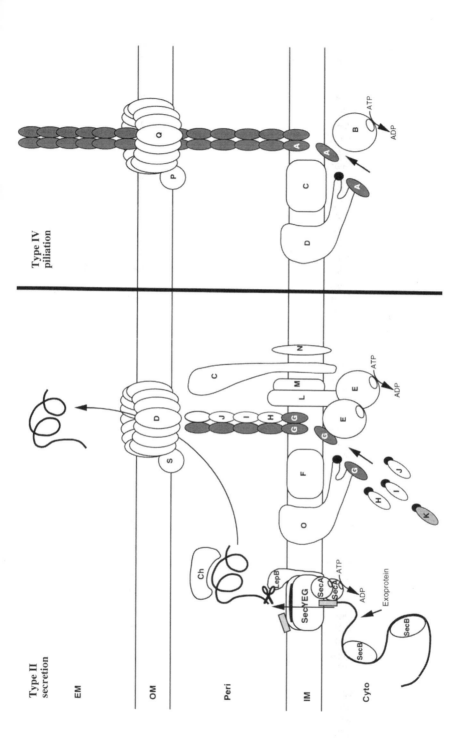

Type II
secretion

Type IV
piliation

EM

OM

Peri

IM

Cyto

secretion of proteins from the periplasm, the assembling structure pushing exoproteins out of the cell (Fig. 5). Surprisingly, the *P. aeruginosa* pilin subunit PilA is also found in association with the *P. aeruginosa* pseudopilin GspG, and it is directly involved in secretion in this organism. However, this raises questions as to the pilin requirement in type II secretion in bacteria that lack type IV pili, and may indicate subtle differences in the mechanism of assembly of the secretory apparatus.

The four pseudopilins, GspG–J, have the G − 1, F + 1 and E + 5 residues conserved (Fig. 4b) (although some differences are permitted at position F + 1, which can be replaced by other hydrophobic residues, such as methionine). It is unclear why four different pseudopilins have been identified in type II secretion systems, whereas only one pilin subunit has been reported for type IV piliation. However, in recent years, five additional genes, the products of which have characteristics in common with Gsp pseudopilins, have been shown to be involved in fimbrial biogenesis in *P. aeruginosa* (Alm & Mattick, 1997). Their sequences suggest that FimT, FimU, PilV and PilW are related to pseudopilins, whereas PilE is more like pilin. Some of these proteins may form a cell envelope complex similar to the speculative 'pseudopilus' involved in type II secretion, whereas PilA is the subunit which forms the pilus structure present at the cell surface (Fig. 5). It has been shown that GspK also belongs to the pseudopilin family (Bleves *et al.*, 1998). It is, however, considered to be atypical because it lacks the E + 5 residue and has a higher relative molecular mass (> 30 kDa) than classical pilins and pseudopilins (15–20 kDa). Such atypical subunits have also been found in the type IV piliation system, with PilX in *P. aeruginosa*. It is unclear whether E + 5, which is required for the methylation of the pilin subunit, is required in the case of pseudopilins because the E + 5V (E11V) substitution in GspG from *K. oxytoca* did not affect N-methylation (Pugsley, 1993). The significance of this difference between the two systems is not understood. The E + 5 residue has also been suggested to play a registration role in assembly of the pilus, by participating in the formation of a salt bridge between the negatively charged glutamate of subunit N and the positively charged methylated N-terminus of subunit N^{+1} (Parge *et al.*, 1995). The absence of E + 5 in the newly incorporated pilin or pseudopilin subunit may prevent incorporation of the subsequent subunit, preventing elongation of the pilus or 'pseudopilus' and initiating disassembly of the structure. This speculation provides clues about the particular role of GspK proteins in protein secretion.

GspE, the putative ATP-binding protein

Protein secretion is an active process that requires energy sources to support the functioning of the secretion machinery. GspE proteins have sequences similar to the Walker A motif and, to a lesser extent, to the Walker B motif of

ATPases (Walker et al., 1982). Despite the presence of these motifs, it was not possible in any case to demonstrate the binding of ATP to these proteins. In contrast, the nucleotide-binding fold seems to be essential for the function of the protein. Mutation of the conserved glycine residue within the Walker A box of GspE from *P. aeruginosa*, *K. oxytoca* or *V. cholerae* causes the bacteria to become secretion-defective (Turner et al., 1993; Possot & Pugsley, 1994; Sandkvist et al., 1995). ATP-binding activity could not be demonstrated for GspE proteins but autokinase activity has been detected in the case of *V. cholerae* GspE (Sandkvist et al., 1995). The protein is labelled with $[\gamma\text{-}^{32}P]ATP$ but not with $[\alpha\text{-}^{32}P]ATP$, suggesting that labelling is due to autophosphorylation. Such labelling requires an intact Walker A box (the K270A substitution within the motif results in one-tenth the amount of labelling in the mutant GspE protein). Mutations in the less well conserved Walker B box have little or no effect on the secretion process (20–30% reduction in pullulanase secretion for mutations in the *K. oxytoca* GspE) (Possot & Pugsley, 1994). The Walker B box is thought to be the determinant for nucleoside recognition, and the weak requirement for this motif may account for the failure to demonstrate ATP-binding activity in GspE proteins. Alternatively, binding may be stimulated when GspE interacts with other components of the type II secretion machinery or other molecules not involved in the type II machinery. ATP binding and hydrolysis are, for example, stimulated when SecA associates with the preprotein (Lill et al., 1989).

Many proteins involved in transport processes have sequence similar to Walker A and B boxes. In some cases, there are additional motifs like the 'LSGGQ' consensus sequence of ABC transporters (Browne et al., 1996). Within the GspE family, there is a highly conserved central region between the Walker A and B boxes, typical of members of this family, consisting of two short aspartate-rich motifs (Possot & Pugsley, 1994). These motifs are called aspartate boxes or T2SP-E, for bacterial type II secretion system protein E signature. They are required for the function of GspE in the secretion process, as shown by the substantial decrease, 80–90%, in pullulanase secretion if the aspartate residues are replaced by asparagine residues in *K. oxytoca* GspE (Possot & Pugsley, 1994). The aspartate residues may be involved in the formation and stabilization of the nucleotide-binding fold by interacting with Mg^{2+}. Another motif, a tetracysteine motif identical to that described for the GspO peptidases, is present in members of the GspE family (Possot & Pugsley, 1997). It appears to be essential for function, because replacement of any of the cysteine residues by a serine within the *K. oxytoca* GspE leads to a large decrease in pullulanase secretion (80%), possibly a decrease of more than 99% if three of the cysteine residues are simultaneously mutated. This similarity between GspO and GspE proteins may be purely coincidental or may reflect, as suggested by Possot and Pugsley, a common function such as coordination of a Zn^{2+} ion. An additional feature of proteins involved in transport processes is their association with the membrane (Higgins, 1992). Some ABC proteins contain a highly hydrophobic domain which anchors them directly in the membrane (Mdr, HlyB). Others, like the bacterial ABC proteins involved in amino acid (HisP) or sugar (MalK) transport, are peripherally bound to the membrane by interaction with integral membrane components of the corresponding permeases. This is also the case for the SecA ATPase, which binds to the SecYE integral membrane domain of the translocase. The deduced amino acid sequences of GspE proteins show that they are mainly hydrophilic and do not possess any hydrophobic domains that could be considered to be transmembrane domains. This observation is consistent with the observation that GspE proteins are present in the cytoplasm if produced in *E. coli* (Possot *et al.*, 1992; Sandkvist *et al.*, 1995; Thomas *et al.*, 1997; Ball *et al.*, 1999). However, these proteins are associated with the cytoplasmic membrane in their original host (Possot & Pugsley, 1994; Sandkvist *et al.*, 1995; Ball *et al.*, 1999). Derivatives of GspE, including proteins with mutations in the Walker A box, have a dominant negative effect on secretion if produced in cells that also contain the wild-type GspE (Turner *et al.*, 1993; Possot & Pugsley, 1994; Sandkvist *et al.*, 1995). This suggests that GspE probably interacts with other Gsp components to form a functional machinery. GspEs are bound to the inner face of the inner membrane thanks to an interaction with GspL. This is demonstrated by the co-expression of the genes encoding GspE and GspL from *V. cholerae* and *P. aeruginosa* in *E. coli*, and by the low level of membrane association of GspE in a corresponding *gspL* mutant strain (Sandkvist *et al.*, 1995; Ball *et al.*, 1999). The interaction of GspE with the membrane via GspL is probably also responsible for its stability. GspL is a bitopic cytoplasmic membrane protein with an N_{in}-C_{out} topology (Bleves *et al.*, 1996), and a large cytoplasmic domain which interacts with GspE proteins (Ball *et al.*, 1999). Finally, traffic ATPases function as dimers in a wide range of transport systems (Higgins, 1992). GspE proteins also appear to associate as homodimers, as shown using a domain of the lambda phage cI repressor as a reporter for dimerization (Turner *et al.*, 1997). It is not yet possible to exclude the possibility that GspE assembles into higher-order multimers.

The exact role of GspE is unknown, but the various observations discussed above suggest that it acts as a traffic ATPase for transport across the cytoplasmic membrane. The substrate for such a transporter has not been identified. Exoproteins are brought into the periplasm via the Sec apparatus and are unlikely to be the substrate for GspE-dependent inner membrane translocation. The most pertinent studies of the function of GspE may be those of Lory and collaborators (Nunn *et al.*, 1990), who showed that the GspE homologue PilB is required for the assembly of type IV pili in *P. aeruginosa*. This suggests that GspE is an ATP-binding protein involved in the assembly of the type II secretory apparatus, and more particularly in the translocation of pseudopilins through the cytoplasmic membrane before their assembly into a 'pseudopilus' structure. Unlike *P. aeruginosa* PilA and PilD (GspO), PilB is exclusively involved in type IV piliation and is not required for type II secretion (Lu *et al.*, 1997). Consistent with this notion, the effects of thermosensitive (*ts*) mutations within the *P. aeruginosa* GspG pseudopilin are suppressed by a secondary mutation within GspE (Kagami *et al.*, 1998). This strongly supports the idea that pseudopilins are substrates for GspE. Moreover, the *ts* mutations affected the periplasmic domain of the pseudopilin, suggesting that the interaction takes place before translocation, or that GspE pushes the pseudopilins through the membrane in a manner similar to the insertion–deinsertion cycles of SecA, pushing the preproteins through the translocase. One additional intriguing result is that it is not possible to rescue pullulanase secretion in *K. oxytoca* cells that have synthesized pullulanase at the same time as expressing all of the *gsp* secretion genes except *gspE*, by subsequent expression of *gspE* alone (Possot *et al.*, 1992). This could indicate that *gspE* must be co-expressed with other *gsp* genes in order to be active, and to participate in the formation of a functional secretion apparatus. If this is indeed the case, we can distinguish two stages in the functioning of the type II secretion machinery: (i) assembly of the core of the machinery, which requires GspE as an 'energizer', and (ii) transport of the exoproteins through the outer membrane, which may require a different energy source. Alternatively, the role of GspE in the secretion process may be to transmit the energy issued from ATP hydrolysis to other Gsp components. This process may involve a series of protein–protein interactions and conformational changes in the Gsp proteins, conserving energy to push the secreted protein through the machinery and the outer membrane.

GspC, F, L, M and N, the inner membrane proteins

Comparison between type II protein secretion and type IV piliation has already suggested that homologous components such as PilD/GspO, PilB/GspE, PilA/GspG–J, PilX/GspK from *P. aeruginosa* are involved in these processes. Another Pil protein, PilC, is essential in type IV piliation. It has a Gsp homologue, GspF. GspF is a polytopic integral cytoplasmic membrane protein with a small periplasmic loop and two larger cytoplasmic domains connected by three transmembrane regions (Thomas *et al.*, 1997). PilB, C, D and A are clustered in most type IV piliated bacteria, suggesting that PilB and C are simultaneously involved in the translocation/assembly of PilA after processing by PilD (Strom & Lory, 1993). The PilC homologue GspF may have a similar function to PilC with respect to pseudopilins. GspL is required for the association of GspE with the membrane. This function could otherwise have been carried out by GspF. The function of GspF is thus unclear. It may strengthen the association of GspE with the membrane (Thomas *et al.*, 1997), or be directly involved in pore formation in the cytoplasmic membrane, allowing pseudopilin translocation. These hypotheses are, however, purely speculative.

Surprisingly, the membrane anchor of GspE, GspL, has no known homologue in the Pil system. This raises questions about the membrane association of PilB and indicates subtle differences in the assembly of these complex machineries. One *P. aeruginosa* protein, PilN, has a low level of sequence similarity (24% identical residues) to the GspL protein from *Xanthomonas campestris* (Martin *et al.*, 1995). The PilN protein is considerably smaller than GspL and the sequence similarity is limited to the C-terminal end of GspL. The extent to which the similarity between PilN and GspL reflects any conservation of function is unknown, but functional conservation is unlikely given that it is the N-terminus of GspL that is involved in GspE membrane association.

No homologues of GspM and GspC specifically involved in type IV piliation have yet been identified. Like GspL, GspM and GspC are bitopic inner membrane proteins with an N_{in}-C_{out} topology (Bleves *et al.*, 1996). However, unlike GspL, the N-terminal domain of these two proteins is very short. Homologues of the *gspC*, *gspL* and *gspM* genes have been identified by analysing the DNA sequence of *E. coli* K-12, close to a *gspO* gene homologue, *pppa*, which is probably a second prepilin peptidase gene in *E. coli* K-12 (Francetic *et al.*, 1998). These genes have yet to be shown to be involved in type IV piliation or any other process.

The precise function of GspM is unknown, but recent studies have suggested that it is crucial for the stability of GspL (Michel *et al.*, 1998). The stabilization process is reciprocal because the abundance of GspM in the cell depends on GspL, indicating that these two Gsp components interact with each other. It is unknown whether and how the cascade of protein–protein interactions is kinetically organized during the assembly of the machinery. The GspM protein may first determine the membrane location of the secretion site, then recruit the GspL component, which in turn brings GspE into association with the membrane.

The function of the GspC protein is also unknown. Its principal feature is the presence of a PDZ domain in the C-terminal region of the protein (Pallen & Ponting, 1997). PDZ domains are named after the three eukaryotic proteins (*p*ost synaptic density protein, *d*isc large and *z*o-1 proteins) in which they were first discovered. They mediate a variety of protein–protein interactions by binding to short sequences (X-T/S-X-V-COO$^-$), usually at the C-termini of target polypeptides, but may also bind homotypically (Ponting *et al.*, 1997). It is widely thought that PDZ-domain-containing proteins mediate the organization of multimolecular complexes at sites of membrane specialization. Some GspCs have no PDZ domain. This is true of the GspCs of *P. aeruginosa* and *P. alcaligenes*, for example. However, in both cases, the PDZ region has been replaced by a coiled-coil structure, a motif also known to be involved in protein–protein interactions. Hence, GspC proteins may form homomultimers or interact with other proteins by one of two mechanisms, one involving a coiled-coil and the other PDZ domains.

Finally, the GspN protein is also a bitopic inner membrane protein. It is absent from many of the known type II systems, and has been identified only in *K. oxytoca*, *Erwinia carotovora* and *Pseudomonas putida* (Table 1). However, GspN appears to be required for protein secretion in *K. oxytoca* and *Erw. carotovora*. The Gsp system of *Erw. chrysanthemi* has no GspN homologue, but can be efficiently reconstituted in an *E. coli* host whereas the *Erw. carotovora* Gsp system cannot. This raises questions about the non-central role of GspN within the machinery.

GspD, a pore-like structure in the outer membrane

The paradox of the type II secretion apparatus concerns the cell envelope distribution of the Gsp components (Fig. 3). The preponderance of inner membrane proteins is puzzling for a transport system that translocates proteins across the outer membrane. However, GspD proteins have a signal peptide and are associated with the outer membrane (Bitter *et al.*, 1998; Hardie *et al.*, 1996a; Hu *et al.*, 1995). Sequence comparisons have shown that in all cases, the C-terminal domain is highly conserved whereas the N-terminal domain is variable (Genin & Boucher, 1994). The conserved C-terminal region covers 200–300 residues and contains a very highly conserved block of about 60 amino acids containing invariant glycine and proline residues that have been shown to be functionally important (Russel, 1994a). The GspD proteins are multimers consisting of 12–15 subunits (Kazmierczak *et al.*, 1994; Chen *et al.*, 1996; Hardie *et al.*, 1996a; Bitter *et al.*, 1998). They are extremely stable and in most cases are heat- and detergent-resistant. Classical outer membrane porins form multimers (trimers) embedded as β-barrels consisting of amphipathic β-strands. They have an invariant C-terminal phenylalanine residue (Struyve *et al.*, 1991). Members of the GspD family lack this phenylalanine, but secondary structure predictions suggest that they form transmembrane β-strands. In the case of the *P. aeruginosa* GspD, 13 putative transmembrane strands were found in the C-terminal domain of the protein (Bitter *et al.*, 1998). This observation suggests that the C-terminal domain of GspD is required for the insertion of the protein into the outer membrane, whereas the N-terminal domain seems to be involved in multimerization and extends into the periplasm to facilitate interactions with other proteins (see next section).

The multimerization of GspDs makes it possible to envisage the formation of a pore-like structure in the outer membrane. Biochemical and electron microscopy studies have shown that the *P. aeruginosa* GspD multimer can indeed adopt a ring-shaped structure with a large central cavity of about 95 nm in diameter (Bitter *et al.*, 1998). The size of this open hole is consistent with type-II-secretion-dependent exoproteins being brought across the outer membrane in an at least partially folded conformation (see next section). For example, the folded extracellular elastase from *P. aeruginosa* is 60 nm across in its wider dimension (Thayer *et al.*, 1991). The presence of such a large pore in the outer membrane could lead to cell death, so the opening of this pore is probably controlled. Other gated outer membrane channels are known, such as the iron–siderophore complex, vitamin B12 and colicin uptake receptors. The mechanism described involves a conformational change in the receptor and an energy-transducing process requiring proton motive force (pmf) (Moeck & Coulton, 1998). The mechanism of energy transduction to the outer membrane is unclear but it involves a bitopic cytoplasmic membrane protein called TonB. TonB has a large periplasmic domain, and its energized form interacts with the outer membrane receptor. On substrate binding, TonB induces a conformational change of the receptor, leading to the entry of the substrate into the periplasm. It has been shown in more than one case that the translocation of proteins across the outer membrane using the type II secretory apparatus is dependent on pmf (Letellier *et al.*, 1997; Possot *et al.*, 1997). Therefore, a system analogous to the TonB system may be involved in protein secretion. In *P. aeruginosa*, *tonB* mutant strains are not secretion-defective (unpublished observation). However, other proteins, particularly bitopic cytoplasmic membrane proteins of the type II secretion machinery, may have a similar function to TonB. GspC proteins are good candidates for this function for several reasons. The topology of these proteins is similar to that of TonB (Bleves *et al.*, 1996), and they probably interact with GspDs (unpublished observation). In addition, in *P. aeruginosa* and *P. alcaligenes*, the genes encoding these two proteins are organized in a separate operon, suggesting their co-ordinated action (Table 1).

A GspD homologue has been identified in *P. aeruginosa*, XqhA, which is responsible for the residual secretion observed in the original *gspD* mutant (Martinez *et al.*, 1998). The role of XqhA is unclear but it does not seem to be essential to Gsp-dependent secretion because no difference in phenotype was observed between an *xqhA* mutant and the wild-type strain. The GspD family of proteins is involved not only in type II secretion but also in several other unrelated membrane translocation systems (Filloux *et al.*, 1998; Hobbs & Mattick, 1993). Indeed, the GspD family is more generally called the secretin family. Its members are involved in type III secretion, type IV pilus assembly, DNA uptake, assembly of S-layers and filamentous phage assembly and extrusion. In phage extrusion (Russel, 1994b), the pIV secretin interacts with the C-terminal periplasmic domain of the cytoplasmic membrane protein pI (Russel, 1993). pI contains a nucleotide-binding site in its N-terminal cytoplasmic domain, and may be involved in the controlled gating of the pIV pore. The energy transduced for the phage secretion process may therefore come partly from ATP hydrolysis, which has been shown to be required together with the pmf for filamentous phage assembly (Feng *et al.*, 1997).

GspS, the secretin-specific chaperone-like protein

In some cases, the insertion of a GspD protein into the outer membrane depends on another protein, GspS (Hardie *et al.*, 1996a). GspS is a small peripheral outer membrane lipoprotein present in the type II secretory systems of *K. oxytoca* and *Erwinia* species (Table 1). It is essential for the outer membrane insertion of GspD and also protects GspD against proteolytic degradation (Hardie *et al.*, 1996b). The high level of GspD degradation in the absence of GspS may be due to the failure of GspD to insert into the outer membrane. However, GspS is not required for GspD multimerization. GspS proteins are not found in the type II secretion systems of other bacteria, but this does not exclude the possibility that the *gspS* gene in these organisms does not belong to the *gsp* gene cluster and may be present at another chromosomal location. In systems containing GspS, GspD proteins have an additional conserved domain of about 65 amino acids at the extreme C-terminus (Shevchik *et al.*, 1997). The protective effect of GspS against proteolytic degradation has been shown using hybrid proteins containing the *K. oxytoca* GspD C-terminus, which is thought to be the GspS-binding site. The addition of this domain to the pIV secretin or to the unrelated maltose-binding protein rendered these proteins dependent on the *K. oxytoca* GspS for stability (Daefler *et al.*, 1997). GspS-devoid type II systems may compensate by efficiently partitioning GspD into the outer membrane. In contrast, if kinetic partitioning is too slow, the binding of GspS to GspD may prevent the degradation of the protein prior to membrane insertion and may accelerate insertion as well. In unrelated transport processes involving secretins, a GspS-like function has been assigned to PilP in type IV piliation in *N. gonorrhoeae* (Drake *et al.*, 1997), and InvH in type III secretion in *Salmonella typhimurium* (Daefler & Russel, 1998). These two proteins are also small outer membrane lipoproteins but they have no clear sequence similarity to known GspS proteins.

GspA and GspB, the alternative 'energizer'

We have seen that there may be small differences in the composition of type II secretory apparatuses. Two additional Gsp proteins, bringing the number of Gsp components up to 16, have been described in *A. hydrophila*. These two proteins, GspA and GspB, are essential to the protein secretion process (Schoenhofen *et al.*, 1998). The GspB protein has a sequence and structure similar to that of TonB, whereas GspA is a membrane protein with a consensus ATP-binding site. GspA and GspB form a complex within the cytoplasmic membrane and it has been suggested that such a complex transduces energy from ATP hydrolysis, and perhaps also pmf energy, to the protein secretion process. If this is indeed the case, then this complex plays a crucial role, and it is not clear why these components have not been found in any other type II system. A GspA homologue but no GspB homologue has been identified in the *E. coli gsp* gene cluster (Francetic & Pugsley, 1996), which has not been shown to produce or make use of functional type II secretion machinery. In *A. hydrophila*, secretion from the periplasm requires pmf and ATP (Letellier *et al.*, 1997), whereas in *K. oxytoca*, it requires only pmf (Possot *et al.*, 1997). This may partly explain the role of GspA and B in *A. hydrophila*.

EXOPROTEIN RECOGNITION

Secretion by the type II system is a two-step process. The first step, translocation across the bacterial inner membrane, is controlled by the Sec machinery, which recognizes signal-peptide-bearing exoproteins. Once it reaches the periplasm, the mature polypeptide becomes the substrate for the Gsp machinery, which should discriminate between exoproteins and the periplasmic or outer membrane proteins. This process may be based on the recognition of a secretion motif by some of the Gsp proteins.

Species-specific recognition of Gsp-dependent exoproteins

The Gsp secretion apparatus is widespread in Gram-negative bacteria and a wide variety of enzymes and toxins use this pathway. However, Gsp-dependent exoproteins, such as *K. oxytoca* pullulanase, are not recognized by the Gsp machinery of *P. aeruginosa*, demonstrating that the process is specific (de Groot *et al.*, 1991). These two bacteria are not closely related and heterologous secretion has been described in more closely related organisms. For example, the *Pseudomonas glumae* lipase and the *A. hydrophila* aerolysin are secreted by *P. aeruginosa* and *Vibrio*, respectively (Frenken *et al.*, 1993a; Wong *et al.*, 1990). In contrast, even though the cellulases EGZ and CelV from *Erw. chrysanthemi* and *Erw. carotovora*, respectively, are very similar in terms of amino acid sequence (40% identity) and domain organization, these two proteins are recognized exclusively by their own secretion machinery and are not exchangeable (Py *et al.*, 1991). Thus, the principle on which the exoprotein recognition process is based is unclear, but it may rely on the presence of a secretion motif within the secreted protein. *P. aeruginosa* secretes enzymes as diverse as lipase, elastase, alkaline phosphatase and the ADP-ribosylating exotoxin A, all of which use the same secretion machinery. These proteins should, therefore, have a common secretion motif. However, sequence analyses of these proteins have not identified a linear motif of residues. This has led researchers in the field to suggest that there is a non-linear motif, a conformational signal constructed by folding.

Translocation of folded proteins across the outer membrane

Two major observations suggest that Gsp-dependent exoproteins are translocated across the outer membrane in a folded conformation. First, the studies with *Erw. chrysanthemi* cellulases and pectate lyases and *K. oxytoca* pullulanase have demonstrated that disulphide bridges are formed in exoproteins during secretion (Bortoli-German *et al.*, 1994; Pugsley, 1992). The cysteine bond is formed in the periplasm, catalysed by disulphide bond isomerases (Dsb), particularly by DsbA (Shevchik *et al.*, 1995). Second, studies with *V. cholerae* and *A. hydrophila* have shown that the B pentamer of the cholera toxin is formed in the periplasm before its secretion (Hirst & Holmgren, 1987), and that proaerolysin is secreted as a dimer (Hardie *et al.*, 1995). Thus, the secretion motif may be formed and presented to the secretory apparatus upon folding of the exoprotein. It is clear that it is not the disulphide bond itself that is recognized by the machinery. CelV, secreted in a Gsp-dependent manner by *Erw. carotovora*, contains no cysteines (Cooper & Salmond, 1993), and an engineered cysteine-free pullulanase is efficiently secreted (Sauvonnet & Pugsley, 1998). Secretion of the cysteine-free pullulanase is DsbA-dependent, raising questions about the role of DsbA in pullulanase secretion. DsbA may act as a chaperone, affecting pullulanase folding, but it may also be important for disulphide bond formation within Gsp components of the secretion machinery. These results of studies on pullulanase conflict with the work of Bortoli-German and collaborators (Bortoli-German *et al.*, 1994), who showed that site-directed mutagenesis of either of the cysteines in EGZ cellulase prevented secretion. The cysteine bond may act as a clip, stabilizing the protein and fixing the protein in its folded conformation. The requirement for the disulphide bond therefore depends on the stability of the structure adopted by the reduced exoprotein.

The folding of exoprotein within the periplasm seems, in some cases, to involve not only general catalysts such as DsbA, but also specifically dedicated chaperones. This is the case, for example, for the propeptide of *P. aeruginosa* elastase (Braun *et al.*, 1996), and for the LipB protein, which is required for the folding and activation of the *P. glumae* lipase (Frenken *et al.*, 1993b). Recent studies proposed that the PlcR protein is required for full haemolytic activity of *P. aeruginosa* phospholipase C, suggesting that it might also act as a helper for secretion (Cota-Gomez *et al.*, 1997). However, it is not clear whether all individual exoproteins are systematically associated with such foldases because these are the only known examples.

The secretion motif

Gsp-dependent exoproteins acquire a highly ordered structure after reaching the periplasm and before outer membrane translocation. In the absence of a linear motif, the secretion signal for the various exoproteins is therefore probably a patch signal involving distal regions brought into close proximity by the folding of the protein. If this is true, the motif should be very sensitive to any changes in the structure of the exoproteins. This is consistent with the observation made by Py and collaborators (Py *et al.*, 1993), who showed that slight modifications in the *Erw. chrysanthemi* EGZ protein result in non-secreted derivatives. In *A. hydrophila*, a single substitution of the W227 residue yields a protein that is not secreted (Wong & Buckley, 1991). In contrast, linker insertion experiments within *K. oxytoca* pullulanase resulted in the construction of 23 derivatives, all of which were efficiently secreted (Sauvonnet *et al.*, 1995). Some of these derivatives were inactive, indicating changes in the enzyme structure, thereby challenging the existence of the conformational motif. However, it is possible that the motif is formed at an intermediate state of folding rather than in the fully active native protein.

The search for the minimal motifs required for protein secretion could also be tackled by fusing domains of exoproteins to a reporter periplasmic protein such as β-lactamase (Bla) or alkaline phosphatase (PhoA) and selecting those hybrids that are efficiently secreted. It is difficult to reconcile this approach with the notion that there is a conformational signal, but two studies have provided information about short domains involved in type II secretion machinery targeting. Two regions in the *K. oxytoca* pullulanase, A and B, which are well separated from each other in the primary protein sequence, together direct the secretion of a PulA–Bla hybrid (Sauvonnet & Pugsley, 1996). These two domains are therefore probably part of a structural motif which is still formed and recognized within such a hybrid. Studies with the *P. aeruginosa* exotoxin A identified a single stretch of 60 amino acids (60–120) which was sufficient to direct Bla secretion (Lu & Lory, 1996). This region appears to be rich in anti-parallel β-sheets but has no apparent sequence similarity with any region of the other type-II-secretion-dependent exoproteins of *P. aeruginosa* or the A and B motifs of pullulanase. Moreover, another study with *P. aeruginosa* exotoxin A has shown that a truncated protein retaining the first 30 and last 305 residues, but not residues 60–120, is still efficiently secreted (McVay & Hamood, 1995). The nature of the secretion signal is therefore unclear. One alternative to a unique structural motif is that a series of successive interactions leads to the secretion of exoproteins. These interactions may involve different secretion signals that are not essential individually but are required simultaneously for optimal secretion. Finally, the secretion motif may not be recognized by a common component of the secretion machinery but by a specific intermediate, such as the previously described dedicated chaperone, which will in turn be recognized by a Gsp component of the secretion machinery.

Recognition by GspD

One of the Gsp components involved in the cascade of recognition events required to direct the exoprotein is the GspD secretin. The pectate lyase PelB of *Erw. chrysanthemi* binds to the N-terminus of GspD (Shevchik *et al.*, 1997). The N-terminus is the most variable region of the protein, consistent with the notion of specificity for recognition of the exoprotein. This observation is also consistent with the fact that, of all the Gsp components of *Erw. chrysanthemi* required for pectate lyase secretion, only two cannot be replaced by Gsp homologues from *Erw. carotovora*, namely GspC and GspD (Lindeberg *et al.*, 1996). This suggests that these two proteins are intimately involved in the species-specific recognition of the exoprotein by the Gsp system. Another possibility, however, is that these proteins are not incorporated into the heterologous system, because they fail to make the specific interactions with the other Gsp components that are required for the assembly of the secretion machinery.

A GSP SECRETION MODEL

With the current state of knowledge of type II secretory mechanisms, it is difficult to clearly envisage the precise molecular mechanisms that specifically drive proteins across the outer membrane. However, based on the observations described in this chapter, it is possible to formulate ideas about several of the main features of the machinery (Fig. 5). Secretion occurs in two steps, as shown by Poquet and collaborators (Poquet *et al.*, 1993), and the Gsp components of the secretory apparatus are involved only in the translocation of the periplasmic intermediate of the exoprotein across the outer membrane. The secretion machinery consists of 12–16 components, and the assembly of the macromolecular membrane complex and the recognition of the secreted proteins can be summarized in several points.

(1) Based on homology with the piliation assembly system, it is possible to imagine the assembly of a so-called 'pseudopilus'. Assembly would require a peptidase, an ATP-binding 'energizer' protein and possibly several inner membrane proteins to achieve the translocation of pseudopilins across the inner membrane. These components, and this first phase of molecular events, may not be directly involved in outer membrane translocation of exoproteins, but rather in assembly of a subcomplex of the secretion machinery.

(2) The final channel, allowing exoproteins to reach the extracellular medium, is made of secretin (GspD), which multimerizes (12–15 subunits) and may form a large hole (95 nm) within the outer membrane. In some cases, GspD outer membrane insertion requires the GspS outer membrane lipoprotein.

(3) The controlled opening of the secretin channel may require energy, from pmf or ATP. Transduction of this energy to the outer membrane may be achieved via large periplasmic domains of cytoplasmic membrane proteins interacting with the secretin. GspC and GspA/GspB complex are good candidates for this function.

(4) The assembly of the 'pseudopilus' (mentioned in 1) may be a key element, helping to push the exoprotein through the secretin for its final release. Alternatively, the pseudopilus may act like a cork, blocking the secretin channel, with secretion occurring only if the structure is retracted. The pseudopilus may also serve as a guide for routing exoproteins to the secretin channel.

(5) Despite the conservation of Gsp systems, most Gsp components are not exchangeable, suggesting that there is specificity within the machinery itself. This specificity, and the probability of producing a functional hybrid machinery, depends on the phylogenetic distance between the organisms involved. As described above, in *Erwinia* species, only GspC and GspD are not exchangeable. More generally, the GspO peptidase may be functionally moved from one species to another (Dupuy *et al.*, 1992; de Groot *et al.*, 1994), probably because it has an enzymic rather than a structural function. This idea is supported by the observation that GspE hybrids may be functional if the variable N-terminus is in a homologous context, whereas the ATP-binding-site-containing C-terminus may be exchanged between species (Sandkvist *et al.*, 1995; de Groot *et al.*, 1996). Apart from that, the most permissive Gsp component is GspG, which can be exchanged, more or less efficiently, in several cases (Francetic & Pugsley, 1996; Lindeberg *et al.*, 1996; Pugsley, 1996; de Groot *et al.*, 1999).

(6) The secreted proteins must have features (secretion motifs) that mediate specific recognition by Gsp components, such as the secretin GspD. The exoproteins are folded in the periplasm and the secretion motif is probably conformational rather than sequential. As they are folded, the exoproteins need the large hole formed by the secretin for their translocation across the outer membrane. This translocation of folded exoproteins contrasts with the 'dogmatic' unfolded competent state for translocation across the bacterial cytoplasmic membrane via the Sec machinery. However, a novel Sec-independent translocation pathway has been shown to translocate metalloenzymes into the periplasm in a stably folded conformation (Santini *et al.*, 1998).

CONCLUDING REMARKS

It is clear that the components of the type II secretion machinery have now been well-established. However, we do not yet understand the precise organization of the machinery and the mechanisms controlling its secretion function. The existence and function of the 'pseudopilus' are among the most controversial features of the type II secretion apparatus, its prediction being mainly based on homology with the type IV pili assembly system. However, the relationship between type III secretion and flagella assembly systems hinged on the structure of part of the secretion machinery, which looked like the base of a flagellum when viewed with an electron microscope (Kubori *et al.*, 1998).

Gsp-like components are also involved in the transport and assembly of various macromolecules through the membranes of Gram-negative bacteria. Therefore, specific mechanisms may be required to adapt regions of the cell envelope, including the peptidoglycan layer, to such transport processes. An important question, which should be addressed in the near future, concerns the location and the distribution of the secretion sites within the bacterial cell envelope. It is possible that these sites are not spread out evenly over the envelope but are rather confined to particular areas, such as the pole of the cell. This distribution would be reminiscent of that for type IV pili, and would limit leakage zones to a part of the bacterial cell in which the periplasm is compartmented (Foley *et al.*, 1989), reducing the effect of the transient opening of large holes in the outer membrane on the integrity of the cell. In this region the peptidoglycan layer may be looser, making the assembly of the machinery and secretion of the exoproteins easier.

ACKNOWLEDGEMENTS

Work cited here that was carried out in the author's laboratory was partly supported by the European Community, E.U. grant Bio4-CT960119.

REFERENCES

Alm, R. A. & Mattick, J. S. (1997). Genes involved in the biogenesis and function of type-4 fimbriae in *Pseudomonas aeruginosa*. *Gene* **192**, 89–98.

Ball, G., Chapon-Hervé, V., Bleves, S., Michel, G. & Bally, M. (1999). Assembly of XcpR in the cytoplasmic membrane is required for extracellular protein secretion in *Pseudomonas aeruginosa*. *Journal of Bacteriology* **181**, 382–388.

Bally, M., Filloux, A., Akrim, M., Ball, G., Lazdunski, A. & Tommassen, J. (1992). Protein secretion in *Pseudomonas aeruginosa*: characterization of seven *xcp* genes and processing of secretory apparatus components by prepilin peptidase [published erratum appears in *Molecular Microbiology* 1992; 6, 2745]. *Molecular Microbiology* **6**, 1121–1131.

Bitter, W., Koster, M., Latijnhouwers, M., de Cock, H. & Tommassen, J. (1998). Formation of oligomeric rings by XcpQ and PilQ, which are involved in protein transport across the outer membrane of *Pseudomonas aeruginosa*. *Molecular Microbiology* **27**, 209–219.

Bleves, S., Lazdunski, A. & Filloux, A. (1996). Membrane topology of three Xcp proteins involved in exoprotein transport by *Pseudomonas aeruginosa*. *Journal of Bacteriology* **178**, 4297–4300.

Bleves, S., Voulhoux, R., Michel, G., Lazdunski, A., Tommassen, J. & Filloux, A.

(1998). The secretion apparatus of *Pseudomonas aeruginosa*: identification of a fifth pseudopilin, XcpX (GspK family). *Molecular Microbiology* **27**, 31–40.

Bortoli-German, I., Brun, E., Py, B., Chippaux, M. & Barras, F. (1994). Periplasmic disulphide bond formation is essential for cellulase secretion by the plant pathogen *Erwinia chrysanthemi*. *Molecular Microbiology* **11**, 545–553.

Braun, P., Tommassen, J. & Filloux, A. (1996). Role of the propeptide in folding and secretion of elastase of *Pseudomonas aeruginosa*. *Molecular Microbiology* **19**, 297–306.

Braun, V., Hobbie, S. & Ondraczek, R. (1992). *Serratia marcescens* forms a new type of cytolysin. *FEMS Microbiology Letters* **79**, 299–305.

Browne, B. L., McClendon, V. & Bedwell, D. M. (1996). Mutations within the first LSGGQ motif of Ste6p cause defects in a-factor transport and mating in *Saccharomyces cerevisiae*. *Journal of Bacteriology* **178**, 1712–1719.

Chen, L. Y., Chen, D. Y., Miaw, J. & Hu, N. T. (1996). XpsD, an outer membrane protein required for protein secretion by *Xanthomonas campestris* pv. *campestris*, forms a multimer. *Journal of Biological Chemistry* **271**, 2703–2708.

Cooper, V. J. & Salmond, G. P. (1993). Molecular analysis of the major cellulase (CelV) of *Erwinia carotovora*: evidence for an evolutionary "mix-and-match" of enzyme domains. *Molecular & General Genetics* **241**, 341–350.

Cota-Gomez, A., Vasil, A. I., Kadurugamuwa, J., Beveridge, T. J., Schweizer, H. P. & Vasil, M. L. (1997). PlcR1 and PlcR2 are putative calcium-binding proteins required for secretion of the hemolytic phospholipase C of *Pseudomonas aeruginosa*. *Infection and Immunity* **65**, 2904–2913.

Daefler, S. & Russel, M. (1998). The *Salmonella typhimurium* InvH protein is an outer membrane lipoprotein required for the proper localization of InvG. *Molecular Microbiology* **28**, 1367–1380.

Daefler, S., Guilvout, I., Hardie, K. R., Pugsley, A. P. & Russel, M. (1997). The C-terminal domain of the secretin PulD contains the binding site for its cognate chaperone, PulS, and confers PulS dependence on pIVf1 function. *Molecular Microbiology* **24**, 465–475.

Drake, S. L., Sandstedt, S. A. & Koomey, M. (1997). PilP, a pilus biogenesis lipoprotein in *Neisseria gonorrhoeae*, affects expression of PilQ as a high-molecular-mass multimer. *Molecular Microbiology* **23**, 657–668.

Duong, F., Eichler, J., Price, A., Leonard, M. R. & Wickner, W. (1997). Biogenesis of the gram-negative bacterial envelope. *Cell* **91**, 567–573.

Dupuy, B., Taha, M. K., Possot, O., Marchal, C. & Pugsley, A. P. (1992). PulO, a component of the pullulanase secretion pathway of *Klebsiella oxytoca*, correctly and efficiently processes gonococcal type IV prepilin in *Escherichia coli*. *Molecular Microbiology* **6**, 1887–1894.

d'Enfert, C., Ryter, A. & Pugsley, A. P. (1987). Cloning and expression in *Escherichia coli* of the *Klebsiella pneumoniae* genes for production, surface localization and secretion of the lipoprotein pullulanase. *EMBO Journal* **6**, 3531–3538.

Feng, J. N., Russel, M. & Model, P. (1997). A permeabilized cell system that assembles filamentous bacteriophage. *Proceedings of the National Academy of Sciences, USA* **94**, 4068–4073.

Filloux, A., Bally, M., Ball, G., Akrim, M., Tommassen, J. & Lazdunski, A. (1990). Protein secretion in gram-negative bacteria: transport across the outer membrane involves common mechanisms in different bacteria. *EMBO Journal* **9**, 4323–4329.

Filloux, A., Michel, G. & Bally, M. (1998). GSP-dependent protein secretion in Gram-negative bacteria: the Xcp system of *Pseudomonas aeruginosa*. *FEMS Microbiology Reviews* **22**, 177–198.

Foley, M., Brass, J. M., Birmingham, J., Cook, W. R., Garland, P. B., Higgins, C. F. & Rothfield, L. I. (1989). Compartmentalization of the periplasm at cell division sites in *Escherichia coli* as shown by fluorescence photobleaching experiments. *Molecular Microbiology* **3**, 1329–1336.

Francetic, O. & Pugsley, A. P. (1996). The cryptic general secretory pathway (*gsp*) operon of *Escherichia coli* K-12 encodes functional proteins. *Journal of Bacteriology* **178**, 3544–3549.

Francetic, O., Lory, S. & Pugsley, A. P. (1998). A second prepilin peptidase gene in *Escherichia coli* K-12. *Molecular Microbiology* **27**, 763–775.

Frenken, L. G., de Groot, A., Tommassen, J. & Verrips, C. T. (1993a). Role of the *lipB* gene product in the folding of the secreted lipase of *Pseudomonas glumae*. *Molecular Microbiology* **9**, 591–599.

Frenken, L. G., Bos, J. W., Visser, C., Muller, W., Tommassen, J. & Verrips, C. T. (1993b). An accessory gene, *lipB*, required for the production of active *Pseudomonas glumae* lipase. *Molecular Microbiology* **9**, 579–589.

Genin, S. & Boucher, C. A. (1994). A superfamily of proteins involved in different secretion pathways in gram-negative bacteria: modular structure and specificity of the N-terminal domain. *Molecular & General Genetics* **243**, 112–118.

Gerritse, G., Ure, R., Bizoullier, F. & Quax, W. J. (1998). The phenotype enhancement method identifies the Xcp outer membrane secretion machinery from *Pseudomonas alcaligenes* as a bottleneck for lipase production. *Journal of Biotechnology* **64**, 23–38.

de Groot, A., Filloux, A. & Tommassen, J. (1991). Conservation of *xcp* genes, involved in the two-step protein secretion process, in different *Pseudomonas* species and other gram-negative bacteria. *Molecular & General Genetics* **229**, 278–284.

de Groot, A., Heijnen, I., de Cock, H., Filloux, A. & Tommassen, J. (1994). Characterization of type IV pilus genes in plant growth-promoting *Pseudomonas putida* WCS358. *Journal of Bacteriology* **176**, 642–650.

de Groot, A., Krijger, J. J., Filloux, A. & Tommassen, J. (1996). Characterization of type II protein secretion (*xcp*) genes in the plant growth-stimulating *Pseudomonas putida*, strain WCS358. *Molecular & General Genetics* **250**, 491–504.

de Groot, A., Gerritse, G., Tommassen, J., Lazdunski, A. & Filloux, A. (1999). Molecular organization of the *xcp* gene cluster in *Pseudomonas putida*: absence of an *xcpX* (*gspK*) homologue. *Gene* **226**, 35–40.

Hahn, H. P. (1997). The type-4 pilus is the major virulence-associated adhesin of *Pseudomonas aeruginosa* – a review. *Gene* **192**, 99–108.

Hardie, K. R., Schulze, A., Parker, M. W. & Buckley, J. T. (1995). *Vibrio* spp. secrete proaerolysin as a folded dimer without the need for disulphide bond formation. *Molecular Microbiology* **17**, 1035–1044.

Hardie, K. R., Lory, S. & Pugsley, A. P. (1996a). Insertion of an outer membrane protein in *Escherichia coli* requires a chaperone-like protein. *EMBO Journal* **15**, 978–988.

Hardie, K. R., Seydel, A., Guilvout, I. & Pugsley, A. P. (1996b). The secretin-specific, chaperone-like protein of the general secretory pathway: separation of proteolytic protection and piloting functions. *Molecular Microbiology* **22**, 967–976.

von Heijne, G. (1985). Signal sequences. The limits of variation. *Journal of Molecular Biology* **184**, 99–105.

Henderson, I. R., Navarro-Garcia, F. & Nataro, J. P. (1998). The great escape: structure and function of the autotransporter proteins. *Trends in Microbiology* **6**, 370–378.

Henrichsen, J. (1983). Twitching motility. *Annual Review of Microbiology* **37**, 81–93.

Higgins, C. F. (1992). ABC transporters: from microorganisms to man. *Annual Review of Cell Biology* **8**, 67–113.

Hirst, T. R. & Holmgren, J. (1987). Conformation of protein secreted across bacterial outer membranes: a study of enterotoxin translocation from *Vibrio cholerae*. *Proceedings of the National Academy of Sciences, USA* **84**, 7418–7422.

Hobbs, M. & Mattick, J. S. (1993). Common components in the assembly of type 4 fimbriae, DNA transfer systems, filamentous phage and protein-secretion apparatus: a general system for the formation of surface-associated protein complexes. *Molecular Microbiology* **10**, 233–243.

Howard, S. P., Critch, J. & Bedi, A. (1993). Isolation and analysis of eight *exe* genes and their involvement in extracellular protein secretion and outer membrane assembly in *Aeromonas hydrophila*. *Journal of Bacteriology* **175**, 6695–6703.

Hu, N. T., Hung, M. N., Liao, C. T. & Lin, M. H. (1995). Subcellular location of XpsD, a protein required for extracellular protein secretion by *Xanthomonas campestris* pv. *campestris*. *Microbiology* **141**, 1395–1406.

Kagami, Y., Ratliff, M., Surber, M., Martinez, A. & Nunn, D. N. (1998). Type II protein secretion by *Pseudomonas aeruginosa*: genetic suppression of a conditional mutation in the pilin-like component XcpT by the cytoplasmic component XcpR. *Molecular Microbiology* **27**, 221–233.

Kazmierczak, B. I., Mielke, D. L., Russel, M. & Model, P. (1994). pIV, a filamentous phage protein that mediates phage export across the bacterial cell envelope, forms a multimer. *Journal of Molecular Biology* **238**, 187–198.

Kubori, T., Matsushima, Y., Nakamura, D., Uralil, J., Lara-Tejero, M., Sukhan, A., Galan, J. E. & Aizawa, S. I. (1998). Supramolecular structure of the *Salmonella typhimurium* type III protein secretion system. *Science* **280**, 602–605.

Letellier, L., Howard, S. P. & Buckley, J. T. (1997). Studies on the energetics of proaerolysin secretion across the outer membrane of *Aeromonas* species. Evidence for a requirement for both the protonmotive force and ATP. *Journal of Biological Chemistry* **272**, 11109–11113.

Lill, R., Cunningham, K., Brundage, L. A., Ito, K., Oliver, D. & Wickner, W. (1989). SecA protein hydrolyzes ATP and is an essential component of the protein translocation ATPase of *Escherichia coli*. *EMBO Journal* **8**, 961–966.

Lindeberg, M., Salmond, G. P. & Collmer, A. (1996). Complementation of deletion mutations in a cloned functional cluster of *Erwinia chrysanthemi out* genes with *Erwinia carotovora out* homologues reveals OutC and OutD as candidate gatekeepers of species-specific secretion of proteins via the type II pathway. *Molecular Microbiology* **20**, 175–190.

Lu, H. M. & Lory, S. (1996). A specific targeting domain in mature exotoxin A is required for its extracellular secretion from *Pseudomonas aeruginosa*. *EMBO Journal* **15**, 429–436.

Lu, H. M., Motley, S. T. & Lory, S. (1997). Interactions of the components of the general secretion pathway: role of *Pseudomonas aeruginosa* type IV pilin subunits in complex formation and extracellular protein secretion. *Molecular Microbiology* **25**, 247–259.

McVay, C. S. & Hamood, A. N. (1995). Toxin A secretion in *Pseudomonas aeruginosa*: the role of the first 30 amino acids of the mature toxin. *Molecular & General Genetics* **249**, 515–525.

Martin, P. R., Watson, A. A., McCaul, T. F. & Mattick, J. S. (1995). Characterization of a five-gene cluster required for the biogenesis of type 4 fimbriae in *Pseudomonas aeruginosa*. *Molecular Microbiology* **16**, 497–508.

Martinez, A., Ostrovsky, P. & Nunn, D. N. (1998). Identification of an additional member of the secretin superfamily of proteins in *Pseudomonas aeruginosa* that is able to function in type II protein secretion. *Molecular Microbiology* **28**, 1235–1246.

Michel, G., Bleves, S., Ball, G., Lazdunski, A. & Filloux, A. (1998). Mutual stabilization of the XcpZ and XcpY components of the secretory apparatus in *Pseudomonas aeruginosa*. *Microbiology* **144**, 3379–3386.

Moeck, G. S. & Coulton, J. W. (1998). TonB-dependent iron acquisition: mechanisms of siderophore-mediated active transport. *Molecular Microbiology* **28**, 675–681.

Nunn, D. N. & Lory, S. (1991). Product of the *Pseudomonas aeruginosa* gene *pilD* is a prepilin leader peptidase. *Proceedings of the National Academy of Sciences, USA* **88**, 3281–3285.

Nunn, D. N. & Lory, S. (1992). Components of the protein-excretion apparatus of *Pseudomonas aeruginosa* are processed by the type IV prepilin peptidase. *Proceedings of the National Academy of Sciences, USA* **89**, 47–51.

Nunn, D. N. & Lory, S. (1993). Cleavage, methylation, and localization of the *Pseudomonas aeruginosa* export proteins XcpT, -U, -V, and -W. *Journal of Bacteriology* **175**, 4375–4382.

Nunn, D., Bergman, S. & Lory, S. (1990). Products of three accessory genes, *pilB*, *pilC*, and *pilD*, are required for biogenesis of *Pseudomonas aeruginosa* pili. *Journal of Bacteriology* **172**, 2911–2919.

Pallen, M. J. & Ponting, C. P. (1997). PDZ domains in bacterial proteins. *Molecular Microbiology* **26**, 411–413.

Parge, H. E., Forest, K. T., Hickey, M. J., Christensen, D. A., Getzoff, E. D. & Tainer, J. A. (1995). Structure of the fibre-forming protein pilin at 2.6 Å resolution. *Nature* **378**, 32–38.

Pepe, C. M., Eklund, M. W. & Strom, M. S. (1996). Cloning of an *Aeromonas hydrophila* type IV pilus biogenesis gene cluster: complementation of pilus assembly functions and characterization of a type IV leader peptidase/N-methyltransferase required for extracellular protein secretion. *Molecular Microbiology* **19**, 857–869.

Ponting, C. P., Phillips, C., Davies, K. E. & Blake, D. J. (1997). PDZ domains: targeting signalling molecules to sub-membranous sites. *Bioessays* **19**, 469–479.

Poquet, I., Faucher, D. & Pugsley, A. P. (1993). Stable periplasmic secretion intermediate in the general secretory pathway of *Escherichia coli*. *EMBO Journal* **12**, 271–278.

Possot, O. & Pugsley, A. P. (1994). Molecular characterization of PulE, a protein required for pullulanase secretion. *Molecular Microbiology* **12**, 287–299.

Possot, O. M. & Pugsley, A. P. (1997). The conserved tetracysteine motif in the general secretory pathway component PulE is required for efficient pullulanase secretion. *Gene* **192**, 45–50.

Possot, O., d'Enfert, C., Reyss, I. & Pugsley, A. P. (1992). Pullulanase secretion in *Escherichia coli* K-12 requires a cytoplasmic protein and a putative polytopic cytoplasmic membrane protein. *Molecular Microbiology* **6**, 95–105.

Possot, O. M., Letellier, L. & Pugsley, A. P. (1997). Energy requirement for pullulanase secretion by the main terminal branch of the general secretory pathway. *Molecular Microbiology* **24**, 457–464.

Pugsley, A. P. (1992). Translocation of a folded protein across the outer membrane in *Escherichia coli*. *Proceedings of the National Academy of Sciences, USA* **89**, 12058–12062.

Pugsley, A. P. (1993). Processing and methylation of PulG, a pilin-like component of the general secretory pathway of *Klebsiella oxytoca*. *Molecular Microbiology* **9**, 295–308.

Pugsley, A. P. (1996). Multimers of the precursor of a type IV pilin-like component of the general secretory pathway are unrelated to pili. *Molecular Microbiology* **20**, 1235–1245.

Pugsley, A. P. & Dupuy, B. (1992). An enzyme with type IV prepilin peptidase activity is required to process components of the general extracellular protein secretion pathway of *Klebsiella oxytoca*. *Molecular Microbiology* **6**, 751–760.

Py, B., Salmond, G. P. C., Chippaux, M. & Barras, F. (1991). Secretion of cellulases in

Erwinia chrysanthemi and *E. carotovora* is species-specific. *FEMS Microbiology Letters* **79**, 315–322.

Py, B., Chippaux, M. & Barras, F. (1993). Mutagenesis of cellulase EGZ for studying the general protein secretory pathway in *Erwinia chrysanthemi*. *Molecular Microbiology* **7**, 785–793.

Reeves, P. J., Douglas, P. & Salmond, G. P. (1994). beta-Lactamase topology probe analysis of the OutO NMePhe peptidase, and six other Out protein components of the *Erwinia carotovora* general secretion pathway apparatus. *Molecular Microbiology* **12**, 445–457.

Russel, M. (1993). Protein-protein interactions during filamentous phage assembly. *Journal of Molecular Biology* **231**, 689–697.

Russel, M. (1994a). Mutants at conserved positions in gene IV, a gene required for assembly and secretion of filamentous phages. *Molecular Microbiology* **14**, 357–369.

Russel, M. (1994b). Phage assembly: a paradigm for bacterial virulence factor export? *Science* **265**, 612–614.

Sandkvist, M., Bagdasarian, M., Howard, S. P. & DiRita, V. J. (1995). Interaction between the autokinase EpsE and EpsL in the cytoplasmic membrane is required for extracellular secretion in *Vibrio cholerae*. *EMBO Journal* **14**, 1664–1673.

Santini, C. L., Ize, B., Chanal, A., Muller, M., Giordano, G. & Wu, L. F. (1998). A novel sec-independent periplasmic protein translocation pathway in *Escherichia coli*. *EMBO Journal* **17**, 101–112.

Sauvonnet, N. & Pugsley, A. P. (1996). Identification of two regions of *Klebsiella oxytoca* pullulanase that together are capable of promoting beta-lactamase secretion by the general secretory pathway. *Molecular Microbiology* **22**, 1–7.

Sauvonnet, N. & Pugsley, A. P. (1998). The requirement for DsbA in pullulanase secretion is independent of disulphide bond formation in the enzyme. *Molecular Microbiology* **27**, 661–667.

Sauvonnet, N., Poquet, I. & Pugsley, A. P. (1995). Extracellular secretion of pullulanase is unaffected by minor sequence changes but is usually prevented by adding reporter proteins to its N- or C-terminal end. *Journal of Bacteriology* **177**, 5238–5246.

Schatz, G. & Dobberstein, B. (1996). Common principles of protein translocation across membranes. *Science* **271**, 1519–1526.

Schoenhofen, I. C., Stratilo, C. & Howard, S. P. (1998). An ExeAB complex in the type II secretion pathway of *Aeromonas hydrophila*: effect of ATP-binding cassette mutations on complex formation and function. *Molecular Microbiology* **29**, 1237–1247.

Shevchik, V. E., Bortoli-German, I., Robert-Baudouy, J., Robinet, S., Barras, F. & Condemine, G. (1995). Differential effect of *dsbA* and *dsbC* mutations on extracellular enzyme secretion in *Erwinia chrysanthemi*. *Molecular Microbiology* **16**, 745–753.

Shevchik, V. E., Robert-Baudouy, J. & Condemine, G. (1997). Specific interaction between OutD, an *Erwinia chrysanthemi* outer membrane protein of the general secretory pathway, and secreted proteins. *EMBO Journal* **16**, 3007–3016.

Strom, M. S. & Lory, S. (1991). Amino acid substitutions in pilin of *Pseudomonas aeruginosa*. Effect on leader peptide cleavage, amino-terminal methylation, and pilus assembly. *Journal of Biological Chemistry* **266**, 1656–1664.

Strom, M. S. & Lory, S. (1993). Structure-function and biogenesis of the type IV pili. *Annual Review of Microbiology* **47**, 565–596.

Struyve, M., Moons, M. & Tommassen, J. (1991). Carboxy-terminal phenylalanine is essential for the correct assembly of a bacterial outer membrane protein. *Journal of Molecular Biology* **218**, 141–148.

Thayer, M. M., Flaherty, K. M. & McKay, D. B. (1991). Three-dimensional structure of the elastase of *Pseudomonas aeruginosa* at 1.5-Å resolution. *Journal of Biological Chemistry* **266**, 2864–2871.

Thomas, J. D., Reeves, P. J. & Salmond, G. P. (1997). The general secretion pathway of *Erwinia carotovora* subsp. *carotovora*: analysis of the membrane topology of OutC and OutF. *Microbiology* **143**, 713–720.

Turner, L. R., Lara, J. C., Nunn, D. N. & Lory, S. (1993). Mutations in the consensus ATP-binding sites of XcpR and PilB eliminate extracellular protein secretion and pilus biogenesis in *Pseudomonas aeruginosa*. *Journal of Bacteriology* **175**, 4962–4969.

Turner, L. R., Olson, J. W. & Lory, S. (1997). The XcpR protein of *Pseudomonas aeruginosa* dimerizes via its N-terminus. *Molecular Microbiology* **26**, 877–887.

Walker, J. E., Saraste, M., Runswick, M. J. & Gay, N. J. (1982). Distantly related sequences in the alpha- and beta-subunits of ATP synthase, myosin, kinases and other ATP-requiring enzymes and a common nucleotide binding fold. *EMBO Journal* **1**, 945–951.

Whitchurch, C. B. & Mattick, J. S. (1994). Characterization of a gene, *pilU*, required for twitching motility but not phage sensitivity in *Pseudomonas aeruginosa*. *Molecular Microbiology* **13**, 1079–1091.

Winans, S. C., Burns, D. L. & Christie, P. J. (1996). Adaptation of a conjugal transfer system for the export of pathogenic macromolecules. *Trends in Microbiology* **4**, 64–68.

Wong, K. R. & Buckley, J. T. (1991). Site-directed mutagenesis of a single tryptophan near the middle of the channel-forming toxin aerolysin inhibits its transfer across the outer membrane of *Aeromonas salmonicida*. *Journal of Biological Chemistry* **266**, 14451–14456.

Wong, K. R., McLean, D. M. & Buckley, J. T. (1990). Cloned aerolysin of *Aeromonas hydrophila* is exported by a wild-type marine *Vibrio* strain but remains periplasmic in pleiotropic export mutants. *Journal of Bacteriology* **172**, 372–376.

Zupan, J. R., Ward, D. & Zambryski, P. (1998). Assembly of the VirB transport complex for DNA transfer from *Agrobacterium tumefaciens* to plant cells. *Current Opinion in Microbiology* **1**, 649–655.

TYPE III SECRETION AND THE PATHOGENESIS OF *YERSINIA* INFECTIONS

DEBORAH M. ANDERSON, LUISA W. CHENG, VINCENT T. LEE, SARKIS MASMANIAN, KUMARAN RAMAMURTHI, CHRISTINA TAM AND OLAF SCHNEEWIND

Department of Microbiology & Immunology, UCLA School of Medicine, 10833 Le Conte Avenue, Los Angeles, CA 90095, USA

PATHOGENESIS OF *YERSINIA* INFECTIONS

Pathogenic yersiniae infect human bodies at different sites. *Yersinia pseudotuberculosis* and *Yersinia enterocolitica* are enteric pathogens that invade the intestinal mucosa via M cells (Autenrieth & Firsching, 1996; Grutzkau *et al.*, 1990; Hanski *et al.*, 1989), a lymphoid organ that samples intestinal contents and delivers them to macrophages (Neutra *et al.*, 1996). The invading yersiniae are drained to the local intestinal lymph nodes (Peyer's patches) where they multiply to form micro-colonies and abscesses, thereby causing acute intestinal infections (Hanski *et al.*, 1989). *Yersinia pestis*, on the other hand, typically invades humans via bites of its insect vector, the flea (Butler, 1983). Plague is a zoonotic infection and rats or other rodents as well as mammals serve as a reservoir for *Y. pestis*. Following inoculation into the skin, *Y. pestis* is drained to and multiplies within local or regional lymph nodes to cause a fulminant lymph adenitis, commonly referred to as bubonic plague (Butler, 1983). If the host cannot clear this infection, *Y. pestis* spreads systemically, eventually colonizing the lungs of infected individuals. Pneumonic plague permits the direct transmission of *Y. pestis* between humans via respiratory droplet infections and the epidemic spread of *Y. pestis* in urban dwellings. Such infections have a rather poor prognosis as pneumonic tissues do not present a formidable defence to the invading pathogen (Butler, 1995).

TYPE III SECRETION OF YOP PROTEINS

To establish disease, all three pathogenic *Yersinia* species, *Y. pestis*, *Y. pseudotuberculosis* and *Y. enterocolitica*, require the presence of a 70 kb virulence plasmid (Gemski *et al.*, 1980b; Zink *et al.*, 1980). Encoded on this genetic element are some 50 genes that specify a type III secretion machinery and its export substrates called Yop proteins (*Yersinia* outer proteins) (Cornelis *et al.*, 1998; Perry *et al.*, 1998; Persson *et al.*, 1995). During infection yersiniae dock on the surface of macrophages or other immune cells and direct Yops into the cytosol of the eukaryotic target cell (Rosqvist *et al.*, 1991, 1994). This process prevents phagocytosis of yersiniae and allows the establishment of bacterial infections within lymphoid tissues (Rosqvist *et al.*, 1988, 1990). Some Yop proteins are also secreted into the extracellular milieu and presumably modulate the cellular immune response at a distance from the site of infection (Lee & Schneewind, 1999; Beuscher *et al.*, 1995).

Higuchi and Smith demonstrated that virulent *Y. pestis* requires calcium in the culture medium for growth at 37 °C (Higuchi *et al.*, 1959; Higuchi & Smith, 1961). Pathogenic *Y. pseudotuberculosis* as well as *Y. enterocolitica* have a similar requirement for calcium in the growth medium, whereas non-pathogenic, virulence-plasmid-less yersiniae do not (Gemski *et al.*, 1980a). Temperature shift to 37 °C and low calcium trigger pathogenic yersiniae to massively secrete Yop proteins into the extracellular milieu (Michiels *et al.*, 1990; Portnoy *et al.*, 1981). Due to the substantial biosynthetic expense of this type III secretion pathway, yersiniae either divide slowly or stop dividing altogether (Higuchi & Smith, 1961). It is not clear whether or not low calcium provides a bona fide signal for type III secretion during *Yersinia* infection of its mammalian hosts. The extracellular fluids of mammals contain approximately 1–2 mM free calcium, a level that suppresses all type III secretion in artificial media. Nonetheless, as observed in live animals or during the infection of tissue cultures, contact with eukaryotic cells at 37 °C provides a signal for the secretion of some Yops into the extracellular milieu (type III secretion) and for the injection of other Yops into eukaryotic target cells (type III targeting) (Lee *et al.*, 1998; Rosqvist *et al.*, 1994). It should be noted that induction of the type III machine via target cell contact does not cause yersiniae to stop or even slow division (Lee *et al.*, 1998).

TYPE III GENES ENCODED ON THE *YERSINIA* VIRULENCE PLASMID

Goguen and Yother sought to identify genes involved in the 'low-calcium response' (*lcr*) and plated populations of *Y. pestis* transposon mutants at 37 °C on low-calcium medium (Goguen *et al.*, 1984). Colonies that formed under these conditions represent mutational variants defective in the secretion of Yop proteins. About half of all transposon mutants mapped to the *Yersinia* virulence plasmid. Most of the insertions prevented expression of genes required for the assembly of the type III machinery, thereby restoring growth under low-calcium conditions (Allaoui *et al.*, 1994, 1995; Bergmann *et al.*, 1991, 1994; Michiels *et al.*, 1991; Plano *et al.*, 1991; Plano & Straley, 1993, 1995; Woestyn *et al.*, 1994). Some of these genes are called *lcr*, while others have recently been referred to as *ysc* for 'Yop secretion' (Michiels *et al.*, 1991). A negative selection scheme was employed to search for *Yersinia* genes involved in sensing calcium (Yother & Goguen, 1985; Yother *et al.*, 1986). Transposon mutants were selected at 37 °C and on low calcium in the presence of penicillin. As β-lactam antibiotics kill only dividing bacteria, the survivors of this selection must be temperature-sensitive for growth or unable to grow altogether. Hence colonies that formed after plating at low temperature in the absence of penicillin were tested for a temperature-sensitive growth phenotype (*ts* or calcium-blind mutants) (Yother & Goguen, 1985; Yother *et al.*, 1986). For example, transposon insertions in *lcrE* answered this selection (Yother *et al.*, 1986) and the gene product, now named YopN, is thought to be involved in sensing the low-calcium signal (Forsberg *et al.*, 1991). Work over the past decade showed that at least 10 genes are involved in the low-calcium signalling pathway [*yopN*, *tyeA*, *sycN*, *yscB*, *yscM1* (*lcrQ*), *yscM2*, *lcrG*, *lcrV*, *lcrH* and *yopD*], most of which are located in two operons, *lcrGVHyopBD* and *yopNtyeAsycN* (Bergmann *et al.*, 1991; Day & Plano, 1998; Forsberg *et al.*, 1991; Iriarte *et al.*, 1998; Jackson *et al.*, 1998; Petterson *et al.*, 1996; Skrzypek & Straley, 1993; Stainier *et al.*, 1997).

Sequencing of the *Yersinia* virulence plasmid helped to reveal other genes involved in the type III secretion pathway (Cornelis *et al.*, 1998; Perry *et al.*, 1998). Current data indicate that some 23 genes are required for type III secretion of Yops and are clustered in four operons, *virA* (*yscXYlcrDR*), *virB* (*yscNOPQRSTU*), *virC* (*yscABCDEFGIJKLM*) and *virF* (*lcrF*). The export substrates of the *Yersinia* type III machinery were identified in concentrated supernatants of cultures that were grown at 37 °C and low calcium. The samples were separated by SDS-PAGE, and the proteins were Coomassie-stained and identified by Edman degradation (Michiels *et al.*, 1990; Straley *et al.*, 1993b). Some 14 different export substrates of the type III machinery are known: YopB, YopD, YopE, YopH, YopM, YopN (LcrE), YopO (YpkA), YopP (YopJ), YopQ (YopK), YopR, YopT, LcrV, YscM1 (LcrQ) and YscM2. Designations in parentheses refer to analogous proteins identified in either *Y. pestis* or *Y. pseudotuberculosis*. Three Yops appear to require dedicated secretion chaperones for their targeting into eukaryotic cells (see below), SycE (YopE), SycH (YopH) and SycT (YopT), the genes of which are located immediately adjacent to that of the cognate *yop* that the chaperone binds to (Michiels *et al.*, 1990; Straley *et al.*, 1993b; Iriarte & Cornelis, 1998; Wattiau & Cornelis, 1993; Wattiau *et al.*, 1994, 1996; Woestyn *et al.*, 1996).

LOCALIZATION OF YOPS DURING *YERSINIA* INFECTION OF TISSUE CULTURE CELLS

Rosqvist and co-workers were the first to show that Yop proteins are injected into the cytosol of eukaryotic cells (Rosqvist *et al.*, 1991, 1994). The investigators employed antibody to YopE to reveal the presence of this polypeptide in the cytosol of infected HeLa cells by immunofluorescence microscopy. The same technique was also employed to reveal the type III injection of YopH and YopO (YpkA) (Hakansson *et al.*, 1996a; Persson *et al.*, 1995, 1997). Sory and Cornelis developed a clever scheme that permitted measurements of type III targeting as an increase in the intracytosolic concentration of cAMP (Sory *et al.*, 1995; Sory & Cornelis, 1994). *Bordetella pertussis* adenylate cyclase (Cya) requires calmodulin in the cytosol of eukaryotic cells for its enzymic activity of converting ATP to cAMP. When fused to the C-terminus of Yop proteins, the targeting of hybrid Yop–Cya polypeptides can be revealed as an increase in cAMP of HeLa cell extracts. Employing this experimental strategy, YopE, YopH, YopM, YopO, YopP and YopT were shown to be targeted into the cytosol of eukaryotic cells (Boland *et al.*, 1996; Boland & Cornelis, 1998; Iriarte & Cornelis, 1998; Mills *et al.*, 1997). Assuming that these polypeptides accomplish the task of manipulating eukaryotic cells to modulate the immune response and to prevent phagocytosis, these polypeptides are also referred to as effector Yops (Cornelis & Wolf-Watz, 1997). YopH is a protein tyrosine phosphatase that acts on p130^{CAP} and focal adhesion kinase (FAK), a mechanism that appears to be essential in preventing phagocytosis (Black & Bliska, 1997; Bliska *et al.*, 1991; Andersson *et al.*, 1996; Fallman *et al.*, 1995; Persson *et al.*, 1997). YopE and YopT are thought to promote actin rearrangements and the establishment of focal adhesions, mechanisms that are also thought to prevent phagocytosis (Iriarte & Cornelis, 1998; Rosqvist *et al.*, 1988, 1990). YopM is directed into the nucleus of eukaryotic cells; however, its precise function is as of yet unknown (Skrzypek *et al.*, 1998). YopO (YpkA) is a protein serine/threonine kinase and is thought to affect signal transduction events once injected into eukaryotic cells (Galyov *et al.*, 1993, 1994). Finally, YopP (YopJ) induces apoptosis in infected tissues; however, the precise mechanism by which this occurs is still unknown (Mills *et al.*, 1997; Monack *et al.*, 1997).

Lee and co-workers applied a unique scheme to analyse infected HeLa cells (Lee *et al.*, 1998). While the experimental approaches described above permit measurements of type III targeting, they cannot detect polypeptides in the extracellular milieu. This can be accomplished by fractionating infected HeLa cells and analysing samples for the presence of Yops via immunoblotting. During this protocol, the medium is first decanted and centrifuged to sediment yersiniae that are non-adherent to eukaryotic cells. Tissue culture cells and adherent bacteria are then extracted with digitonin, a detergent that disrupts the cholesterol-containing plasma membrane of eukaryotic cells but not the bacterial envelope. The extract is again centrifuged to sediment bacteria as well as eukaryotic organelles and separate them from the supernatant containing the cytosolic extract. This scheme revealed that YopB, YopD and YopR are secreted into the extracellular milieu while YopE, YopH, YopM and YopN are injected into eukaryotic cells (Lee *et al.*, 1998). Injection proved to be specific, i.e. YopE, YopH, YopM and YopN were found in cytosolic extracts but not in the tissue culture medium. Small amounts of YopR, YopB and YopD were found in the digitonin extract supernatant and pellet; however, this appears to be a contamination of extracellular protein binding to the eukaryotic cell surface. In protease protection experiments, YopB, YopD and YopR were determined to be mostly extracellular (>95%) (Lee & Schneewind, 1999). Some type III export substrates remain associated with the bacteria during the infection of HeLa cells, e.g. YopQ, LcrV and YscM1. Together these results suggest that during the infection of HeLa cells, Yop proteins are directed to specific locations, as if contact with eukaryotic cells triggered a unique program of type III export. This could be explained by the existence of different types of type III machines that direct proteins to different locations. Alternatively, a defined program of protein export may recognize subsets of Yops as secretion substrates which are exported as the type III machine assembles to generate an injection device that ultimately leads to the deployment of effector Yops into the cytosol of eukaryotic cells (see below).

SIGNALS FOR YOP SECRETION AND TARGETING

Because Yop proteins do not appear to share similarities in their amino acid sequences, the mode by which they are recognized by the type III machinery has been studied in detail. Translational fusions of many different reporter proteins to the 3' end of the *yopE* coding sequence can be exported by the type III machinery, e.g. the *lacZ* alpha peptide, *Escherichia coli* alkaline phosphatase, TEM β-lactamase, neomycin phosphotransferase, *B. pertussis* adenylate cyclase and jellyfish green fluorescent protein (Jacobi *et al.*, 1998; Michiels & Cornelis, 1991; Sory & Cornelis, 1994; Sory *et al.*, 1995; Anderson & Schneewind, 1997, 1999; Cheng *et al.*, 1997). The minimal signal sufficient for the secretion of reporter fusions to the C-terminus of YopE is located within the first 15 codons (Sory *et al.*, 1995; Schesser *et al.*, 1996; Anderson & Schneewind, 1997, 1999). This signal can be mutated extensively without loss of function (Anderson & Schneewind, 1997). For example, when frameshift mutations were introduced immediately following the AUG start codon and the reading frame was restored by appropriate suppressor mutations at the fusion site with the reporter gene, the mutational changes did not interfere with signalling secretion of hybrid polypeptides (Anderson & Schneewind, 1997). These observations led to the proposal that the secretion signal may be encoded within the nucleotide sequence of *yop* mRNA (Anderson & Schneewind, 1997). Indeed, all Yops examined thus far (YopE, YopN, YopH and YopQ) have a common secretion signal located within the first 15 codons that is tolerant of at least some frameshift mutations. The RNA signal hypothesis implies that *yop* mRNA can not be translated in the cytoplasm of bacteria unless ribosomal polypeptide synthesis is tethered to the type III secretion machinery. It is not yet clear whether repression of *yop* mRNA translation is a property of the nucleic acid's sequence alone or whether it requires other factors. Translational repression may be relieved once ribosomes charged with *yop* mRNA dock onto the type III machinery (Anderson & Schneewind, 1997).

A second mode of substrate recognition of Yop proteins has been described (Cheng *et al.*, 1997). Deletion of the first 15 codons of *yopE* mRNA gives rise to a truncated polypeptide that can be exported by low-calcium-induced yersiniae. This signal is also sufficient for the secretion of fused reporter proteins and was mapped to amino acid residues 15–100 of YopE. However, secretion absolutely requires binding of the SycE secretion chaperone to amino acids 15–100 of YopE (Cheng *et al.*, 1997; Woestyn *et al.*, 1996; Schesser *et al.*, 1996). To determine which of the two signals is required for the injection of YopE into eukaryotic cells, the targeting of YopE–Npt fusion proteins was measured by digitonin fractionation (Lee *et al.*, 1998). The first 100 amino acids of YopE are necessary and sufficient for the targeting of hybrid reporter constructs by a mechanism that absolutely required the presence of the SycE chaperone. Unfortunately, these experiments could not provide a definitive answer whether the mRNA signal is required for type III targeting. Nevertheless, the first 15 codons (amino acids) of YopE are necessary for the injection of reporter proteins by yersiniae.

YopE and YopH appear to harbour similar secretion and targeting signals (Sory *et al.*, 1995). Other Yops, e.g. YopQ, are exported by only one (mRNA) secretion signal without the need for a cytoplasmic Syc chaperone (Anderson & Schneewind, 1999). During infection, YopQ remains associated with the bacteria and it is thus not possible to measure its export under these conditions. However, in *Yersinia* cultures induced by temperature shift to 37 °C and low calcium, the first 15 codons of YopQ are both necessary and sufficient for the secretion of fused Npt reporter proteins. YopQ is found quantitatively in the culture medium of low-calcium-induced yersiniae. *yopQ* mRNA is not translated in the presence of calcium and YopQ polypeptide can only be secreted co- but not post-translationally (Anderson & Schneewind, 1999). Combined with mutational analysis of its secretion signal, these data indicate that YopQ is exported exclusively by an mRNA signal (Anderson & Schneewind, 1999).

REGULATION OF *YOP* GENE EXPRESSION

Many *yop* genes are regulated by VirF, a transcriptional activator of the AraC family that is thought to activate expression of type III genes in a temperature-dependent manner (Wattiau & Cornelis, 1994). Mutations in genes specifying components of the type III secretion machinery either abrogate or significantly reduce the expression of *yop* genes (Allaoui *et al.*, 1995; Straley *et al.*, 1993a). Clearly, this is not only due to transcriptional regulation. Recent work on the expression of *yopQ* showed that its expression is controlled at the level of translation, in that the removal of calcium ions from the extracellular milieu stimulated the translation of *yopQ* transcripts about 10-fold (Anderson & Schneewind, 1999). Mutations that interfere with calcium signalling also abolish the translational repression of *yopQ* expression. The detailed molecular mechanisms by which transcription and translation of *yop* genes might be controlled are as of yet unclear.

RECOGNITION OF mRNA SIGNALS BY COMPONENTS OF THE TYPE III MACHINE

Pathogenic *Yersinia* species require 23 genes [*yscABCDEFGIJKL, yscNOPQRSTU* and *yscV(lcrD)yscWXY*] for the low-calcium-induced secretion of Yops (Cornelis *et al.*, 1998), whereas plant-pathogenic *Erwinia chrysanthemi* needs only 11 genes to deliver Avr proteins and cause plant disease (called *hrc* for hypersensitive response conserved gene) (Dangl & Holub, 1997; Ham *et al.*, 1998). The *Erw. chrysanthemi* type III machine is functional for the secretion of heterologous type III substrates when cloned in *E. coli* (Ham *et al.*, 1998). Anderson *et al.* (1999) asked whether this cloned type III machine recognizes mRNA signals of Yops. YopQ and YopE can be exported by the recombinant *E. coli*, even in the absence of the SycE secretion chaperone, suggesting that the *Erwinia* type III machine recognizes secretion substrates via mRNA signals (Anderson *et al.*, 1999). Furthermore, plant-pathogenic secretion substrates that are secreted by the *Erwinia* type III machine, *Pseudomonas* AvrB and AvrPto (Ham *et al.*, 1998), can also be secreted in *Y. enterocolitica* induced via low calcium (Anderson *et al.*, 1999). The secretion signals of *Pseudomonas syringae* AvrB and AvrPto were mapped to the first 15 codons and could be altered by frameshift mutation without loss of function, suggesting that it is recognized at the RNA level (Anderson *et al.*, 1999). Comparison of the genes required for type III secretion of other Gram-negative pathogens reveals that homologues of nine genes, *yscC*, *yscJ*, *yscN*, *yscQRSTU* and *yscV* (*lcrD*), are found in all of these machines (including the *Erw. chrysanthemi hrp* genes) and may be involved in recognizing mRNA secretion signals and translocating poly-peptide across the bacterial envelope. Eight of the nine conserved genes are homologous to the genes needed for assembly and export of the flagellar basal body hook complex (Hueck, 1998). The *yscC* gene, specifying an outer-membrane secretin (Koster *et al.*, 1997; Russel, 1994), is not found in the flagellar gene cluster (Macnab, 1992). Secretins form multimeric, gated channels in the outer membrane and function to translocate proteins as well as bacteriophages (Linderoth *et al.*, 1997). This can be explained by the function of the flagellar basal body to export and polymerize hook and flagellar components, whereas both the type II and type III secretion path-ways appear to release virulence factors into the extracellular milieu via their secretins (Daefler *et al.*, 1997).

What then is the function of the remaining 14 *ysc* genes? Perhaps the remaining proteins of the *Yersinia* machine can recognize secretion substrates each with a certain specificity and thereby function to regulate the export of Yops. Our notion that type III machines recognize heterologous secretion substrate via mRNA signals is in conflict with a previous observation that salmonellae require SycE chaperone to export YopE polypeptide (Rosqvist *et al.*, 1995).

MECHANISMS OF YOP TARGETING INTO EUKARYOTIC CELLS

The injection of Yop proteins into eukaryotic cells requires a mechanism of protein translocation across three membranes: the bacterial inner and outer membranes, as well as the eukaryotic plasma membrane. While it is clear that the type III secretion machine transports proteins across the bacterial envelope, translocation across the plasma membrane may require other polypeptides. The existence of such translocator proteins was first postulated by Wolf-Watz and co-workers (Hakansson *et al.*, 1993, 1996b). These authors observed that *Yersinia* strains carrying mutations in the *yopB* and *yopD* genes are unable to inject effector Yops into the cytosol of eukaryotic cells. Purified YopB is reported to form aqueous channels in the plasma membrane and YopD has been proposed to participate in the formation of a pore or to shuttle proteins into eukaryotic cells (Hakansson *et al.*, 1996b; Francis & Wolf-Watz, 1998). The mechanism of Yop protein targeting was presumed to occur in two steps (Cornelis & Wolf-Watz, 1997). Effector Yops, for example YopE, are first secreted across the bacterial envelope by a mechanism requiring the signal encoded in the first 15 codons. Amino acids 15–100 of YopE, called the translocation signal, are then responsible for interaction with the translocator pore YopB/YopD to permit import of YopE into eukaryotic cells (Cornelis & Wolf-Watz, 1997; Sory *et al.*, 1995). This model accounts for the observations that YopE may be exported by a secretion signal encoded within the first 15 codons yet requires the first 100 amino acids for proper targeting. Nonetheless, the model implies the existence of a translocation intermediate as well as a recognition event for effector Yops by the YopB/YopD translocator pore. The latter is difficult to explain as effector Yops such as YopE, YopH, YopM, YopO, YopP and YopT do not share common sequence or physical similarity that would permit their recognition by the translocator.

Lee and others proposed a different mechanism, in which the type III machine is extended to form an injection device that permits the simultaneous translocation of effector Yops across three membranes into eukaryotic cells (Lee *et al.*, 1998). Substrate recognition of effector Yops appears to require binding of Syc chaperone to the unfolded polypeptide and this interaction is a prerequisite for translocation via the type III channel. This model makes do without the need for an extracellular translocation intermediate or recognition event at the eukaryotic plasma membrane. In keeping with the observations described above, YopB/YopD could serve as an extension of the type III channel that leads into the eukaryotic cytosol. However, there now is debate about the role of YopB/YopD in type III targeting. Recent work showed that *yopB* mutant *Yersinia* can inject effector Yops just like wild-type bacteria (Lee & Schneewind, 1999). Although *yopD* mutations abrogate type III targeting, Gst–YopD fusions that can not be exported by the type III machinery complement this defect (Lee & Schneewind, 1999). Thus, it appears that YopD plays an intra-bacterial role that is essential for type III targeting, whereas the secreted, extracellular species does not seem to be necessary for this mechanism. If so, the factors required for the injection of effector Yops are still unknown. If these components form an extension of the type III channel, it is conceivable that such polypeptides may be exported in a distinct manner that allows their assembly from the inside out, similar to the synthesis of bacterial flagellar filaments.

ACKNOWLEDGEMENTS

Work in the laboratory of O. Schneewind is supported by grants from the US Public Health Service AI38897 and AI42797.

REFERENCES

Allaoui, A., Woestyn, S., Sluiters, C. & Cornelis, G. (1994). YscU, a *Yersinia enterocolitica* inner membrane protein involved in Yop secretion. *Journal of Bacteriology* **176**, 4534–4542.

Allaoui, A., Schulte, R. & Cornelis, G. R. (1995). Mutational analysis of the *Yersinia enterocolitica virC* operon: characterization of *yscE, F, G, I, J, K* required for Yop secretion and *yscH* encoding YopR. *Molecular Microbiology* **18**, 343–355.

Anderson, D. M. & Schneewind, O. (1997). A mRNA signal for the type III secretion of Yop proteins by *Yersinia enterocolitica. Science* **278**, 1140–1143.

Anderson, D. M. & Schneewind, O. (1999). *Yersinia enterocolitica* Type III secretion: an mRNA signal that couples translation and secretion of YopQ. *Molecular Microbiology* **31**, 1139–1148.

Anderson, D. M., Fouts, D., Collmer, A. & Schneewind, O. (1999). *Molecular Microbiology* (in press).

Andersson, K., Carballeira, N., Magnusson, K., Persson, C., Stendahl, O., Wolf-Watz, H. & Fallman, M. (1996). YopH of *Yersinia pseudotuberculosis* interrupts early phosphotyrosine signalling associated with phagocytosis. *Molecular Microbiology* **20**, 1057–1069.

Autenrieth, I. B. & Firsching, R. (1996). Penetration of M cells and destruction of Peyer's patches by *Yersinia enterocolitica*: an ultrastructural and histological study. *Journal of Medical Microbiology* **44**, 285–294.

Bergman, T., Hakansson, S., Forsberg, A., Norlander, L., Macellaro, A., Backman, A., Bolin, I. & Wolf-Watz, H. (1991). Analysis of the V antigen *lcrGVH-yopBD* operon of *Yersinia pseudotuberculosis*: evidence for a regulatory role of LcrH and LcrV. *Journal of Bacteriology* **173**, 1607–1616.

Bergman, T., Erickson, K., Galyov, E., Persson, C. & Wolf-Watz, H. (1994). The *lcrB* (*yscN/U*) gene cluster of *Yersinia pseudotuberculosis* is involved in Yop secretion and shows high homology to the spa gene clusters of *Shigella flexneri* and *Salmonella typhimurium. Journal of Bacteriology* **176**, 2619–2626.

Beuscher, H. U., Rodel, F., Forsberg, A. & Rollinghoff, M. (1995). Bacterial evasion of host immune defense: *Yersinia enterocolitica* encodes a suppressor for tumor necrosis factor alpha expression. *Infection and Immunity* **63**, 1270–1277.

Black, D. S. & Bliska, J. B. (1997). Identification of p130[Cas] as a substrate of *Yersinia* YopH (Yop51), a bacterial protein tyrosine phosphatase that translocates into mammalian cells and targets focal adhesions. *EMBO Journal* **16**, 2730–2744.

Bliska, J. B., Guan, K., Dixon, J. E. & Falkow, S. (1991). Tyrosine phosphate hydrolysis of host proteins by an essential *Yersinia* virulence determinant. *Proceedings of the National Academy of Sciences, USA* **88**, 1187–1191.

Boland, A. & Cornelis, G. R. (1998). Role of YopP in suppression of tumor necrosis factor alpha release by macrophages during *Yersinia* infection. *Infection and Immunity* **66**, 1878–1884.

Boland, A., Sory, M.-P., Iriarte, M., Kerbourch, C., Wattiau, P. & Cornelis, G. R. (1996). Status of YopM and YopN in the *Yersinia yop* virulon: YopM of *Y. enterocolitica* is internalized inside the cytosol of PU5-1.8 macrophages by the YopB, D, N delivery apparatus. *EMBO Journal* **15**, 5191–5201.

Butler, T. (1983). *Plague and Other Yersinia Infections.* New York: Plenum.

Butler, T. (1995). *Yersinia* species. In *Infectious Diseases*, 5th edn, pp. 1748–1756. Edited by G. Mandell. New York: Churchill Livingstone.

Cheng, L. W., Anderson, D. M. & Schneewind, O. (1997). Two independent type III secretion mechanisms for YopE in *Yersinia enterocolitica. Molecular Microbiology* **24**, 757–765.

Cornelis, G. R. & Wolf-Watz, H. (1997). The *Yersinia* Yop virulon: a bacterial system for subverting eukaryotic cells. *Molecular Microbiology* **23**, 861–867.

Cornelis, G. R., Boland, A., Boyd, A. P., Geuijen, C., Iriarte, M., Neyt, C., Sory, M.-P. & Stainier, I. (1998). The virulence plasmid of *Yersinia*, an antihost genome. *Microbiology and Molecular Biology Reviews* **62**, 1315–1352.

Daefler, S., Guilvout, I., Hardie, K. R., Pugsley, A. P. & Russel, M. (1997). The C-terminal domain of the secretin PulD contains the binding site for its cognate chaperone, PulS, and confers PulS dependence on pIVf1 function. *Molecular Microbiology* **24**, 465–475.

Dangl, J. & Holub, E. (1997). La dolce vita: a molecular feast in plant-pathogen interactions. *Cell* **91**, 17–24.

Day, J. B. & Plano, G. V. (1998). A complex composed of SycN and YscB functions as a specific chaperone for YopN in *Yersinia pestis. Molecular Microbiology* **30**, 777–789.

Fallman, M., Andersson, K., Hakansson, S., Magnusson, K. E., Stendahl, O. & Wolf-Watz, H. (1995). *Yersinia pseudotuberculosis* inhibits Fc receptor mediated phagocytosis in J774 cells. *Infection and Immunity* **63**, 3117–3124.

Forsberg, A., Viitanen, A.-M., Skunik, M. & Wolf-Watz, H. (1991). The surface-located YopN protein is involved in calcium signal transduction in *Yersinia pseudotuberculosis. Molecular Microbiology* **5**, 977–986.

Francis, M. S. & Wolf-Watz, H. (1998). YopD of *Yersinia pseudotuberculosis* is translocated into the cytosol of HeLa epithelial cells: evidence of a structural domain necessary for translocation. *Molecular Microbiology* **29**, 799–814.

Galyov, E. E., Hakansson, S., Forsberg, A. & Wolf-Watz, H. (1993). A secreted protein kinase of *Yersinia pseudotuberculosis* is an indispensable virulence determinant. *Nature* **361**, 730–732.

Galyov, E. E., Hakansson, S. & Wolf-Watz, H. (1994). Characterization of the operon encoding the YpkA Ser/Thr protein kinase and the YopJ protein of *Yersinia pseudotuberculosis. Journal of Bacteriology* **176**, 4543–4548.

Gemski, P., Lazere, J. R. & Casey, T. (1980a). Plasmid associated with pathogenicity and calcium dependency of *Yersinia enterocolitica*. *Infection and Immunity* **27**, 682–685.

Gemski, P., Lazere, J. R., Casey, T. & Wohlhieter, J. A. (1980b). Presence of a virulence-associated plasmid in *Yersinia pseudotuberculosis*. *Infection and Immunity* **28**, 1044–1047.

Goguen, J. D., Yother, J. & Straley, S. C. (1984). Genetic analysis of the low calcium response in *Yersinia pestis* Mud1(Ap *lac*) insertion mutants. *Journal of Bacteriology* **160**, 842–848.

Grutzkau, A., Hanski, C., Hahn, H. & Riecken, E. O. (1990). Involvement of M cells in the bacterial invasion of Peyer's patches: a common mechanism shared by *Yersinia enterocolitica* and other enteroinvasive bacteria. *Gut* **31**, 1011–1015.

Hakansson, S., Bergman, T., Vanooteghem, J.-C., Cornelis, G. & Wolf-Watz, H. (1993). YopB and YopD constitute a novel class of *Yersinia* Yop proteins. *Infection and Immunity* **61**, 71–80.

Hakansson, S., Galyov, E., Rosqvist, R. & Wolf-Watz, H. (1996a). The *Yersinia* YpkA Ser/Thr kinase is translocated and subsequently targeted to the inner surface of the HeLa cell plasma membrane. *Molecular Microbiology* **20**, 593–603.

Hakansson, S., Schesser, K., Persson, C., Galyov, E. E., Rosqvist, R., Homble, F. & Wolf-Watz, H. (1996b). The YopB protein of *Yersinia pseudotuberculosis* is essential for the translocation of Yop effector proteins across the target cell plasma membrane and displays a contact-dependent membrane disrupting activity. *EMBO Journal* **15**, 5812–5823.

Ham, J. H., Bauer, D. W., Fouts, D. E. & Collmer, A. (1998). A cloned *Erwinia chrysanthemi* Hrp (type III protein secretion) system functions in *Escherichia coli* to deliver *Pseudomonas syringae* Avr signals to plant cells and to secrete Avr proteins in culture. *Proceedings of the National Academy of Sciences, USA* **95**, 10206–10211.

Hanski, C., Kutschka, U., Schmoranzer, H. P., Naumann, M., Stallmach, A., Hahn, H., Menge, H. & Riecken, E. O. (1989). Immunohistochemical and electron microscopic study of the interaction of *Yersinia enterocolitica* serotype O:8 with intestinal mucosa during experimental enteritis. *Infection and Immunity* **57**, 673–678.

Higuchi, K. & Smith, J. L. (1961). Studies on the nutrition and physiology of *Pasteurella pestis*. *Journal of Bacteriology* **81**, 605–608.

Higuchi, K., Kupferberg, L. L. & Smith, J. L. (1959). Studies on the nutrition and physiology of *Pasteurella pestis*. *Journal of Bacteriology* **77**, 317–321.

Hueck, C. J. (1998). Type III protein secretion in bacterial pathogens of animals and plants. *Microbiology and Molecular Biology Reviews* **62**, 379–433.

Iriarte, M. & Cornelis, G. R. (1998). YopT, a new *Yersinia* effector protein, affects the cytoskeleton of host cells. *Molecular Microbiology* **29**, 915–929.

Iriarte, M., Sory, M.-P., Boland, A., Boyd, A. P., Mills, S. D., Lambermont, I. & Cornelis, G. R. (1998). TyeA, a protein involved in control of Yop release and in translocation of *Yersinia* Yop effectors. *EMBO Journal* **17**, 1907–1918.

Jackson, M. W., Day, J. B. & Plano, G. V. (1998). YscB of *Yersinia pestis* functions as a specific chaperone for YopN. *Journal of Bacteriology* **180**, 4912–4921.

Jacobi, C. A., Roggenkamp, A., Rakin, A., Zumbihl, R., Leitritz, L. & Heesemann, J. (1998). *In vitro* and *in vivo* expression studies of *yopE* from *Yersinia enterocolitica* using the *gfp* reporter gene. *Molecular Microbiology* **30**, 865–882.

Koster, M., Bitter, W., de Cock, H., Allaoui, A., Cornelis, G. R. & Tommassen, J. (1997). The outer membrane component, YscC, of the Yop secretion machinery of *Yersinia enterocolitica* forms a ring-shaped multimeric complex. *Molecular Microbiology* **26**, 789–797.

Lee, V. T. & Schneewind, O. (1999). Multiple pathogenic strategies of *Yersiniae*: the role of extra-cellular and cytosolic Yops. *Molecular Microbiology* **31**, (in press).

Lee, V. T., Anderson, D. M. & Schneewind, O. (1998). Targeting of *Yersinia* Yop proteins into the cytosol of HeLa cells: one-step translocation of YopE across bacterial and eukaryotic membranes is dependent on SycE chaperone. *Molecular Microbiology* **28**, 593–601.

Linderoth, N. A., Simon, M. N. & Russel, M. (1997). The filamentous phage pIV multimer visualized by scanning transmission electron microscopy. *Science* **278**, 1635–1638.

Macnab, R. M. (1992). Genetics and biogenesis of bacterial flagella. *Annual Review of Genetics* **26**, 131–158.

Michiels, T. & Cornelis, G. R. (1991). Secretion of hybrid proteins by the *Yersinia* Yop export system. *Journal of Bacteriology* **173**, 1677–1685.

Michiels, T., Wattiau, P., Brasseur, R., Ruysschaert, J.-M. & Cornelis, G. (1990). Secretion of Yop proteins by *Yersiniae*. *Infection and Immunity* **58**, 2840–2849.

Michiels, T., Vanooteghem, J.-C., Lambert de Rouvroit, C., China, B., Gustin, A., Boudry, P. & Cornelis, G. R. (1991). Analysis of *virC*, an operon involved in the secretion of Yop proteins by *Yersinia enterocolitica*. *Journal of Bacteriology* **173**, 4994–5009.

Mills, S. D., Boland, A., Sory, M.-P., van der Smissen, P., Kerbouch, C., Finlay, B. B. & Cornelis, G. R. (1997). *Yersinia enterocolitica* induces apoptosis in macrophages by a process requiring functional type III secretion and translocation mechanisms and involving YopP, presumably acting as an effector protein. *Proceedings of the National Academy of Sciences, USA* **94**, 12638–12643.

Monack, D. M., Mecsas, J., Ghori, N. & Falkow, S. (1997). *Yersinia* signals macrophages to undergo apoptosis and YopJ is necessary for this cell death. *Proceedings of the National Academy of Sciences, USA* **94**, 10385–10390.

Neutra, M. R., Frey, A. & Kraehenbuhl, J. P. (1996). Epithelial M cells: gateways for mucosal infection and immunization. *Cell* **86**, 345–348.

Perry, R. D., Straley, S. C., Fetherston, J. D., Rose, D. J., Gregor, J. & Blattner, F. R. (1998). DNA sequencing and analysis of the low-Ca^{2+} response plasmid pCD1 of *Yersinia pestis* KIM5. *Infection and Immunity* **66**, 4611–4623.

Persson, C., Nordfelth, R., Holmstrom, A., Hakansson, S., Rosqvist, R. & Wolf-Watz, H. (1995). Cell-surface-bound *Yersinia* translocate the protein tyrosine phosphatase YopH by a polarized mechanism into the target cell. *Molecular Microbiology* **18**, 135–150.

Persson, C., Carballeira, N., Wolf-Watz, H. & Fallman, M. (1997). The PTPase YopH inhibits uptake of *Yersinia*, tyrosine phosphorylation of p130^{cas} and FAK, and the associated accumulation of these proteins in peripheral focal adhesion. *EMBO Journal* **16**, 2307–2318.

Petterson, J., Nordfelth, R., Dubinina, E., Bergman, T., Gustafsson, M., Magnusson, K. E. & Wolf-Watz, H. (1996). Modulation of virulence factor expression by pathogen target cell contact. *Science* **273**, 1231–1233.

Plano, G. V. & Straley, S. C. (1993). Multiple effects of *lcrD* mutations in *Yersinia pestis*. *Journal of Bacteriology* **175**, 3536–3545.

Plano, G. V. & Straley, S. C. (1995). Mutations in *yscC*, *yscD*, and *yscG* prevent high level expression and secretion of V antigen and Yops in *Yersinia pestis*. *Journal of Bacteriology* **177**, 3843–3854.

Plano, G. V., Barve, S. S. & Straley, S. C. (1991). LcrD, a membrane-bound regulator of the *Yersinia pestis* low-calcium response. *Journal of Bacteriology* **173**, 7293–7303.

Portnoy, D. A., Moseley, S. L. & Falkow, S. (1981). Characterization of plasmids and plasmid-associated determinants of *Yersinia enterocolitica* pathogenesis. *Infection and Immunity* **31**, 775–782.

Rosqvist, R., Bolin, I. & Wolf-Watz, H. (1988). Inhibition of phagocytosis in *Yersinia pseudotuberculosis*: a virulence plasmid-encoded ability involving the Yop2b protein. *Infection and Immunity* **56**, 2139–2143.

Rosqvist, R., Forsberg, A., Rimpilainen, M., Bergman, T. & Wolf-Watz, H. (1990). The cytotoxic protein YopE of *Yersinia* obstructs the primary host defence. *Molecular Microbiology* **4**, 657–667.

Rosqvist, R., Forsberg, A. & Wolf-Watz, H. (1991). Intracellular targeting of the *Yersinia* YopE cytotoxin in mammalian cells induces actin microfilament disruption. *Infection and Immunity* **59**, 4562–4569.

Rosqvist, R., Magnusson, K.-E. & Wolf-Watz, H. (1994). Target cell contact triggers expression and polarized transfer of *Yersinia* YopE cytotoxin into mammalian cells. *EMBO Journal* **13**, 964–972.

Rosqvist, R., Hakansson, S., Forsberg, A. & Wolf-Watz, H. (1995). Functional conservation of the secretion and translocation machinery for virulence proteins of *yersiniae*, *salmonellae* and *shigellae*. *EMBO Journal* **14**, 4187–4195.

Russel, M. (1994). Phage assembly: a paradigm for bacterial virulence factor export? *Science* **265**, 612–614.

Schesser, K., Frithz-Lindsten, E. & Wolf-Watz, H. (1996). Delineation and mutational analysis of the *Yersinia pseudotuberculosis* YopE domains which mediate translocation across bacterial and eukaryotic cellular membranes. *Journal of Bacteriology* **178**, 7227–7233.

Skrzypek, E. & Straley, S. C. (1993). LcrG, a secreted protein involved in negative regulation of the low-calcium response in *Yersinia pestis*. *Journal of Bacteriology* **175**, 3520–3528.

Skrzypek, E., Cowan, C. & Straley, S. C. (1998). Targeting of the *Yersinia pestis* YopM protein into HeLa cells and intra-cellular trafficking to the nucleus. *Molecular Microbiology* **30**, 1051–1065.

Sory, M.-P. & Cornelis, G. R. (1994). Translocation of a hybrid YopE-adenylate cyclase from *Yersinia enterocolitica* into HeLa cells. *Molecular Microbiology* **14**, 583–594.

Sory, M.-P., Boland, A., Lambermont, I. & Cornelis, G. R. (1995). Identification of the YopE and YopH domains required for secretion and internalization into the cytosol of macrophages, using the *cyaA* gene fusion approach. *Proceedings of the National Academy of Sciences, USA* **92**, 11998–12002.

Stainier, I., Iriarte, M. & Cornelis, G. R. (1997). YscM1 and YscM2, two *Yersinia enterocolitica* proteins causing downregulation of *yop* transcription. *Molecular Microbiology* **26**, 833–843.

Straley, S. C., Plano, G. V., Skrzypek, E., Haddix, P. L. & Fields, K. A. (1993a). Regulation by Ca^{2+} in the *Yersinia* low-Ca^{2+} response. *Molecular Microbiology* **8**, 1005–1010.

Straley, S. C., Skrzypek, E., Plano, G. V. & Bliska, J. B. (1993b). Yops of *Yersinia spp.* pathogenic for humans. *Infection and Immunity* **61**, 3105–3110.

Wattiau, P. & Cornelis, G. R. (1993). SycE, a chaperone-like protein of *Yersinia enterocolitica* involved in the secretion of YopE. *Molecular Microbiology* **8**, 123–131.

Wattiau, P. & Cornelis, G. R. (1994). Identification of DNA sequences recognized by VirF, the transcriptional activator of the *Yersinia* yop regulon. *Journal of Bacteriology* **176**, 3878–3884.

Wattiau, P., Bernier, B., Deslee, P., Michiels, T. & Cornelis, G. R. (1994). Individual chaperones required for Yop secretion by *Yersinia*. *Proceedings of the National Academy of Sciences, USA* **91**, 10493–10497.

Wattiau, P., Woestyn, S. & Cornelis, G. R. (1996). Customized secretion chaperones in pathogenic bacteria. *Molecular Microbiology* **20**, 255–262.

Woestyn, S., Allaoui, A., Wattiau, P. & Cornelis, G. (1994). YscN, the putative energizer of the *Yersinia* Yop secretion machinery. *Journal of Bacteriology* **176**, 1561–1569.

Woestyn, S., Sory, M.-P., Boland, A., Lequenne, O. & Cornelis, G. R. (1996). The cytosolic SycE and SycH chaperones of *Yersinia* protect the region of YopE and YopH involved in translocation across eukaryotic cell membranes. *Molecular Microbiology* **20**, 1261–1271.

Yother, J. & Goguen, J. D. (1985). Isolation and characterization of Ca^{2+}-blind mutants of *Yersinia pestis*. *Journal of Bacteriology* **164**, 704–711.

Yother, J., Chamness, T. W. & Goguen, J. D. (1986). Temperature controlled plasmid regulon associated with low calcium response in *Yersinia pestis*. *Journal of Bacteriology* **165**, 443–447.

Zink, D. L., Feeley, J. C., Wells, J. G., Vanderzant, C., Vickery, J. C., Roof, W. D. & O'Donovan, G. A. (1980). Plasmid-mediated tissue invasiveness in *Yersinia enterocolitica*. *Nature* **283**, 224–226.

ASSEMBLY OF BACTERIAL ADHESINS ACROSS THE OUTER MEMBRANE VIA THE CHAPERONE– USHER PATHWAY

GABRIEL E. SOTO AND SCOTT J. HULTGREN

Department of Molecular Microbiology, Washington University School of Medicine, Saint Louis, MO 63110, USA

INTRODUCTION

The chaperone–usher pathway of Gram-negative bacteria is involved in the assembly of over 25 architecturally diverse organelles of attachment including many different kinds of pili (Table 1). Pili mediate binding to specific receptors on host cells, and therefore play a crucial role in the early events of many bacterial infections. Bacterial attachment can lead to the activation of innate host defences or the subversion of cellular processes facilitating bacterial colonization or invasion. In addition, the binding event may also activate the expression of new genes in the microbe that are important in the pathogenic process.

The best-characterized systems to date are P pili and type 1 pili. P pili are expressed on the surfaces of uropathogenic strains of *Escherichia coli* associated with acute pyelonephritis (Kallenius *et al.*, 1981). Type 1 pili are important virulence determinants expressed in *E. coli* as well as in most of the *Enterobacteriaceae* family that mediate binding to mannose-oligosaccharides (Krogfelt *et al.*, 1990). The study of the assembly of these organelles has provided a framework for understanding how complex hetero-oligomeric interactions are orchestrated within the bacterial cell. In this review, we focus on the assembly of P pili and type 1 as models for the assembly of adhesins across the outer membrane (OM) via the chaperone–usher pathway.

STRUCTURE AND ASSEMBLY OF P AND TYPE 1 PILI

Eleven genes organized in the *pap* gene cluster are required for the expression and assembly of P pili (Hull *et al.*, 1981; Hultgren & Normark, 1991; Hultgren *et al.*, 1991; Marklund *et al.*, 1992). Quick-freeze, deep-etch electron microscopy revealed that these are composite fibres consisting of flexible fibrillae joined end-to-end to pilus rods (Kuehn *et al.*, 1992). The tip fibrillae are comprised predominantly of PapE subunits, while the rod is composed of repeating PapA subunits packed into a right-handed helical cylinder (3.28 subunits per turn), with an external diameter of 68 Å and an axial hole of 15 Å (Bullitt & Makowski, 1995; Gong & Makowski, 1992).

PapG is the adhesin that mediates binding to the Gal-α(1,4)-Gal digalacto-side present in the globoseries of glycolipids on uroepithelial cells and erythrocytes (Leffler & Svanborg-Eden, 1980; Striker *et al.*, 1995). PapG is joined to the distal end of the PapE fibrillum via a specialized adaptor protein, PapF. The PapK adaptor protein joins the tip fibrillum to the PapA rod. Another minor component, PapH, is located at the base of the PapA rod and is thought to terminate assembly.

The expression and assembly of type 1 pili require at least nine genes that are present in the type 1 gene cluster (Hull *et al.*, 1981; Hultgren *et al.*, 1991). Like P pili, type 1 pili are also composite structures. The tip fibrillum is short and stubby. It contains FimG and the FimH adhesin and possibly FimF, which may join the tip to a rod comprised predominantly of FimA subunits (Jones *et al.*, 1995). The overall structure of the type 1 rod is similar to that of the P pilus rod comprised of PapA. The type 1 subunits are arranged in a helix with an external diameter of 6–7 nm and an axial hole of 20–25 Å, with 3.125 subunits per turn (Brinton, 1965).

The assembly of P and type 1 pili involves the folding and subsequent transport of pilus subunits in a non-polymerized state across the periplasmic compartment to the OM. Two specialized classes of proteins constitute the assembly machinery in each case: a periplasmic immunoglobulin-like cha-perone and an OM usher. The crystal structure of the PapD chaperone involved in the assembly of P pili has been solved to 2.0 Å resolution. PapD has two immunoglobulin-like domains oriented towards one another in an L shape (Fig. 1) (Holmgren & Brändén, 1989). All periplasmic chaperones identified to date can be organized into two structurally and functionally distinct subfamilies based on conserved amino acid differences in the chaperone cleft and the length of the loop that connects the F1 and G1 β-strands of domain 1 (Hung *et al.*, 1996). The two subfamilies are designated FGS (F1–G1 Short) and FGL (F1–G1 Long), corresponding to loop lengths of 20 or fewer amino acids, or 21 or greater amino acids, respectively. Interestingly, these two subfamilies assemble fibres with distinct molecular architectures. FGS chaperones, of which PapD is a member, are involved in assembling pili with rod-like architectures. FGL chaperones, on the other hand, mediate the assembly of very thin or afimbrial adhesive structures on the surface of bacteria.

PapD binds to and caps interactive surfaces on each of the pilus subunits, forming chaperone–subunit complexes. Chaperone–subunit complex forma-tion involves the binding of PapD to a highly conserved motif present in the C-termini of all subunits via a β-zippering interaction (Kuehn *et al.*, 1993; Soto *et al.*, 1998). This motif is characterized by a series of alternating hydrophobic residues flanked by a glycine located 14 residues upstream from the C-terminus and by a penultimate tyrosine. Pilus subunits are initially translocated across the cytoplasmic membrane via the Sec (general secretion system) machinery, although this pathway itself is not sufficient for the

Table 1. *Adhesive structures on the bacterial cell surface assembled by the chaperone–usher pathway*

Structure	Assembly gene products	Organism	Disease(s) associated with pilus expression	Reference
Thick, rigid pili	*FGS chaperone/usher*			
P pili	PapD/PapC	*E. coli*	Pyelonephritis/cystitis	Hull *et al.* (1984); Hull & Hull (1994); Hultgren *et al.* (1991)
Type 1 pili	FimC/FimD	*E. coli* Salmonella sp. *Klebsiella pneumoniae*	Cystitis	Jones *et al.* (1995); Ofek *et al.* (1977); Sokurenko *et al.* (1992)
Prs pili	PrsD/PrsC	*E. coli*	Cystitis	Lund *et al.* (1988); Stromberg *et al.* (1990)
S pili	SfaE/SfaF	*E. coli*	UTI Newborn meningitis	Moch *et al.* (1987); Schmoll *et al.* (1990)
F1C pili	FocC/FocD	*E. coli*	Cystitis	Riegman *et al.* (1990)
Haemophilus influenzae fimbriae	HifB/HifC	*Haemophilus influenzae*	Otitis media Meningitis	Guerina *et al.* (1982); van Alphen *et al.* (1991)
Haemophilus influenzae biogroup aegyptius fimbriae	HafB/HafE	*Haemophilus influenzae*	Brazilian purpuric fever	Read *et al.* (1996)
Type 2 and 3 pili	FimB/FimC	*Bordetella pertussis*	Whooping cough	Locht *et al.* (1992); Mooi *et al.* (1992); Willems *et al.* (1992)
MR/P pili	MrpD/MrpC	*Proteus mirabilis*	Nosocomial UTI	Bahrani *et al.* (1991)
PMF pili	PmfC/PmfD	*Proteus mirabilis*	Nosocomial UTI	Bahrani *et al.* (1993)
Long polar fimbriae	LpfB/LpfC	*Salmonella typhimurium*	Gastroenteritis	Baumler & Heffron (1995)
Pef pili	PefD/PefC	*Salmonella typhimurium*	Gastroenteritis	Friedich *et al.* (1993)
Ambient-temperature fimbriae	AftB/AftC	*Proteus mirabilis*	UTI	Massad *et al.* (1994, 1996)
987P fimbriae	FasB/FasD	*E. coli*	Diarrhoea in piglets	Cao *et al.* (1995); Edwards *et al.* (1996)

Thin, flexible pili

K99 pili	FaeE/FaeD	*E. coli*	Neonatal diarrhoea in calves, lambs, piglets	Bakker *et al.* (1991); Smit *et al.* (1984)
K88 pili	FanE/FanD	*E. coli*	Neonatal diarrhoea in piglets	Bakker *et al.* (1991); Erickson *et al.* (1992)
F17 pili	F17D/F17papC	*E. coli*	Diarrhoea	Lintermans *et al.* (1988)
MR/K pili	MrkB/MrkC	*Klebsiella pneumoniae*	Pneumonia	Allen *et al.* (1991); Gerlach *et al.* (1988)
REPEC fimbriae	RalE/RalD	*E. coli*	Diarrhoea in rabbits	Adams *et al.* (1997)

Atypical structures

FGL chaperone/usher

CS31A capsule-like protein	ClpE/ClpD	*E. coli*	Diarrhoea	Girdeau *et al.* (1988)
Antigen CS6	CssC/CssD	*E. coli*	Diarrhoea	Knutton *et al.* (1989); Wolf *et al.* (1989)
Myf fimbriae	MyfB/MyfC	*Yersinia enterocolitica*	Enterocolitis	Iriarte *et al.* (1993)
pH 6 antigen	PsaB/PsaC	*Yersinia pestis*	Plague	Lindler & Tall (1993)
CS3 pili	CS3-1/CS3-2	*E. coli*	Diarrhoea	Jalajakumari *et al.* (1989); Levine *et al.* (1984)
Envelope antigen F1	Caf1M/Caf1A	*Yersinia pestis*	Plague	Galyov *et al.* (1991); Karlyshev *et al.* (1992)
Non-fimbrial adhesins I	NfaE/NfaC	*E. coli*	UTI / Newborn meningitis	Ahrens *et al.* (1993); Goldhar *et al.* (1987)
SEF14 fimbriae	SefB/SefC	*Salmonella enteritidis*	Gastroenteritis	Clouthier *et al.* (1993); Muller *et al.* (1991)
Aggregative adherence fimbriae I	AggD/AggC	*E. coli*	Diarrhoea	Savarino *et al.* (1994)
AFA-III	AfaB/AfaC	*E. coli*	Pyelonephritis	Garcia *et al.* (1994); Le Bouguenec *et al.* (1993)

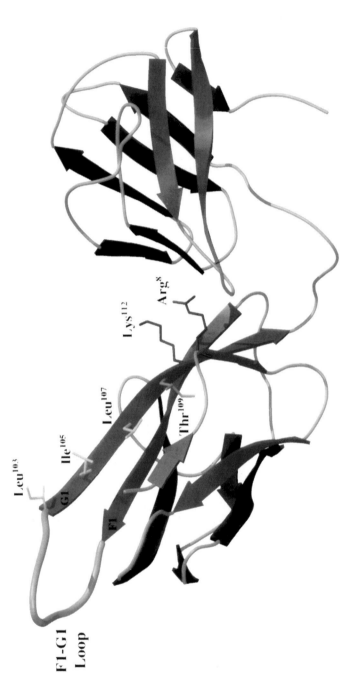

Fig. 1. Ribbon model of the three-dimensional structure of the PapD chaperone. PapD consists of two globular domains orientated in the shape of a boomerang. Each domain is a β-barrel structure formed by two antiparallel β-pleated sheets that have a topology similar to an immunoglobulin fold. The hypervariable F1–G1 loop is shown on the left. Also shown are the invariant Arg[8] and Lys[112] residues, together with the highly conserved hydrophobic residues at positions 103, 105 and 107 of the G1 β-strand.

efficient release of subunits into the periplasm (Jones *et al.*, 1997). Nascent subunits are retained in the inner cytoplasmic membrane through an interaction mediated by their hydrophobic C-termini. In the presence of chaperone, the subunits are partitioned into the periplasmic space as chaperone–subunit preassembly complexes following formation of the chaperone–subunit β-zipper (Fig. 2). Based on available crystallographic data, the G1 β-strand as well as invariant cleft residues are thought to participate in the β-zippering interaction with a subunit. Mutations in these invariant cleft residues of the chaperone abolish the ability of the chaperone to import subunits and form chaperone–subunit complexes (Jones *et al.*, 1997; Kuehn *et al.*, 1993; Soto *et al.*, 1998). Release of the subunits from the inner membrane is a prerequisite for their folding into an assembly-competent conformation, and there is evidence that folding of the subunits occurs directly on the chaperone template (Jones *et al.*, 1997; Soto *et al.*, 1998). Furthermore, it has been shown that the C-terminal chaperone recognition motif of pilus subunits is also involved in mediating quaternary subunit–subunit interactions in the mature pilus (Soto *et al.*, 1998). Thus, periplasmic chaperones may function by coupling the folding of pilus subunits with the capping of their assembly surfaces.

After their formation, the chaperone–subunit complexes are targeted to the OM usher for assembly. The PapC usher is predicted to possess predominantly β-sheet secondary structure, typical of bacterial OM pore-forming proteins. PapC likely exposes a large surface to the periplasm for interaction with chaperone–subunit complexes. To facilitate pilus assembly, the usher must be able to translocate pilus subunits across the OM. The PapC usher was purified and shown to form a pore when reconstituted into liposomes (Thanassi *et al.*, 1998). Using high-resolution electron microscopy, it was shown that PapC assembled into ring-shaped complexes containing central pores of 2–3 nm in diameter, each consisting of at least six subunits (Thanassi *et al.*, 1998). This diameter is not large enough to accommodate the 6.8-nm-wide helical pilus rod. However, experiments have shown that P pilus rods can be unravelled into linear fibres held together by subunits interacting in a head-to-tail fashion (Thanassi *et al.*, 1998). These unravelled rods measure 2 nm in diameter and would be narrow enough to pass through the usher pore. Therefore it has been proposed that the pilus rod is assembled across the OM in this linear form, and only adopts its final helical conformation upon reaching the external surface. This may be part of a mechanism that drives the outward growth of the organelle.

Ushers have a more active role in pilus assembly than simply functioning as diffusion pores for the translocation of pili across the OM. PapC was shown to recognize chaperone–subunit complexes with differential affinities depending on their final positions in the pilus (Dodson *et al.*, 1993). Later studies examining the 'real-time' kinetics of the chaperone–subunit–usher interactions demonstrated that chaperone–adhesin complexes from both the

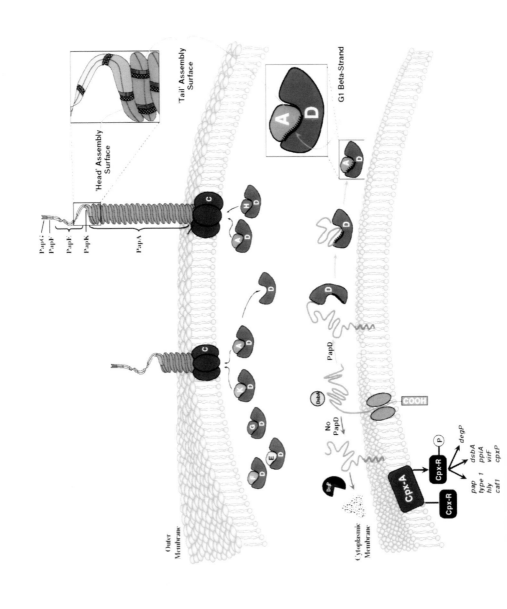

'Head' Assembly Surface

'Tail' Assembly Surface

G1 Beta-Strand

PapC
PapF
PapE
PapK
PapA

Outer Membrane

No PapD

PapD

COOH

Cytoplasmic Membrane

Cpx-A

Cpx-R

Cpx-R P

degP
dsbA
ppiA
virF
cpxP

pap
type 1
hly
caf1

P and type 1 pilus systems bound tightest and fastest to their respective ushers (Saulino *et al.*, 1998). These results suggested that kinetic partitioning of chaperone–adhesin complexes to the usher is a key factor in the tip localization of the adhesin. In addition, the dissociation rates for all of the chaperone–subunit complexes from the usher were slow, suggesting that after association of a complex with the usher, the subunit is destined for assembly into the pilus. The FimC–FimH complex was shown to form a stable complex with FimD. Expression of other combinations of chaperone–subunit complexes with the usher did not result in formation of a stable ternary complex. In addition, the interaction of the FimCH complex with FimD induced a protease-resistant conformation in FimD (Saulino *et al.*, 1998). Furthermore, this conformational change was maintained during pilus assembly, suggesting that interaction of the FimCH complex with FimD stabilizes the usher in an assembly-competent conformation. These data argue that interaction of the chaperone–adhesin complex with the usher is critical to initiate pilus biogenesis. Consistent with this hypothesis was the observation that mutations in the FimH adhesin greatly reduce pilus biogenesis in several clinical strains arguing that FimCHD ternary complex formation initiates pilus assembly (Saulino *et al.*, 1998).

In addition to preferential interactions of different chaperone–subunit complexes with the usher, subunit–subunit surface complementarity is another factor that plays a key role in dictating the relative order of subunit incorporation into the growing organelle. Jacob-Dubuisson *et al.* (1993) demonstrated that the PapF and PapK adaptor proteins were required for the efficient initiation of tip fibrillae and pilus rods, respectively. Deletion of both the *papF* and *papK* genes abolished piliation altogether, suggesting that other pilus subunits do not possess the structural determinants necessary to initiate the formation of tip fibrillae and pilus rods. The highly conserved C-terminal region of pilus subunits, together with another highly conserved region near the N-terminus, have recently been identified as serving as the primary assembly surfaces that mediate subunit–subunit interactions in the quaternary structure of the mature pilus (Soto *et al.*, 1998). Subtle differences in these primary assembly regions from one subunit to another may be responsible for controlling the order of incorporation of pilus subunits.

Fig. 2. Assembly of P pili from *E. coli* via the chaperone–usher pathway. Chaperone-mediated extraction of subunits from the inner cytoplasmic membrane is coupled with their folding into an assembly-competent state. The G1 β-strand of the immunoglobulin-like chaperones, which may serve as a template in the subunit-folding pathway, protects nascently folded subunits from premature oligomerization in the periplasmic space by directly capping the newly formed assembly surfaces. These interactive surfaces remain protected by the chaperone until delivery of the preassembly complex to the OM assembly site comprising the usher. Expression of pilus subunits in the absence of PapD leads to the activation of the Cpx pathway and the expression of several virulence genes.

A LINK BETWEEN PILUS BIOGENESIS AND PATHOGENESIS

Studies addressing the fate of pilus subunits when not folded properly unveiled a possible link between the assembly pathway and a signalling pathway that may be important in pathogenesis. Expression of pilus subunits in the absence of the chaperone leads to an OFF pathway of misfolding that is sensed by the CpxA–CpxR two-component system (Jones *et al.*, 1997). CpxA is the membrane-bound sensor/kinase and CpxR is the DNA-binding response/regulator (Raivio & Silhavy, 1997). This signalling pathway up-regulates *degP* transcription as well as a number of other chaperone-like proteins such as the disulfide isomerase DsbA and *cis trans* prolyl isomerases (Danese & Silhavy, 1997; De Las Penas *et al.*, 1997). These factors facilitate subunit folding: DsbA is required for pilus biogenesis (Jacob-Dubuisson *et al.*, 1994). These studies suggested that Cpx monitors pilus biogenesis and responds by controlling the expression of factors that facilitate pilus bio-genesis (Jones *et al.*, 1997).

Colonization is a dynamic process that involves a complex cascade of molecular cross-talk at the host–pathogen interface. Mulvey *et al.* (1998) have recently investigated the consequences of microbial colonization in a murine cystitis model using scanning and high-resolution transmission electron microscopy. These studies revealed that type 1 pilus tips interacted directly with the lumenal surface of the bladder, which is embedded with hexagonal arrays of integral membrane glycoproteins known as uroplakins. The attached pili were shortened and facilitated a more intimate contact of the bacteria with the uroplakin-coated host cells. The mechanism by which this apparent shortening occurs remains unknown, but mechanisms based on retraction of pili or impeded growth of nascent pili have been suggested (Mulvey *et al.*, 1998). Either of these mechanisms would likely result in a build-up of unassembled pilin subunits in the periplasm. It is intriguing to speculate that activation of the CpxA–CpxR pathway in response to pilus-mediated attachment leads to the expression of an array of virulence genes necessary for establishing an infection. Hung & Hultgren (1998) have referred to this state as the 'attached' phenotype. Bacterial attachment in the bladder was shown to trigger an apoptotic-like response in the host bladder epithelial cells which led to their exfoliation. Bacteria resisted this innate exfoliation mechanism by invading into the epithelium.

CONCLUSIONS

The assembly of adhesins on their outer surfaces poses many challenges to Gram-negative organisms. These include the folding of intrinsically aggrega-tion-prone subunits and their transport through the periplasmic compart-ment in a non-polymerized state to an OM assembly site where they are incorporated into a growing organelle in a defined order. The chaperone–

usher pathway has evolved to meet these challenges and to ensure that the bacterium is able to exert tight control over these protein–protein interactions not only during the assembly of these organelles, but also during the events that may follow their attachment to target receptors. Its study provides a model system for understanding the biological processes underlying the control of complex hetero-oligomeric interactions and organelle assembly across biological membranes as well as the molecular basis of crosstalk at the host–pathogen interface.

REFERENCES

Adams, L. M., Simmons, C. P., Rezmann, L., Strugnell, R. A. & Robins-Browne, R. M. (1997). Identification and characterization of a K88- and CS31A-like operon of a rabbit enteropathogenic *Escherichia coli* strain which encodes fimbriae involved in the colonization of rabbit intestine. *Infection and Immunity* **65**, 5222–5230.

Ahrens, R., Ott, M., Ritter, A., Hoschutzkky, H., Buhler, T., Lottspeich, F., Boulnois, G. J., Jann, K. & Hacker, J. (1993). Genetic analysis of the gene cluster encoding nonfimbrial adhesin I from an *Escherichia coli* uropathogen. *Infection and Immunity* **61**, 2505–2512.

Allen, B. L., Gerlach, G. F. & Clegg, S. (1991). Nucleotide sequence and functions of *mrk* determinants necessary for expression of type 3 fimbriae in *Klebsiella pneumoniae*. *Journal of Bacteriology* **173**, 916–920.

van Alphen, L., Geelen-van Den Broek, L., Blaas, L., van Ham, M. & Dankert, J. (1991). Blocking of fimbria-mediated adherence of *Haemophilus influenzae* by sialyl gangliosides. *Infection and Immunity* **59**, 4473–4477.

Bahrani, F. K., Johnson, D. E., Robbins, D. & Mobley, H. L. T. (1991). *Proteus mirabilis* flagella and MR/P fimbriae: isolation, purification, N-terminal analysis, and serum antibody response following experimental urinary tract infection. *Infection and Immunity* **59**, 3574–3580.

Bahrani, F. K., Cook, S., Hull, R., Massad, G. & Mobley, H. L. T. (1993). *Proteus mirabilis* fimbriae: N-terminal amino acid sequence of a major fimbrial subunit and nucleotide sequences of the genes from two strains. *Infection and Immunity* **61**, 884–891.

Bakker, D., Vader, C. E., Roosendaal, B., Mooi, F. R., Oudega, B. & de Graff, F. K. (1991). Structure and function of periplasmic chaperone-like proteins involved in the biosynthesis of K88 and K99 fimbriae in enterotoxigenic *Escherichia coli*. *Molecular Microbiology* **5**, 875–886.

Baumler, A. J. & Heffron, F. (1995). Identification and sequence analysis of *lpfABCDE*, a putative fimbrial operon of *Salmonella typhimurium*. *Journal of Bacteriology* **177**, 2087–2097.

Brinton, C. C., Jr (1965). The structure, function, synthesis, and genetic control of bacterial pili and a model for DNA and RNA transport in gram negative bacteria. *Transactions of the New York Academy of Science* **27**, 1003–1065.

Bullitt, E. & Makowski, L. (1995). Structural polymorphism of bacterial adhesion pili. *Nature* **373**, 164–167.

Cao, J., Chan, A. S., Bayer, M. E. & Schifferli, D. M. (1995). Ordered translocation of 987P fimbrial subunits through the outer membrane of *Escherichia coli*. *Journal of Bacteriology* **177**, 3704–3713.

Clouthier, S. C., Muller, K. H., Doran, J. L., Collinson, S. K. & Kay, W. W. (1993). Characterization of three fimbrial genes, *sefABC*, of *Salmonella enteritidis. Journal of Bacteriology* **175**, 2523–2533.

Danese, P. N. & Silhavy, T. J. (1997). The σE and the Cpx signal transduction systems control the synthesis of periplasmic protein-folding systems in *Escherichia coli. Genes & Development* **11**, 1183–1193.

De Las Penas, A., Connolly, L. & Gross, C. A. (1997). The σE-mediated response to extracytoplasmic stress in *Escherichia coli* is transduced by RseA and RseB, two negative regulators of σE. *Molecular Microbiology* **24**, 373–385.

Dodson, K. W., Jacob-Dubuisson, F., Striker, R. T. & Hultgren, S. J. (1993). Outer-membrane PapC molecular usher discriminately recognizes periplasmic chaperone-pilus subunit complexes. *Proceedings of the National Academy of Sciences, USA* **90**, 3670–3674.

Edwards, R. A., Cao, J. & Schifferli, D. M. (1996). Identification of major and minor chaperone proteins involved in the export of 987P fimbriae. *Journal of Bacteriology* **178**, 3426–3433.

Erickson, A. K., Willgohs, J. A., McFarland, S. Y., Benfield, D. A. & Francis, D. H. (1992). Identification of two porcine brush border glycoproteins that bind the K88ac adhesin of *Escherichia coli* and correlation of these glycoproteins with the adhesive phenotype. *Infection and Immunity* **60**, 983–988.

Friedich, M. J., Kinsey, N. E., Vila, J. & Kadner, R. J. (1993). Nucleotide sequence of a 13.9 kb segment of the 90 kb virulence plasmid of *Salmonella typhimurium*: the presence of a fimbrial biosynthetic gene. *Molecular Microbiology* **8**, 543–558.

Galyov, E. E., Karlishev, A. V., Chernovskaya, T. V., Dolgikh, D. A., Smirnov, O. Y., Volkovoy, K. I., Abramov, V. M. & Zav'yalov, V. P. (1991). Expression of the envelope antigen F1 of *Yersinia pestis* is mediated by the product of *caf1M* gene having homology with the chaperone protein PapD of *Escherichia coli. FEBS Letters* **286**, 79–82.

Garcia, M. I., Labigne, A. & Le Bouguenec, C. (1994). Nucleotide sequence of the afimbrial-adhesin-encoding *afa-3* gene cluster and its translocation via flanking IS1 insertion sequences. *Journal of Bacteriology* **176**, 7601–7613.

Gerlach, G. F., Allen, B. L. & Clegg, S. (1988). Molecular characterization of the type 3 (MR/K) fimbriae of *Klebsiella pneumoniae. Journal of Bacteriology* **170**, 3547–3553.

Girdeau, J. P., Vartanian, M. D., Ollier, J. L. & Contrepois, M. (1988). CS31A, a new K88-related fimbrial antigen on bovine enteropathogenic and septicemic *Escherichia coli* strains. *Infection and Immunity* **56**, 2180–2188.

Goldhar, J., Perry, R., Golecki, J. R., Hoschutzky, H., Jann, B. & Jann, K. (1987). Nonfimbrial, mannose-resistant adhesins from uropathogenic *Escherichia coli* 083:K1:H4 and 014:K?:H11. *Infection and Immunity* **55**, 1837–1842.

Gong, M. & Makowski, L. (1992). Helical structure of P pili from *Escherichia coli. Journal of Molecular Biology* **228**, 735–742.

Guerina, N. G., Langerman, S., Clegg, H. W., Kessler, T. W., Goldman, D. A. & Gilsdorf, J. R. (1982). Adherence of piliated *Haemophilus influenzae* to human oropharyngeal cells. *Journal of Infectious Diseases* **146**, 564.

Holmgren, A. & Brändén, C. I. (1989). Crystal structure of chaperone protein PapD reveals an immunoglobulin fold. *Nature* **342**, 248–251.

Hull, R. A. & Hull, S. I. (1994). Adherence mechanisms in urinary tract infections. In *Molecular Genetics of Bacterial Pathogenesis*, pp. 79–90. Edited by V. L. Miller, J. B. Kaper, D. A. Portnoy & R. R. Isberg. Washington, DC: American Society for Microbiology.

Hull, R. A., Gill, R. E., Hsu, P., Minshaw, B. H. & Falkow, S. (1981). Construction and expression of recombinant plasmids encoding type 1 and D-mannose-resistant pili from a urinary tract infection *Escherichia coli* isolate. *Infection and Immunity* **33**, 933–938.

Hull, R. A., Hull, S. I. & Falkow, S. (1984). Frequency of gene sequences necessary for pyelonephritis-associated pili expression among isolates of *Enterobacteriaceae* from human extraintestinal infections. *Infection and Immunity* **43**, 1064–1067.

Hultgren, S. J. & Normark, S. (1991). Biogenesis of the bacterial pilus. *Current Opinion in Genetics & Development* **1**, 313–318.

Hultgren, S. J., Normark, S. & Abraham, S. N. (1991). Chaperone-assisted assembly and molecular architecture of adhesive pili. *Annual Review of Microbiology* **45**, 383–415.

Hung, D. L. & Hultgren, S. J. (1998). Pilus biogenesis via the chaperone/usher pathway: an integration of structure and function. *Journal of Structural Biology* **124**, 201–220.

Hung, D. L., Knight, S. D., Woods, R. M., Pinkner, J. S. & Hultgren, S. J. (1996). Molecular basis of two subfamilies of immunoglobulin-like chaperones. *EMBO Journal* **15**, 3792–3805.

Iriarte, M., Vanooteghem, C., Delor, I., Diaz, R., Knutton, S. & Cornelis, G. R. (1993). The Myf fibrillae of *Yersinia enterocolitica*. *Molecular Microbiology* **9**, 507–520.

Jacob-Dubuisson, F., Heuser, J., Dodson, K., Normark, S. & Hultgren, S. (1993). Initiation of assembly and association of the structural elements of a bacterial pilus depend on two specialized tip proteins. *EMBO Journal* **12**, 837–847.

Jacob-Dubuisson, F., Pinkner, J., Xu, Z., Striker, R., Padmanhaban, A. & Hultgren, S. J. (1994). PapD chaperone function in pilus biogenesis depends on oxidant and chaperone-like activities of DsbA. *Proceedings of the National Academy of Sciences, USA* **91**, 11552–11556.

Jalajakumari, M. B., Thomas, C. J., Halter, R. & Manning, P. A. (1989). Genes for biosynthesis and assembly of CS3 pili of CFA/II enterotoxigenic *Escherichia coli*: novel regulation of pilus production by bypassing an amber codon. *Molecular Microbiology* **3**, 1685–1695.

Jones, C. H., Pinkner, J. S., Roth, R., Heuser, J., Nicholes, A. V., Abraham, S. N. & Hultgren, S. J. (1995). FimH adhesin of type 1 pili is assembled into a fibrillar tip structure in the *Enterobacteriaceae*. *Proceedings of the National Academy of Sciences, USA* **92**, 2081–2085.

Jones, C. H., Danese, P. N., Pinkner, J. S., Silhavy, T. J. & Hultgren, S. J. (1997). The chaperone-assisted membrane release and folding pathway is sensed by two signal transduction systems. *EMBO Journal* **16**, 6394–6406.

Kallenius, G., Svenson, S. B., Hultberg, H., Molby, R., Helin, I., Cedergen, B. & Windberg, J. (1981). Occurrence of P fimbriated *Escherichia coli* in urinary tract infection. *Lancet* **2**, 1369–1372.

Karlyshev, A., Galyov, E., Smirnov, O., Guzayev, A., Abramov, V. & Zav'yalov, V. (1992). A new gene of the *f1* operon of *Y. pestis* involved in the capsule biogenesis. *FEBS Letters* **297**, 77–80.

Knutton, S., McConnel, M. M., Rowe, B. & McNeish, A. S. (1989). Adhesin and ultrastructural properties of human enterotoxigenic *Escherichia coli* producing colonization factor antigens III and IV. *Infection and Immunity* **57**, 3364–3371.

Krogfelt, K. A., Bergmans, H. & Klemm, P. (1990). Direct evidence that the FimH protein is the mannose specific adhesin of *Escherichia coli* type 1 fimbriae. *Infection and Immunity* **58**, 1995–1999.

Kuehn, M. J., Heuser, J., Normark, S. & Hultgren, S. J. (1992). P pili in uropathogenic *E. coli* are composite fibers with distinct fibrillar adhesive tips. *Nature* **356**, 252–255.

Kuehn, M. J., Ogg, D. J., Kihlberg, J., Slonim, L. N., Flemmer, K., Bergfors, T. & Hultgren, S. J. (1993). Structural basis of pilus subunit recognition by the PapD chaperone. *Science* **262**, 1234–1241.

Le Bouguenec, C., Garcia, M. I., Oulin, V., Desperrier, J.-M., Gounon, P. & Labigne, A. (1993). Characterization of plasmid-borne *afa-3* gene clusters encoding afimbrial adhesins expressed by *Escherichia coli* strains associated with intestinal or urinary tract infections. *Infection and Immunity* **61**, 5106–5114.

Leffler, H. & Svanborg-Eden, C. (1980). Chemical identification of a glycosphingo-lipid receptor for *Escherichia coli* attaching to human urinary tract epithelial cells and agglutinating human erythrocytes. *FEMS Microbiology Letters* **8**, 127–134.

Levine, M. M., Ristaino, P., Marley, G. & 8 other authors (1984). Coli surface antigens 1 and 3 of colonization factor antigen II-positive enterotoxigenic *Escherichia coli*: morphology, purification, and immune responses in humans. *Infection and Immunity* **44**, 409–429.

Lindler, L. E. & Tall, B. D. (1993). *Yersinia pestis* pH 6 antigen forms fimbriae and is induced by intracellular association with macrophages. *Molecular Microbiology* **8**, 311–324.

Lintermans, P. F., Pohl, P., Bertels, A. & 8 other authors (1988). Characterization and purification of the F17 adhesin on the surface of bovine enteropathogenic and septicemic *Escherichia coli*. *American Journal of Veterinary Research* **49**, 1794–1799.

Locht, C., Geoffroy, M.-C. & Renauld, G. (1992). Common accessory genes for the *Bordetella pertussis* filamentous hemagglutinin and fimbriae share sequence similarities with the *papC* and *papD* gene families. *EMBO Journal* **11**, 3175–3183.

Lund, B., Marklund, B. I., Stromberg, N., Lindberg, F., Karlsson, K. A. & Normark, S. (1988). Uropathogenic *Escherichia coli* can express serologically identical pili of different receptor binding specificities. *Molecular Microbiology* **2**, 255–263.

Marklund, B. I., Tennent, J. M., Garcia, E., Hamers, A., Baga, M., Lindberg, F., Gaastra, W. & Normark, S. (1992). Horizontal gene transfer of the *Escherichia coli* *pap* and *prs* pili operons as a mechanism for the development of tissue-specific adhesive properties. *Molecular Microbiology* **6**, 2225–2242.

Massad, G., Bahrani, F. K. & Mobley, H. L. T. (1994). *Proteus mirabilis* fimbriae: identification, isolation, and characterization of a new ambient-temperature fimbria. *Infection and Immunity* **62**, 1989–1994.

Massad, G., Fulkerson, J. F., Jr, Watson, D. C. & Mobley, H. L. T. (1996). *Proteus mirabilis* ambient-temperature fimbriae: cloning and nucleotide sequence of the *aft* gene cluster. *Infection and Immunity* **64**, 4390–4395.

Moch, T., Hoschutzky, H., Hacker, J., Kroncke, K.-D. & Jann, K. (1987). Isolation and characterization of the α-sialyl-β-2,3-galactosyl-specific adhesin from fim-briated *Escherichia coli*. *Proceedings of the National Academy of Sciences, USA* **84**, 3462–3466.

Mooi, F. R., Jansen, W. H., Brunings, H., Gielen, H., van der Heide, H. G. J., Walvoort, H. C. & Guniee, P. A. M. (1992). Construction and analysis of *Bordetella pertussis* mutants defective in the production of fimbriae. *Microbial Pathogenesis* **12**, 127–135.

Muller, K. H., Collinson, S. K., Trust, T. J. & Kay, W. W. (1991). Type 1 fimbriae of *Salmonella enteritidis*. *Journal of Bacteriology* **173**, 454–457.

Mulvey, M. A., Lopez-Boado, Y. S., Wilson, C. L., Roth, R., Parks, W. C., Heuser, J. & Hultgren, S. J. (1998). Induction and evasion of host defenses by type 1-piliated uropathogenic *Escherichia coli*. *Science* **282**, 1494–1497.

Ofek, I., Mirelman, D. & Sharon, S. (1977). Adherence of *Escherichia coli* to human mucosal cells mediated by mannose receptors. *Nature* **265**, 623–625.

Raivio, T. L. & Silhavy, T. J. (1997). Transduction of envelope stress in *Escherichia coli* by the Cpx two-component system. *Journal of Bacteriology* **179**, 7724–7733.

Read, T. D., Dowdell, M., Satola, S. W. & Farley, M. M. (1996). Duplication of pilus gene complexes of *Haemophilus influenzae* biogroup aegyptius. *Journal of Bacteriology* **178**, 6564–6570.

Riegman, N., Kusters, R., van Veggel, H., Bergmans, H., Van Bergen en Henegouwen, P., Hacker, J. & Van Die, I. (1990). F1C fimbriae of a uropathogenic *Escherichia coli* strain: genetic and functional organization of the *foc* gene cluster and identification of minor subunits. *Journal of Bacteriology* **172**, 1114–1120.

Saulino, E. T., Thanassi, D. G., Pinkner, J. & Hultgren, S. J. (1998). Ramifications of kinetic partitioning on usher-mediated pilus biogenesis. *EMBO Journal* **17**, 2177–2185.

Savarino, S., Fox, P., Yikang, D. & Nataro, J. P. (1994). Identification and characterization of a gene cluster mediating enteroaggregative *Escherichia coli* aggregative adherence fimbriae I biogenesis. *Journal of Bacteriology* **176**, 4949–4957.

Schmoll, T., Morschhauser, J., Ott, M., Luwig, B., van Die, I. & Hacker, J. (1990). Complete genetic organization and functional aspects of the *Escherichia coli* S fimbrial adhesin determinant: nucleotide sequence of the genes *sfa* B, C, D, E, F. *Microbial Pathogenesis* **9**, 331–343.

Smit, H., Gaastra, W., Kamerling, J. P., Vliegenthart, J. F. G. & de Graaf, F. K. (1984). Isolation and structural characterization of the equine erythrocyte receptor for enterotoxigenic *Escherichia coli* K99 fimbrial adhesin. *Infection and Immunity* **46**, 578–584.

Sokurenko, E. V., Courtney, H. S., Abraham, S. N., Klemm, P. & Hasty, D. L. (1992). Functional heterogeneity of type 1 fimbriae of *Escherichia coli*. *Infection and Immunity* **60**, 4709–4719.

Soto, G. E., Dodson, K. W., Ogg, D., Liu, C., Heuser, J., Knight, S., Kihlberg, J., Jones, C. H. & Hultgren, S. J. (1998). Periplasmic chaperone recognition motif of subunits mediates quaternary interactions in the pilus. *EMBO Journal* **17**, 6155–6167.

Striker, R., Nilsson, U., Stonecipher, A., Magnusson, G. & Hultgren, S. J. (1995). Structural requirements for the glycolipid receptor of human uropathogenic *E. coli*. *Molecular Microbiology* **16**, 1021–1029.

Stromberg, N., Marklund, B. I., Lund, B., Ilver, D., Hamers, A., Gaastra, W., Karlsson, K. A. & Normark, S. (1990). Host-specificity of uropathogenic *Escherichia coli* depends on differences in binding specificity to Galα(1–4)Gal-containing isoreceptors. *EMBO Journal* **9**, 2001–2010.

Thanassi, D. G., Saulino, E. T., Lombardo, M. J., Roth, R., Heuser, J. & Hultgren, S. J. (1998). The PapC usher forms an oligomeric channel: implications for pilus biogenesis across the outer membrane. *Proceedings of the National Academy of Sciences, USA* **95**, 3146–3151.

Willems, R. J. L., van der Heide, H. G. J. & Mooi, F. R. (1992). Characterization of a *Bordetella pertussis* fimbrial gene cluster which is located directly downstream of the filamentous haemagglutinin gene. *Molecular Microbiology* **6**, 2661–2671.

Wolf, M. K., Andrews, G. P., Tall, B. D., McConnel, M. M., Levine, M. M. & Boedeker, E. C. (1989). Characterization of CS4 and CS6 antigenic components of PCF8775, a putative colonization factor complex from enterotoxigenic *Escherichia coli* E8775. *Infection and Immunity* **57**, 164–173.

DNA UPTAKE BY TRANSFORMABLE BACTERIA

SANFORD A. LACKS

Biology Department, Brookhaven National Laboratory, Upton, NY 11973, USA

HISTORICAL INTRODUCTION

The first dim realization that hereditary material was transferred from male to female animals probably arose among early pastoralists some 10 000 years ago, if not before. However, it was not until 1944 that the chemical substance of heredity, DNA, was directly shown to effect genetic change by demonstration of its role in bacterial transformation (Avery *et al.*, 1944). The work of Avery and colleagues indicated that cells of *Streptococcus pneumoniae* could take up naked DNA. In the next decade, conjugative transfer of genetic markers, requiring cellular contact, was observed in bacteria (Lederberg & Tatum, 1946), and the introduction of viral DNA was shown to accompany bacteriophage infection (Hershey & Chase, 1952).

By the time that the double-stranded, helical structure of DNA was proposed in 1953, a number of animal and plant viruses had been purified and analysed and found to contain either DNA or RNA. It was realized that all viruses contain nucleic acid as their genetic material, and that the nucleic acid must be introduced into the cell to infect it. A natural ability to be transformed genetically by free DNA was discovered in additional bacterial species including *Haemophilus influenzae* and *Bacillus subtilis*. During the 1950s extrachromosomal DNA elements called plasmids were discovered in bacteria, and many plasmids were found to carry the ability to be transferred by conjugation. Thus, mechanisms for DNA transfer appeared to be widespread.

Beginning in the late 1950s, the use of isotopically labelled material showed that DNA was physically introduced into recipient cells during transformation (Fox, 1957; Lerman & Tolmach, 1957). It was shown that transforming DNA is converted to single-stranded segments during entry (Lacks, 1962) and that these lengthy segments are incorporated into the chromosomal DNA of the recipient cell (Fox & Allen, 1964; Notani & Goodgal, 1966). This molecular processing of transforming DNA and the realization that the development of competence for DNA uptake was a transient property of the cell, itself under elaborate control (Tomasz & Hotchkiss, 1964), that required synthesis of a dozen proteins (Morrison & Baker, 1979) pointed to the complexity of the natural transformation systems. Meanwhile, conjugation was also shown to result in transfer of single-stranded DNA (Cohen *et al.*,

1968), and this process, too, required the function of at least a dozen genes (see review by Wilkins, 1995).

In the 1970s various methods for artificially introducing DNA into cells began to be elaborated. These artificial mechanisms will be discussed here only briefly. Also, the mechanisms of viral transfer and conjugation will not be presented in much detail. The emphasis of this review will be on DNA uptake by the natural mechanisms of bacterial transformation. Research up to the mid-1970s on this topic was previously reviewed (Lacks, 1977b). Several fine recent reviews cover DNA uptake more broadly or from different perspectives (Dreiseikelmann, 1994; Dubnau, 1991, 1997; Sabelnikov, 1994; Lorenz & Wackernagel, 1994; Lunsford, 1998; Palmen *et al.*, 1994). In the case of transformation, discovery of the mechanisms of competence regulation during the past decade has facilitated the identification of genes involved in DNA uptake. Therefore, most of the components of the uptake systems are now known, although their arrangements and precise functions are not. No case of DNA uptake is yet understood in detail, but it should be instructive at this time to assess the state of our knowledge.

OVERVIEW OF DNA UPTAKE MECHANISMS

The various processes of DNA uptake by cells can be categorized as viral DNA entry, conjugation or transformation. Within each category, a variety of mechanisms have been found. However, considerable similarities occur among the different mechanisms of conjugation and, especially, transformation. All of these natural mechanisms of DNA transfer are quite elaborate and involve multiple protein components, as the case may be, of the virus, the donor cell and the recipient cell. The mechanisms of viral infection and conjugation will be discussed mainly with respect to their relevance to transformation.

Infection by viruses

Introduction of DNA in viral infection always involves interaction of viral proteins with receptors on the cell surface. Some viruses have elaborate structures and appendages composed of multiple proteins. For example, *Escherichia coli* T4 phage contains a tail-like appendage implicated in DNA uptake. As illustrated in Fig. 1(a), tail fibres bind to surface lipopolysaccharide (Cerritelli *et al.*, 1996), after which the tail core penetrates the outer membrane, fusing it with the inner cell membrane of the Gram-negative bacterium, to allow passage of phage DNA into the cell (Tarahovsky *et al.*, 1991). The phage protein gp2 may serve as a 'pilot' that attaches to an end of incoming DNA to protect it from cellular exonucleases (Lipinska *et al.*, 1989) and perhaps to assist its entry. Such pilot proteins have been proposed in various DNA uptake systems but with little direct evidence,

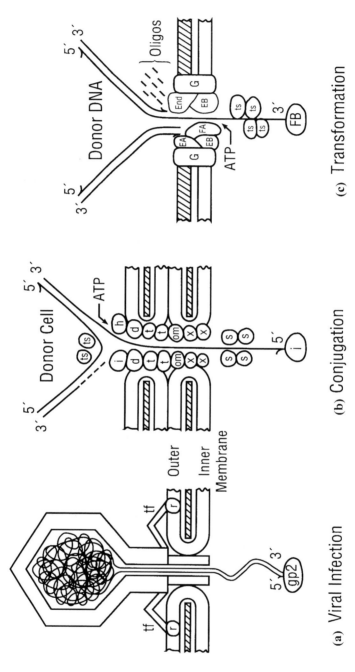

(a) Viral Infection   **(b)** Conjugation   **(c)** Transformation

Fig. 1. Modes of DNA uptake. (a) Viral infection: gp2, phage-encoded protein; r, cell membrane receptor; tf, phage tail fibre. (b) Conjugation: d, TraD; i, TraI; t, other plasmid *tra* gene products; om, OmpA; x, unknown, host proteins; h, donor cell helicase; s and ts, single-stranded DNA-binding proteins. (c) Transformation: End, membrane-bound endonuclease releases oligonucleotides (Oligos); other proteins, EA, EB, FA, FB and G, are induced during competence development and have structural and functional roles in uptake; ts, competence-induced single-stranded DNA-binding protein. The mechanisms shown are in part hypothetical.

possibly because a single protein molecule per viral genome is difficult to detect.

A simpler virus, such as the filamentous phage fd, contains only a few coat proteins. This virus infects 'male' cells of *E. coli*, i.e. cells which contain the sex-factor plasmid F and which, therefore, extrude fibrils, called pili, from their surface. A minor coat protein, gp3, located at one end of the filamentous virus, binds to a pilus (Gray *et al.*, 1981), and the virus is drawn to the cell surface as the pilus retracts. Oligomers of this same gp3 protein then form a transmembrane channel (Glaser-Wuttke *et al.*, 1989) through which the single-stranded fd DNA enters the cell. In addition to the pilin proteins forming the pilus, entry may require other host proteins such as TolA, mutants of which cannot be infected by fd (Sun & Webster, 1986). TolA is a membrane protein, and it may contribute to the entry channel (Levengood & Webster, 1989).

Viral transfer mechanisms can introduce either single- or double-stranded DNA. The other natural mechanisms of DNA transfer, however, introduce only single-stranded DNA. This may reflect their function, which is to transfer genetic information from one cell to another, unlike the virus, which acts as an independent entity and takes over the cell machinery.

Conjugation – cell to cell transfer

The key feature of conjugative DNA transfer is its requirement for close contact between cells, so that the DNA passes directly from donor cell to recipient without external exposure. Close contact is achieved by the action of donor cell pili, which first bind to the recipient cell surface. Upon retraction of the pili, the two cells are juxtaposed, and a channel through which single-stranded DNA can pass is formed (reviewed by Wilkins, 1995).

Conjugative transfer systems are often present on plasmids in both Gram-positive and Gram-negative bacteria. They allow transmission of the plasmid to other cells independent of gene transfer mechanisms in the host. Typically, up to two dozen *tra* genes in the plasmid may encode components of the transfer system (Wilkins, 1995). In addition to the pilins and proteins that process and export them, these genes encode proteins that make up the transmembrane channel, and proteins that act on the plasmid DNA prior to its transfer, and even, by accompanying the DNA, after its transfer to the recipient cell (Rees & Wilkins, 1990).

Once the cells are in contact, a bridge-like channel must be formed between them for passage of the DNA. Many of the *tra* genes presumably encode proteins for this structure. Recipient proteins, such as OmpA, may play roles in cell contact and channel formation (Manoil & Rosenbusch, 1982; Sugawara & Nikaido, 1992). Several *tra* gene products process the donor DNA. In the case of F-factor, the protein TraI, a DNA-dependent ATPase with helicase activity, makes a single-strand break at the *oriT* site on the plasmid, binds covalently to the newly formed 5'-end, and unwinds the strand (Matson *et al.*, 1993). It is not clear whether TraI leads the donor strand into the recipient cell, as depicted in Fig. 1(b), or whether it binds to the intercellular pore so that donor DNA loops into the cell (Wilkins, 1995). A single-stranded DNA-binding protein is encoded by the F-factor; it may bind to the complementary strand, which is used as a template to resynthesize in the donor cell the strand corresponding to the one transferred.

Related to conjugative transfer is the interkingdom transmission of bacterial plasmid DNA from *Agrobacterium tumefaciens* to plant cells to form crown gall tumours (for reviews, see Kado, 1991; Citovsky & Zambryski, 1993). In this case, the Ti plasmid, which is also capable of conjugative transfer between bacteria, carries a set of approximately two dozen *vir* genes that, when activated, remove a single-strand segment of plasmid DNA, called T-DNA, and transfer it through a conjugative bridge structure through the membrane of a plant cell, where it finds its way into the nucleus and becomes incorporated into the plant genome. The *virD1* and *virD2* products form an endonuclease that cuts at the T-DNA borders to release the strand segment. The VirD2 protein is bound to the 5'-end of T-DNA, probably covalently, where it may act as a pilot protein to shepherd the strand into the host cell nucleus, protecting it from exonucleolytic digestion, and/or conserving the original bond energy for ligation to plant DNA (Dürrenberger *et al.*, 1989). Some of the *vir* genes are homologous to *tra* genes, for example *virB2* and *traA*, the pilin-encoding gene of the F-factor (Shirasu & Kado, 1993). Another gene, *virB11*, shares sequence similarity with *comGA*, which is required for transformation (Albano *et al.*, 1989).

Genetic transformation by free DNA

Natural bacterial transformation

Genetic transformation by purified DNA has now been demonstrated in over 40 species of bacteria (Lorenz & Wackernagel, 1994), but it is by no means a universal capability. As in conjugation, many specific proteins are required for transformation. However, since donor DNA is not accompanied by proteins, probably even under natural conditions, the proteins required for DNA uptake are produced in the recipient cell, rather than in the donor cell, as is the case for conjugation. It is not known whether under natural circumstances donor DNA is extruded from or leaks out of the cell or is released on lysis of cells, but extracellular DNA can be detected in cultures (reviewed by Lorenz & Wackernagel, 1994).

The process of natural transformation allows large amounts of DNA – as much as 10 % of their DNA content – to be taken up by cells (for a review, see Lacks, 1988). Chromosomal DNA markers routinely transform more than 1 % of the recipient population; homogeneous, plasmid DNA containing a chromosomal marker can transform more than half of the cells. In most species all of the cells become competent to take up DNA; in *B. subtilis*, however, only ~10% of the population becomes competent. The DNA that is taken up by the recipient cell can be integrated with very high efficiency, with estimates ranging between 0.5 and 1.0.

Artificial transformation – electroporation

In contrast to natural transformation, artificial methods introduce relatively small amounts of DNA into cells. Two types of methods have been used. A cell–DNA mixture is subjected to abrupt shifts in divalent cation concentration and temperature in one method (Mandel & Higa, 1970) and to high-voltage electrical pulses in the other (Harlander, 1987). The former was more important historically, but the latter method, called electroporation, has become the method of choice because of its simplicity and general applicability. Although the precise mechanisms by which these agents act are not known, it is presumed that they form transient pores in the cell membrane(s), through which DNA can pass (for reviews see Chang *et al.*, 1992; Sabelnikov, 1994).

Despite the small quantities of DNA introduced by artificial transformation, it has proven extremely useful in working with the many species that are not naturally transformable. The DNA is taken up in the form presented to the cells without degradative processing, which is advantageous for establishing a plasmid. This compensates, in part, for the relatively poor uptake.

MOLECULAR FATE OF DNA IN TRANSFORMATION

Initial attachment of double-stranded DNA

In all known systems of natural transformation, DNA must be double-stranded, which is its native state, to be efficiently taken up. Although denatured (Miao & Guild, 1970) or single-stranded (Barany & Boeke, 1983) DNA can enter cells, its ability to do so is less than 1% of double-stranded DNA. Kinetic studies in which the concentration of donor DNA is varied suggest the existence on the surface of Gram-positive bacteria of ~50 sites that can reversibly bind double-stranded DNA. Very quickly, however, DNA becomes irreversibly bound, so that it can no longer be washed away, although its sensitivity to external agents, like DNase, shows it to be still outside the cell. With *S. pneumoniae*, donor DNA at this stage can be examined in mutants lacking the EndA nuclease required for entry (Lacks *et al.*, 1974) or in wild-type cells treated with chelating agents (Seto & Tomasz, 1974), which block the nuclease action. With *B. subtilis*, the entry process is slower, and the external state of DNA can be investigated without inhibiting its entry. These Gram-positive species and others are able to take up DNA indiscriminately with regard to its nucleotide sequence.

At least some Gram-negative species, typified by *H. influenzae*, require for binding a 9–11 nt sequence found specifically in their own DNA (Danner *et al.*, 1980). Cells of *Neisseria gonorrhoeae* also require a specific sequence, but one that is different from *H. influenzae* (Goodman & Scocca, 1988). In addition to the sequence requirement, DNA bound by these species becomes resistant to extracellular agents. In the case of *H. influenzae* it has been shown that such bound DNA is protected by cell membranes in a vesicle (Kahn *et al.*, 1982) called a transformasome. Within the transformasome, donor DNA remains double-stranded, but as in the case of Gram-positive cells, it is converted to single strands on entry into the cell proper (Barany *et al.*, 1983). According to one model (Sabelnikov, 1994), DNA binds to receptors on the outer membrane that recognize the specific sequence. This triggers curvature of the membrane, which envelopes the DNA to form a vesicle within the outer membrane. The vesicle, or transformasome, then contacts the inner membrane, fusing with it at the location of a competence-induced structure for DNA entry.

DNA degradation and entry of single strands

External strand breaks

It is likely that DNA binds initially to a cellular receptor in a reversible manner. However, in the transformation of *S. pneumoniae* and *B. subtilis*, a stage of irreversible binding in which the DNA is still external can be identified. Such binding is an active process with its own ionic and energy requirements. In *S. pneumoniae* it has been possible to isolate this stage in mutants lacking the membrane nuclease EndA (Lacks *et al.*, 1974) and to show that concomitant with this binding is the appearance of single-strand breaks (nicks) in the DNA. These breaks apparently occur randomly along the DNA, every 2.5 kb on average, to give single-strand segments of 5 kb (weight average; Lacks & Greenberg, 1976); they may result from enzymic action at entry sites. Proteins with strand-nicking activity have been isolated from the cell surface of competent streptococci (Fujii *et al.*, 1987; Lunsford *et al.*, 1996), but they have not been well characterized. That the breaks are required for binding is indicated by the occurrence of at least one break on every molecule bound, as demonstrated for a covalently closed circular DNA donor (Lacks, 1979b). The reduction by half of DNA strand length upon uptake by cells (Morrison & Guild, 1972) also indicates that a strand break is required for DNA uptake.

In leaky mutants of *endA* that produce some nuclease, the nicks become double-strand breaks (Lacks & Greenberg, 1976); therefore, it can be presumed that in wild-type cells, the single-strand breaks are quickly converted to double-strand breaks as the opposite strand is degraded and the initially nicked strand begins to enter the cell. In *B. subtilis*, donor DNA can be detected on the surface of wild-type cells, irreversibly bound, and nicked to give single-strand fragments of ~5 kb (weight average; Dubnau & Cirigliano, 1972). Although the molecular processing of DNA during binding and entry has been investigated more fully in *S. pneumoniae*, these stages of uptake are considered to be similar in *B. subtilis* (Dubnau, 1991).

Binding is not sequence-specific in these Gram-positive species, but it is specific for unencumbered double-stranded DNA. Neither single-stranded DNA, RNA–DNA hybrids nor glycosylated phage T4 DNA show appreciable binding or ability to compete with native DNA (Lacks, 1977a). Since the single-strand breaks render the binding irreversible, it is possible that a covalent link is formed with a surface protein during breakage. This protein might serve as a pilot, leading the DNA strand into the cell, or it might have another function, perhaps as the putative helicase proposed to assist DNA entry in *B. subtilis* (Londoño-Vallejo & Dubnau, 1993). The TraI protein of the F-factor is reported to nick the transferred DNA, bind to it, and act as a helicase (Matson *et al.*, 1993). In the conjugative plasmid RP4, its TraI protein also accompanies single-stranded DNA into the recipient cell (Rees & Wilkins, 1990).

Single-strand entry

Only single strands of donor DNA enter into *S. pneumoniae* cells (Lacks *et al.*, 1967). Similarly, newly entered DNA in *B. subtilis* is detectable as single strands (Piechowska & Fox, 1971). An amount of DNA approximately equal to the amount taken up is released in an acid-soluble form outside the cells (Lacks & Greenberg, 1973; Morrison & Guild, 1973). The released product consists of oligonucleotides of size distribution identical to that produced by the action of the EndA membrane nuclease of *S. pneumoniae* (Lacks *et al.*, 1974). It appears that one strand of DNA enters the cell as the complementary strand is degraded (Fig. 1c). Selective end-labelling of donor DNA showed that the 3'-end of the strand must enter first, whereas the opposite strand is degraded from the 5'-end (Mejean & Claverys, 1993). A reasonable view of events during entry is that a DNA strand, with its 3'-end formed by the nick made on binding, enters the cell at the same time that the EndA nuclease makes an endonucleolytic break on the opposite strand and continues to degrade that strand from 5' to 3' as its complement enters 3' to 5'. A rate of 100 nt s^{-1} was calculated for DNA entry in *S. pneumoniae* (Mejean & Claverys, 1993).

Null mutants of EndA are reduced in transformability to <0.1% of the normal level (Puyet *et al.*, 1990), and they are correspondingly reduced in DNA uptake (Lacks *et al.*, 1975). This accords with their role in DNA entry. The residual transformation, which is low but real, and the low level of transformation demonstrable with single-stranded DNA (Barany & Boeke,

Table 1. *Genes implicated in DNA uptake*[a]

B. subtilis	S. pneumoniae	H. influenzae[b]	Possible functions
	coiA		Cell wall degradation
	murA[c]		Cell wall resynthesis
comGA	cglA	[HI0298] pilB	Prepilin membrane transport pore
comGB	cglB	[HI0297] pilC	Prepilin transport ATPase
comGC	cglC	[HI0299] pilA	Pilin-like wall structure
comGD	cglD		Pilin-like wall structure
comGE	cglE		Pilin-like wall structure
comGF	cglF		Pilin-like wall structure
comGG	cglG		Pilin-like wall structure
comC	cilC	[HI0296] pilD	Prepilin processing
comEA	celA	[HI1018][c]	Donor dsDNA binding
comEC	celB	[HI0061] rec2	DNA membrane transport pore
comFA	cflA		Nickase, translocase, helicase?
comFC	cflB	[HI0434] com101A (comF)	Pilot protein?
bsn[c] (endA)	endA		Degrades donor DNA strand
nucA[c]	–		Alternative degradative nuclease
ssbB[c]	cilA[c]		ssDNA binding protein
	cilB[c]	[HI0985] dprA	Late DNA processing

[a] See text for references. All genes except *murA*, *bsn* and *endA* are competence-inducible.
[b] Genome designations in brackets; additional gene names in parentheses.
[c] No experimental evidence for uptake.

1983) may result from an alternative mechanism for entry in the absence of EndA activity. A possible candidate for catalysing such entry is the putative helicase encoded by *comFA* in *B. subtilis* (Londoño-Vallejo & Dubnau, 1993) and its homologue, *cflA*, in *S. pneumoniae* (Table 1). An interesting set of mutants found in *S. pneumoniae* do not take up DNA into the cell but still bind DNA and degrade the same amount as wild-type cells (Morrison *et al.*, 1983). They are presumably normal for EndA but lack another factor necessary for entry, which might be the helicase, a pilot protein, or a protein forming the entry pore.

   In Gram-negative species, DNA entry into the cell proper from the transformasome may occur by a mechanism similar to the Gram-positive species. DNA apparently enters the cell as a single strand, with its 3′-end entering first (Barany *et al.*, 1983).

Ionic and energy requirements

Irreversible, external binding of DNA to *S. pneumoniae* cells requires an energy source (Lacks *et al.*, 1974) and potassium ions (Lacks, 1979a). The energy is required, possibly, for covalent linking of the DNA to a surface protein at the nick site. No divalent cations are necessary for binding. Potassium ions are necessary, perhaps for orienting the binding/entry complex in the membrane by maintaining the electrostatic potential across it.

   When cells were preloaded with bound DNA in the presence of a chelating agent, uptake into the cells was observed upon its removal, and that uptake was partially inhibited by a sugar analogue (Seto & Tomasz, 1974). So it appears that energy is needed for entry too, perhaps to power a helicase that unwinds the entering DNA or a translocase that draws it into the cell. Calcium and magnesium ions are both required for entry (Seto & Tomasz, 1976; Lacks, 1977a). The role of calcium ions is unknown, but magnesium ions are required for the membrane nuclease activity (Lacks *et al.*, 1974).

The contribution to DNA uptake of the proton motive force at the membrane, both with respect to electrostatic potential and pH gradient across the membrane, has been examined in both *S. pneumoniae* and *B. subtilis*. The data were reviewed recently (Palmen *et al.*, 1994), but no firm conclusions were drawn with respect to a requirement or function in uptake. However, some data with *S. pneumoniae* suggest that uptake depends on the ATP level in the cells (Clavé & Trombe, 1989). Possible roles for ATP are to activate a protein that is linked covalently to donor DNA during binding or to power helicase or translocase activity during entry.

Consequences of the entry mechanism

Nonhomologous DNA

DNA that enters into *S. pneumoniae* is coated with a single-stranded DNA-binding protein that is induced during competence (Morrison, 1978). Nonhomologous DNA that is taken up persists in the cell with a half-life of ~ 30 min at $30\,°C$ (Lacks *et al.*, 1967). It is apparently degraded slowly by exonucleases, and its breakdown products are reused for DNA synthesis.

Homologous recombination

DNA that is homologous to the recipient genome has a much shorter half-life as single strands, ~ 6 min at $30\,°C$. This corresponds to the time required for integration of donor DNA into the chromosome (Ghei & Lacks, 1967). The single-stranded form of the donor DNA facilitates its interaction with the duplex DNA of the chromosome, and recombination occurs with an efficiency of at least 50% (Fox, 1957; Lerman & Tolmach, 1957).

Plasmid establishment

Because strand breaks occur on DNA binding and one strand is degraded during entry, the most plasmid material that can be introduced in a single entry event is a linearized single strand. Inasmuch as such a strand cannot circularize to form a replicon, establishment of a plasmid requires the interaction of complementary strand segments introduced in two separate events (Saunders & Guild, 1981). However, if a plasmid carries a DNA segment homologous to the chromosome, pairing can occur in this segment between plasmid and chromosome to enable a single plasmid strand to establish the plasmid by a process called chromosomal facilitation (Lopez *et al.*, 1982).

Restriction of donor DNA

Restriction endonucleases generally do not act on either single-stranded or hemimethylated DNA. Therefore, chromosomal transformation by unmodified DNA is not affected by restriction enzymes in the host (Trautner *et al.*, 1974; Lacks & Springhorn, 1984). However, plasmid transfer requires interaction of two donor strands and is susceptible to restriction. In the case of *S. pneumoniae*, its *Dpn*II restriction system circumvents the plasmid susceptibility by expressing an accessory methyltransferase, DpnA, that can methylate single-stranded DNA, so that plasmid DNA is already methylated at potential restriction sites by the time that it becomes a double-stranded replicon (Cerritelli *et al.*, 1989).

REGULATION OF COMPETENCE FOR DNA UPTAKE

Streptococcal species

Competence-stimulating polypeptides

Early investigators of streptococcal transformation realized that the bacteria were not always transformable. They became competent for transformation only when grown in particular media and then only at a certain stage in the growth cycle, namely late-exponential growth. Under these conditions an extracellular factor that stimulated competence in noncompetent cultures was isolated from cultures of *Streptococcus gordonii* (Pakula & Walczak, 1963) and *S. pneumoniae* (Tomasz & Hotchkiss, 1964). The factor from *S. pneumoniae* appeared to be a small protein (Tomasz & Mosser, 1966). This competence-stimulating polypeptide (CSP) was recently shown to be a 17-mer polypeptide, EMRLSKFFRDFILQRKK (Håverstein *et al.*, 1995). Other strains of *S. pneumoniae* and related streptococcal species have CSPs varying slightly in size and composition (Morrison, 1997).

Quorum-sensing mechanism

It was realized that a factor accumulating extracellularly could act as a signal of sufficient bacterial culture density to render it likely that DNA leaking out of one cell would reach another cell of the species (Tomasz, 1965). The molecular mechanism of such quorum sensing is now understood (reviewed by Morrison, 1997). The precursor of CSP is encoded by *comC*, a small gene in an operon containing also *comD* and *comE*. (An operon is a cluster of genes transcribed on the same messenger RNA.) Another operon contains *comA* and *comB*, the membrane-located products of which act to process the *comC* protein and export CSP out of the cell. When CSP accumulates, it signals ComD, which is a membrane-located histidine protein kinase, to phosphorylate ComE. The phosphorylated ComE activates transcription, possibly after another stage in the cascade, of competence-specific proteins.

Induction of competence genes

At least a dozen proteins are preferentially synthesized during the development of competence. Several operons encoding these proteins were found to have the 'combox' sequence TACGAATA (Campbell *et al.*, 1998; Pestova & Morrison, 1998), in place of the typical pneumococcal SigA promoter (TTGACAN$_{12}$TNTGNTATAAT; Sabelnikov *et al.*, 1995). Thus, induction of these late competence genes may depend on a competence-specific sigma factor ultimately elicited by CSP. In a *trt* mutant strain of *S. pneumoniae*, which is transformable in the presence of trypsin and does not require CSP for competence (Lacks & Greenberg, 1973), combox genes are constitutively expressed (S. Lacks & B. Greenberg, unpublished data), perhaps because the putative sigma factor is no longer regulated.

Bacillus species

Competence-stimulating polypeptides

Cells of *B. subtilis* also become competent late in the culture cycle. In this species, two different polypeptides are released by the cell (reviewed by Solomon & Grossman, 1996). The ComX pheromone is a polypeptide of 9 or 10 aa, apparently containing a modified tryptophan residue, that is processed from a 55 aa precursor. Similarly to CSP of *S. pneumoniae*, it acts on a histidine protein kinase (ComP) in the cell membrane, which in turn phosphorylates a transcription factor, ComA. Another competence-stimulating factor, which appears to be a small peptide, may act by blocking a phosphatase that restores ComA to an unphosphorylated and inactive state.

Regulation of gene expression

Phosphorylated ComA increases transcription of *srfA* (*comS*), which leads eventually to the accumulation of ComK (van Sinderen *et al.*, 1995b). ComK acts to induce the late competence genes, not as a sigma factor, but as a transcription activator. Operons that are induced have a consensus SigA promoter, but they also have a characteristic pattern of A- and T-rich sequences upstream of the promoter, which apparently allow binding and activation by ComK (Hamoen *et al.*, 1998).

Gram-negative species

Regulated competence in H. influenzae

Competence appears in cells of *H. influenzae* when a culture is nutritionally depleted but can still carry on protein synthesis (Goodgal & Herriott, 1961).

It appears to be mediated by the accumulation of cAMP in the cell, which triggers the expression of a gene called *sxy* (Dorocicz *et al.*, 1993) or *tfoX* (Zulty & Barcak, 1995), the product of which, in turn, induces late competence genes (Karudapuram & Barcak, 1997). Null mutations in *sxy/tfoX* prevent competence, but certain missense mutants are constitutively competent, but at a lower level than normally attained (Redfield, 1991; Karudapuram & Barcak, 1997). The *sxy/tfoX* product acts as a transcription activator for late competence genes, which generally have a 26-nt dyad symmetry element upstream from a SigA-like promoter (Karudapuram & Barcak, 1997). In the latter respect the induction system is more similar to *B. subtilis* than to *S. pneumoniae*.

Perpetual competence

A number of bacterial species, including the Gram-negative species *N. gonorrhoeae*, *Acinetobacter calcoaceticus*, *Pseudomonas stutzeri*, the cyanobacterium *Synechococcus* and the Gram-positive *Deinococcus radiodurans*, are competent throughout the culture cycle (see review by Lorenz & Wackernagel, 1994). This constitutive ability to take up DNA is mimicked by the *trt* mutant of *S. pneumoniae* (Lacks & Greenberg, 1973) and by some *sxy* mutants of *H. influenzae* (Redfield, 1991), which no longer regulate the late functions of competence.

COMPETENCE-INDUCIBLE GENES

Genes relating to competence have generally been classified as having early or late functions. Genes required for the *development* of competence, such as those involved in quorum sensing and signalling, are the early genes. In *S. pneumoniae*, this class includes *comA*, *B*, *C*, *D* and *E*. There may be additional genes mediating between these early genes and late genes. Late genes appear to constitute a regulon, i.e. a set of coordinately induced operons. The late-gene functions are needed both for DNA uptake and for allowing the introduced donor strands to persist in the cell and recombine with host DNA. The three best studied inducible systems, those of *S. pneumoniae*, *B. subtilis* and *H. influenzae*, differ markedly in their early genes and their regulatory systems. However, their late genes show many similarities. The fundamental mechanisms of uptake, therefore, may be quite alike.

Genes unique to the uptake mechanism

Inasmuch as the earliest and most complete characterization of late genes has been in *B. subtilis* (reviewed by Dubnau, 1997), the nomenclature originally used for them (i.e. *comGA*, *comGB*, etc.) will be taken to describe the general type. Their homologues in other species will derive their name from the

B. subtilis gene, e.g. *comG*–like: *cglA*, *cglB*, etc. Some exceptions based on the historical record, such as *cilC* for a *comC* homologue in *S. pneumoniae* (Campbell *et al.*, 1998), will be made. Since this review is centred on *S. pneumoniae*, reference will frequently be made to *cgl*, *cel* and *cfl* genes of this species. Uptake genes are listed in Table 1.

The cgl family and cilC

The *comG* operon of *B. subtilis* contains seven genes, which encode proteins related to type IV pilins of Gram-negative bacteria (Albano *et al.*, 1989). Similar operons are found in other transformable species, and defects in any of the genes prevent binding and entry of DNA. The *cgl* operon of *S. pneumoniae* (Pestova & Morrison, 1998; see TIGR database for full sequence) also contains seven genes, which are homologous to those of *B. subtilis* (partly shown in Fig. 2). The *comG* proteins have been characterized with respect to their primary sequences, processing and membrane location (Chung *et al.*, 1998). Based on those findings, the *cglA* and *cglB* products, which are larger proteins, 313 and 347 residues in length, are, respectively, an energy-transducing protein and a membrane-spanning protein responsible for transporting through the cell membrane the products of the five downstream *cgl* genes, which are all smaller polypeptides with hydrophobic segments at their N-termini. In *B. subtilis*, ComGC, ComGD and ComGE contain the residues KGFT preceding the hydrophobic segment; in *S. pneumoniae*, a similar sequence, KAFT, precedes the segment in CglC, CglD and CglF (Fig. 2). These N-terminal sequences are characteristic of prepilin-like proteins, typified by the type IV pilus protein of *Pseudomonas aeruginosa* (Sastry *et al.*, 1985). The earliest evidence connecting pilins to transformation was from *N. gonorrhoeae*, where strains unable to make pili were defective in transformation (Biswas *et al.*, 1977).

The *B. subtilis comC* gene (Dubnau, 1991) and its *S. pneumoniae* homologue, *cilC* (Campbell *et al.*, 1998; see TIGR database for full sequence), are both stand-alone genes unlinked to the *comG* or *cgl* locus. They encode a peptidase-methyltransferase that processes prepilins by cleaving the polypeptide at the F residue and methylating the new N-terminus (Strom *et al.*, 1993). In *H. influenzae*, the *cgl* locus contains only four genes. They encode proteins HI0299, HI0298, HI0297 and HI0296 (Fleischmann *et al.*, 1995), which are homologues of the *P. aeruginosa* prepilin (Fig. 2b), ComGA, ComGB and ComC, respectively.

In the three species with different patterns of competence regulation, the *cgl* operon is under late-gene control. In *S. pneumoniae* the operon is preceded by the TACGAATA combox (Pestova & Morrison, 1998), in *B. subtilis* by the ComK recognition site (Hamoen *et al.*, 1998) and in *H. influenzae* by the dyad symmetry element (Karudapuram & Barcak, 1997). *S. gordonii* uses a similar regulatory mechanism to *S. pneumoniae*, and a TACGAATA combox precedes its *comY* operon, which contains genes

(a)

```
SpnCglC    ----------------MKKMMTFLKKAKVKAFTLVEMLVVLLIISVLFLLFVPNLTKQKEAVNDKGKAAVVKVVESQAELYSLE   68
SgoComYC   -------------------MNKLKKLRVKAFTLVEMLVVLLIISVLMLLFVPNLTKQKEAVSDTGNAAVVKVVESQAELYELK   64
SpyCglC    ----------------MINQWNNLRHKKLKGFTLLEMLLVILVISVLMLLFVPNLSKQKDRVTETGNAAVVKLVENQAELYELS   67
LlaCglC    MKIENITLIMIKALSLIKIHGRKLWQKKQKAFTLIEMLIVLAIISILILLFVPNLIKEKAQVQKTGEAAVVKVVESQAQLYELD   84
BsuComGC   ------------------------MNEKGFTLVEMLIVLFIISILLLITIPNVTKHNQTIQKKGCEGLQNMVKAQMTAFELD    58
                .*.***;***;*;  ;**;*;*; ;**; *.;  ; ..*  .;  ;;*;  *   ;.*.
```

```
SpnCglC    KN-EDASLRKLQADGRITEEQAKAYKEYNDKNGGANRKVND   108
SgoComYC   NTGDQATLSKLVAAGNISQKQADSYKAYYGKNNSETQAVAN   105
SpyCglC    QG-SKPSLSQLKADGSITEKQEKAYQDYYDKHKNEKARLSN   108
LlaCglC    HDNDKPNLSELLSAGMITQKQVTAYDDYYDQNKNEQRNFDD   125
BsuComGC   HEGQTPSLADLQSEGYVKKDAVCPNGKRIIITGGEVK-VEH    98
              ;  ...*  .*  ;  *  .;;.        .
```

(b)

```
HI0299     -----MKLTTLQTLKKGFTLIELMIVIAIIAILATIAIPSYQNYTKKAAVSELLQASAPYKADVELCVYS------TNETTSC   72
PaePilA    -----MKA---Q---KGFTLIELMIVVAIIGILAAIAIPQYQNYVARSEGASALATINPLKTTVEESLSRGIAGSKIKIGTTAS   73
                **    *   ********** ;*** ;*** ;****.****.; ;;  ;; * ;  * *; **  .;      . **;;
```

```
HI0299     TGGKN--GIAADIKTAKGYVASVITQSG-G-ITVK---GNGTLANMEYILQAKGNAAAGVTWTTTCKGTDASLFPANFCGSVTK   149
PaePilA    TATETYVGVEPDAN-KLGVIAVAIEDSGAGDITFTFQTGTSSPKNATKVITLN-RTADGV-WA--CKSTQDPMFTPKGCDN---   149
                *.;;.  *;.*;  ; * ;*.*.;** ** **..  *..;  *  ;;  ;.;* ** *; **.*;  .;*.;; *..
```

(c)

```
SpnCglD    MDASRKNRLKLIKNTMIKMEEQIVKSMIKAFTMLESLLVLGLVSILALGLSGSVQSTFSAVEEQIFFMEFEELYRETQKRSVAS   84
SgoComYD   MVRTIVRLRQLPIKAFTVLESLLVLMISSFILLALSSSVQATFEQIQAKIFFLEFEHFYQESQKLSVSS   69
SpyCglD    MVVSLRNKKKLIKTIMTNIKMKKPVLAIKAFTLLETLLSLSVMSFIILGLSVPVTKSYQKVEEHLFFSHFEHLYRHQQKLAILQ   84
LlaCglD    -----------MMITMTRIKMNSAISMIRAFTLLESLLVLLIVSFIILFFSAELTQTVHLFKGRLFVLQFENLYKISQE-NAAL   73
BsuComGD   -------------------MNIKLNEEKGFTLLESLLVLSLASILLVAVFTTLPPAYDNTAVRQAASQLKNDIMLTQQTAISR   64
                              ;  ;**;**;**  * ; *;;   *  ;  ;;;.       ; ;;; ;;
```

```
SpnCglD    QQKTSLNLDGQT--LSNGSQKLPVPKGIQAPSGQSTTFDRAGGNSSLAKVEFQT--SKGAIRY--QLYLGNGKIKRIKETKN--   160
SgoComYD   QRKLVLEISSQE--ISNGYARL------------------------------------------------------------    89
SpyCglD    QKQRVLDISSTK--IVTEGNSLTVPKSITVNHPYRLVIDQMGGNHSLAKIIFDM--TDRRFKY--QFYLGSGNYQKTSQSLHSP   162
LlaCglD    QSSSNLGSENGK--LIYENKEIDIPKEVEMA-EFLIIFDEDGGNSSLQKIKVYLPYEKKTILY--QMEMGSGKYKKKIN-----   146
BsuComGD   QQRTKILFHKKEYQLVIGDTVIERPYATGLSIELLTLKDRLEFNEKGHPNAGGKIRVKGHAVYDITVYLGSGRVNVERK-----   143
               *    ;      ;  *       ;; ; *.  *        *   ; *  ;*  *;;
```

(d)

```
SpnCglF    ---------MVQNSCWQSKSHKVKAFTLLESLLALIVISGGLLLFQAMSQLLISEVRYQQQSEQKEWLLFVDQLEVELDRSQFE   75
SpyCglF    --------------MSKQLSNIKAFTLLEALIALLVISGSLLVYQGLRTRTLLKHSHYLARHDQDNWLLFSHQLREELSGARFY   69
LlaCglF    --------MTMERKFCDLKLKIRAFTLLECLVALLAISGSVLVISGLTRMIEEQMKISQNDSRKDWQIFCEQMRSELSGAKLD   75
BsuComGF   MKTYVSKPQLIKKNHFASAFCRQNGYTLLNVLFSLSVFLLISGSLAAIIHLFLSRQQEHDGFTQQEWMISIEQMMNECKESQAV   84
                         ;;***;.*;.* ;;.          ;;.*  ;     ;  *
```

```
SpnCglF    KVEGN-RLYMKQ--DGKDIAIGKSKSDDFRKTNARGRGYQPMVYGLKSVRITEDNQLVRFHFQFQKG--LEREF-IYRVEKEKS   153
SpyCglF    KVADN-KLYVEK--GKKVLAFGQFKSHDFRKSASNGKGYQPMLFGISRSHIHIEQSQICITLKWKSG--LERTF-YYAFQD---   144
LlaCglF    NVNQN-FLYVTK--DKK-LRFG-LVGDDFRKSDDKGQGYQPMLYDLKGAKIQAEENLIKITIDFDNG--GERVF-IYRFTDTK-   150
BsuComGF   KTAEHGSVLICTNLSGQDIRFD-IYHSMIRKRVD-GKGHVPILDHITAMKADIENGVVLLKIESEDQKVYQTAFPVYSYLGGG-   165
                       ;**  *;*;.*;     ;;;; ;;  ;;  .;    ;;   .
```

Fig. 2. Comparison of homologous ComG-like (Cgl) proteins related to type IV pilins. (a) ComGC-like (CglC) proteins. (b) *H. influenzae* HI0299 (PilA) compared to PaePilA, the type IV prepilin from *P. aeruginosa*. (c) ComGD-like (CglD) proteins. Only the N-terminal half of SgoComYD was cloned. LlaCglD was translated from the reported sequence data after insertion of a single nucleotide before position 1509. (d) ComGF-like (CglF) proteins. Key to species: Spn, *S. pneumoniae*; Sgo, *S. gordonii*; Spy, *S. pyogenes*; Lla, *L. lactis*; Bsu, *B. subtilis*. Key to symbols indicating similarity of residues: *, identical; :, very similar; ., similar. This figure and Figs 3, 4 and 5 were constructed with the aid of the CLUSTAL W program (Thompson *et al.*, 1994).

closely homologous to at least four *cgl* genes (Fig. 2; Lunsford & Roble, 1997). Although *S. gordonii* is transformable, *Streptococcus pyogenes* has never been shown to be transformable, yet it has a homologous *cgl* locus (Fig. 2), preceded by a combox (Håverstein & Morrison, 1999). *Lactococcus lactis* also has a *cgl* operon (Fig. 2; sequence from A. Breunder & K. Hammer, GenBank accession no. Y15043, 1997), but it is apparently transcribed from a typical SigA promoter. These observations raise the possibility that both species are transformable. Perhaps *S. pyogenes*, which lacks genes homologous to *comA–E* in *S. pneumoniae*, has a different system for signalling the development of competence than other streptococci, and maybe in *L. lactis* the entire regulatory mechanism is different, as in *B. subtilis*. Alternatively, the *cgl* pathway may have functions other than enabling DNA uptake; for example, in bacterial adhesion to surfaces.

How *cgl* products act in DNA uptake is unknown, but several possibilities can be envisioned (Chung *et al.*, 1998). By analogy to pilins, the products may form an appendage outside the cell membrane. This structure may act as a scaffold on which other proteins that bind and process DNA for entry are arranged. This scaffold may penetrate the peptidoglycan layer of the cell wall to form a passage for external DNA. In this connection some remodelling of the cell wall may occur, as discussed below. It is possible that *cgl* proteins are directly involved in binding or processing DNA, but no evidence for such functions has been adduced. An intriguing hypothesis is that the *cgl* system, in a manner akin to pilus extrusion and retraction, serves as a motor for bringing DNA into the cell (Dubnau, 1997).

cel (comE-like) genes

Two genes, *comEA* and *comEC*, of the three in the *comE* operon of *B. subtilis* are essential for DNA uptake (Hahn *et al.*, 1993). Homologues of these two genes, *celA* and *celB*, comprise a late competence operon in *S. pneumoniae* (Pestova & Morrison, 1998). In *H. influenzae*, corresponding homologues, HI1008 (Fleischmann *et al.*, 1995) and *rec2* (Clifton *et al.*, 1994), are transcribed separately, each under control of the dyad symmetry element. Relationships between products of the *celA* genes are shown in Fig. 3.

ComEA and CelA are proteins of ∼20 kDa with a long stretch of hydrophobic residues at the N-terminus. The 12 kDa *H. influenzae* homo-

```
SpnCelA  MEAIIEKIKEYKIIVICTGLGLLVGGFFLLKPAPQTPVKETNLQAEVAA---VSKDLVSEKEVNKEEKEEPLEQDL  73
BsuComEA MNWLNQHKKA--IILAASAAVFTAIMIFLATGKNKEPVKQA-----VPT---ETENTVVKQEANNDESNE-----T  61
EcoUvrC  KLPEVKRMECFDISHTMGEQTVASCVVFDANGPLRAEYRRYNITGITPGDDYAAMNQVLRRRYGKAIDDSKIP-DV 462
                     :  *      .  .*  .  . :.        :        *  :::. :  :

SpnCelA  ITVD-VKGAVKSPGIYDLPVGSRINDAVQKAGGLTEQADSKSLNLAQKVSDEA--LVYVPTKGEEAVSQQTGLGTA 146
BsuComEA IVID-IKGAVQHPGVYEMRTGDRVSQAIEKAGGTSEQADEAQVNLAEILQDGT--VVYIPKKGEETAVQQGGGGSV 134
Hin1008  --------------------------MKTLFTSVVLCGALVVSSSFAEEK--ATXQTAQSVVTTQAEAQVAPA  45
EcoUvrC  ILIDGGKGQLAQAKNVFAELDVSWDKNHPLLLGVAKGADRKAGLETLFFEPEGEGFSLPPDSPALHVIQHIRDESH 538
         * :*  **  :  :   .    :      .:.:  .    .       .     .   .   .  .    .

SpnCelA  SSIS-KEKKVNLNKASLEELKQ-VKGLGGKRAQDIIDHR--EANGKFKSVDELKKVSGIGGKTIEKLKDYVTVD 216
BsuComEA QSDGGKGALVNINTATLEELQG-ISGVGPSKAEAIIAYR--EENGRFQTIEDITKVSGIGEKSFEKIKSSITVK 205
Hin1008  VVSD----KLNINTATASEIQKSLTGIGAKKAEAIVQYR--EKHGNFXNAEQLLEVQGIGKATLEKNRDRIIF- 112
EcoUvrC  DHAIGGHRKKRAKVKNTSSLET-IEGVGPKRRQMLLKYMGGLQGLRNASVEEIAKVPGISQGLAEKIFWSLKH- 610
                .  :  .  .. :  :  *: *  .:  :  :  ::  :        .   .  .::  :*  **.    **    :
```

Fig. 3. Comparison of *S. pneumoniae* CelA and *B. subtilis* ComEA to each other and to the carboxyl half of *E. coli* UvrC. Symbols as in Fig. 2.

logue has a similar hydrophobic stretch but otherwise matches only the C-terminal half of the two other proteins. A resemblance between the C-terminus of ComEA and that of *E. coli* UvrC was previously noted (Dubnau, 1997). Mutations in *comEA* prevent binding of DNA to the cell, and it was shown recently that the ComEA protein *in vitro* binds tightly to DNA in a nonspecific fashion, but does not nick DNA (Provvedi & Dubnau, 1999). This protein, therefore, is very likely responsible for the initial, reversible binding of DNA to the cell.

ComEC, CelB and Rec2 are 80 kDa proteins with multiple hydrophobic stretches that could be transmembrane segments. ComEC is not required for DNA binding, but it is necessary for DNA entry (Inamine & Dubnau, 1995). These proteins may form channels in the membrane for passage of DNA (Dubnau, 1997).

cfl (comF-like) genes

The *comF* operon of *B. subtilis* consists of three genes, two of which are essential for transformation and have counterparts in the two-gene *cfl* operon of *S. pneumoniae* (sequence from TIGR). ComFA mutants are reduced in transformability ∼1000-fold (Londoño-Vallejo & Dubnau, 1994). CflA mutations are similarly defective, and they fail to take up DNA (D. Morrison, personal communication). These 50 kDa proteins show sequence similarity to ATP-dependent DNA helicases (Londoño-Vallejo & Dubnau, 1993). Although they lack hydrophobic regions, ComFA was associated with the cytoplasmic face of the cell membrane (Londoño-Vallejo & Dubnau, 1994).

With respect to function, it was suggested that ComFA acts as a helicase to help separate the DNA strands and propel one strand into the cell. It may also have a nicking activity, inasmuch as TraI of the F-factor conjugative system acts as a helicase after nicking the donor DNA (Matson *et al.*, 1993). Comparison of protein sequences indicates some resemblance of ComFA and CflA to UvrB of *E. coli* (Fig. 4a). UvrB contains a latent ATPase that is activated when it interacts with UvrC, and the UvrB–UvrC complex is able to nick DNA (Orren & Sancar, 1989). Conceivably, CflA, either by itself or together with CelA (which resembles UvrC), has the dual functions of nicking and unwinding donor DNA. If it binds covalently to the nicked DNA, as TraI is thought to do, it might also assist its entry into the cell.

ComFC, CflB and the *com101A* [=[HI0434] or *comF* (Tomb *et al.*, 1991)] gene product of *H. influenzae* (Larson & Goodgal, 1991) are homologous 25 kDa proteins. Mutations in *comFC* reduced transformation only ∼10-fold, but the reduction was much greater in *comF*. Mutants in *comF* bound normal amounts of DNA, but whether these proteins function in entry or in a later stage of transformation is unknown. The proteins contain two putative zinc fingers at their N-termini (Fig. 4), which may enable them to bind other components of the entry complex.

(a)

```
SpnCflA   -MKVNLDYLGRLFTENELTEEERQLA--------EKLPAMRKE------------KGKLFCQRCNST-----ILEEWYLP   54
BsuComFA  -MNVPVEKNSSFSKELQQTLRSRHLLRTELSFSDEMIEWHIKNGYITAENSISINKRRYRCNRCGQTDQ---RYFSFYHS   76
EcoUvrB   YVPSSDTFIEKDASVNEHIEQMRLSAT------KAMLERRDVVVVASVSAIYGLGDPDLYLKMMLHLTVGMIIDQRAILR  174
                  :         .   :  .*                          .                .  :     *

SpnCflA   IG--AYYCR--ECLLMK--RVRSDQTLYYFPQE--DFP--KQDVLKWRGQLTPFQEKVSEGLLQVVDKQKPTLVHAVTGA  124
BsuComFA  SGKNKLYCR--SCVMMG--RVSEEVPLYSWKEE--NESNWKSIKLTWDGKLSSGQQKAANVLIEAISKKEELLIWAVCGA  150
EcoUvrB   RLAELQYARNDQAFQRGTFRVRGEV-IDIFPAESDDIALRVELFDEEVERLSLFDPLTGQIVSTIPRFTIYPKTHYVTPR  253
          *.*   ...    **    :.     :*:  .. :.:  :          *
```

```
SpnCflA   GKTEMIYQVVAKVINAGGAVCLASPRIDVCLELYKRLQQDFS-CGIALLHGESEPYFR-TPLVVA--TTHQLLKFYQAFD  200
BsuComFA  GKTEMLFPGIESALNQGLRVCIATPRTDVVLELAPRLKAAFQGADISALYGGSDDKGRLSPLMIS--TTHQLLRYKDAID  228
EcoUvrB   ERIVQAMEEIKEELAARRKVLLENNKLLEEQRLTQRTQFDLEMMNELGYCSGIENYSRFLSGRGPGEPPPTLFDYLPADG  333
           :       :  . :     *    : .  .* *   :. .        *   .  . *     ...  *: :  * .
```

```
SpnCflA   LLIVDEVDAFPYVDNPMLY--HAVKNSVKENGLRIF--LTATSTNELDKKVRLGELKRLN-LPRRFHGNP---LIIPKPI  272
BsuComFA  VMIIDEVDAFPYSADQTLQ--FAVQKARKKNSTLVY--LSATPPKELKRKALNGQLHSVR-IPARHHRKP---LPEPRFV  300
EcoUvrB   LLVVDESHVTIPQIGGMYRGDRARKETLVEYGFRLPSALDNRPLKFEEFEALAPQTIYVSATPGNYELEKSGGDVVDQVV  413
          ::::**  ..       .    *  ::  :. :      *   .  : :. :            *  ...:      : :
```

```
SpnCflA   WLSDFNRYLDKNR-LSPKLKSYIE--KQRKTAYPLLIFASEIK-KGEQLAEILQEQFPNEKIGFVSS--VTEDRLEQVQA  346
BsuComFA  WCGNWKKKLNRNK-IPPAVKRWIE--FHVKEGRPVFLFVPSVS-ILEKAAACFKG--VHCRTASVHA--EDKHRKEKVQQ  372
EcoUvrB   RPTGLLDPIIEVRPVATQVDDLLSEIRQRAAINERVLVTTLTKRMAEDLTEYLEE--HGERVRYLHSDIDTVERMEIIRD  491
            .  :  : :.       :  .     .:.:       *:. :        : :         :        .* * .:
```

```
SpnCflA   FRDGELTILISTTILERGVTFPCV-DVFVVEANHRLFTKS--SLIQIGGRVGR---------------SMDRPTGDLLFF  408
BsuComFA  FRDGQLDLLITTTILERGVTVPKV-QTGVLGAESSIFTES--ALVQIAGRTGR---------------HKEYADGDVIYF  434
EcoUvrB   LRLGEFDVLVGINLLREGLDMPEVSLVAILDADKEGFLRSERSLIQTIGRAARVNVNGKAILYGDKITPSMAKAIGETERR  471
          :*  *::  :*:   .:*..*: .*  *    .  ::  *:  .*  :*:*  **..*               . *:
```

```
SpnCflA   ----------H----DGLNASIKKAIKEIQMMNKEAGL----------------------------------- 432
BsuComFA  ----------H----FGKTKSMLDARKHIKEMNELAAKVECTD----------------------------- 463
EcoUvrB   REKQQKYNEEHGITPQGLNKKVVDILALGQNIAKTKAGRGKSRPIVEPDNVPMDMSPKALQQKIHELEGLMMQHA  647
                   *     *   .  .: .      :  : :  .
```

(b)

```
SpnCflB   MKCLLCGQTMKTVLTFSSLLLLRNDDSCLCSDCDSTFERIGEENCPNCM-KTE-LSTKCQDCQLW----CKEGVEVS-HRAI   75
BsuComFC  MICLLCDSQFSQDVTWRALFLLKPDE-KVCYSCRSKLKKITGHICPLCG-RPQSVHAVCRDCEVWRTRIRDSLLLRQNRSV   79
HI0434    MMNFFN---FRCIHCRGNLHIAKNG---LCSGCQKQIKSFP--YCGHCGSELQYYAQHCGNCLKQ----EPSWDK-MVII   67
                        *    :  .   :*.,*...*   *   :      . :  .   * *:  :
```

```
SpnCflB   FTYNQAMKDFFSRYKFDGDFLLRKVFAS-FLSEELKKYKEYQF----VVIP--LSPDRYANRGFNQVEGLVEAAG-PEYLD  148
BsuComFC  YTYNDMMKETLSRFKFRGDAEIINAFKSDFSSTFSKVYPDKHF----VLVPIPLSKEREEERGFNQAHLLAECLDRPSHHP  156
HI0434    GHYIEPLSILIQRFKFQNQFWIDRTLARLLYLAVRDAKRTHQLKLPEAIIPVPLYHFRQWRRGYNQADLLSQQLSRWLDIP  158
           * : :.  :.*;** .: : ...:  :           .  ::    .::*  *    *  .**;**..* .:
```

```
SpnCflB   LLE--------KREERASSSKNRSERLGTELPFFIKSGVTIPKKILLIDDIYTTGATINRVKKLLEEAG-AKDVKTFSLVR-  220
BsuComFC  LIR--------LNNEKQSK-KKKTERLLSECIFDTKNNSAEGMNIILIDDLYTTGATLHFAARCLLEKGKAASVSSFTLIRS  229
HI0434    NLNNIVKRVKHTYTQRGLSAKDRRQNLKNAFSLAVSKNEFPYRRVALVDDVITTGSTLNEISKLLRKLG-VEEIQVWGLARA  229
          :.           ::  *.:: *..*  . .       *:**: ***:*:: .*  . .
```

Fig. 4. Comparison of homologous ComF-like (Cfl) proteins. (a) *S. pneumoniae* CflA and *B. subtilis* ComFA proteins compared to each other and to *E. coli* UvrB. (b) Comparison of *S. pneumoniae* CflB, *B. subtilis* ComFC and *H. influenzae* HI0434 proteins. Symbols as in Fig. 2.

Post-entry genes

Examination of their genomes indicates that both *S. pneumoniae* and *B. subtilis* harbour two *ssb* loci that encode proteins homologous to the single-stranded DNA-binding protein of *E. coli*. One is expressed as a late competence gene, and the other is constitutively expressed and presumably functions in DNA replication. In *S. pneumoniae*, it is apparently the competence-induced protein that is found bound to newly entered donor strands (Morrison, 1978). Whether such binding assists the uptake process is unknown, inasmuch as no mutants in the competence-induced *ssbB* gene [= *cilA* (Campbell *et al.*, 1998)] have been tested.

There is only a single *recA* gene in *S. pneumoniae*, but its expression is increased ~fivefold during competence (Martin *et al.*, 1995), as a result of the presence of a combox (Campbell *et al.*, 1998) preceding its upstream neighbour, *exp10* [Pearce *et al.*, 1995; = *cinA* (Martin *et al.*, 1995)]. DNA appears to enter normally in *recA* mutants.

Genes of unknown or accessory function

Three late competence genes in *S. pneumoniae* of unknown significance for DNA uptake are *exp10/cinA*, *coiA* and *cilB*. Mutations in the first two genes reduce transformation ~10-fold (Pearce *et al.*, 1995; Pestova & Morrison, 1998). Homologues of *exp10/cinA* are found in many bacteria, but aside from a hydrophobic N-terminal region indicating a membrane location, nothing is known about the function of its protein product. As deduced from the genomic sequence of *S. pneumoniae*, the *coiA* gene is the first in an operon containing four genes. The products of *coiA* and *coiB* both appear to be peptidases, *coiC* encodes a putative methyltransferase, and *coiD* encodes a putative cell wall serine proteinase. It may be speculated that these enzymes play a role in cell wall remodelling during the development of competence.

cilB is homologous to the *H. influenzae* gene *dprA*, in which mutations reduce chromosomal transformation 10000-fold, with no effect on plasmid transfer (Karudapuram *et al.*, 1995). Mutations in *dprA* may affect only chromosomal DNA entry because plasmids enter *H. influenzae* cells by a different mechanism (Pifer, 1986). Otherwise, the gene may be required only for chromosomal integration, as appears to be the case for *recP* in *S. pneumoniae* (Morrison *et al.*, 1983).

The case of *dpnA* is unusual in that it is regulated as a late competence gene (S. Lacks & B. Greenberg, unpublished), but it is not normally required for either chromosomal or plasmid transformation. It encodes a DNA methyl-transferase that protects unmodified incoming plasmid DNA by methylating it while it is in a single-stranded form, thereby allowing plasmid establishment by unmodified donor DNA in a cell containing the *Dpn*II restriction system (Cerritelli *et al.*, 1989).

ROLE OF A MEMBRANE NUCLEASE IN DNA UPTAKE

The endA gene

The only gene known to function in DNA uptake that is not induced during the development of competence is *endA* of *S. pneumoniae*. This constitutive gene is expressed from several upstream promoters, all with a SigA consensus sequence (Puyet *et al.*, 1990). The strongest promoter precedes the first gene of the operon in which *endA* is the third and final gene. EndA is a 30 kDa polypeptide with a hydrophobic segment near its N-terminus (Puyet *et al.*, 1990), and it is located in the cell membrane from which it can be solubilized by mild detergents (Lacks & Neuberger, 1975). *In vitro*, it acts as a general endonuclease, producing oligonucleotide fragments from both single- and double-stranded DNA and from RNA (Rosenthal & Lacks, 1977). When isolated from membranes, EndA is found in a larger structure of ∼250 kDa (Rosenthal & Lacks, 1980).

Mutants of *endA* that retain at least 5% of wild-type nuclease activity are fully transformable, but those with less than 1% activity are defective. Null mutants bind DNA but do not transport it into the cell; they are reduced in transformability more than 1000-fold (Lacks *et al.*, 1975; Puyet *et al.*, 1990). It is interesting that neither *endA* nor *cflA* null mutants reduce transformability as much as defects in *cgl* genes or those required for development of competence. Perhaps both EndA, which hydrolyses the complementary strand, and CflA, which may unwind donor DNA and/or translocate a donor strand into the cell, are required for efficient DNA strand entry, but in the absence of one, the other activity still manages to introduce some DNA.

No-entry degraders

Normally, one donor strand is degraded as the other enters the cell. In certain mutants, however, the expected amount of donor DNA is degraded and found outside the cell, but no DNA enters the cell (Morrison *et al.*, 1983). The single strand produced by EndA action presumably remains outside the cell. The mutations that give rise to these 'no-entry degraders' occur in both the *cel* and *cfl* loci (D. Morrison, personal communication). If CelB composes the pore for DNA entry through the cell membrane and CflA is needed for DNA passage through the pore, then defects in either could block entry without blocking degradation.

B. subtilis nucleases

Cells of *B. subtilis* contain at least 10 enzymes with DNase activity (Rosenthal & Lacks, 1977). One of them, a 16 kDa protein encoded by *nucA*, is membrane-located (Smith *et al.*, 1983). This gene is under competence control, but mutations of it do not prevent transformation (van Sinderen *et al.*, 1995a). However, the plethora of DNases in *B. subtilis* might allow redundancy of the entry nuclease. A possible candidate for this redundant function is the 30 kDa product of the gene *yurI* of *B. subtilis* (Kunst *et al.*, 1997), which has a potential membrane-binding segment and is homologous (28% identity; Fig. 5b) to the periplasmic nuclease EndA of *E. coli*, a protein with enzymic activities very similar to EndA of *S. pneumoniae*. A form of the *yurI* gene, called *bsn*, was cloned from a strain of *B. subtilis*,

(a)

```
SpnEndA  MNKKTRQTLIGLLVLLLLSTGSYYIKQMPSAPNSPKTNLSQKKQASEAPSQALAESVLTDAVKSQIKGSLEWNGSGAF  78
SpyEndB  MSKSNRRTWQGLVVILIAILTTFTTSTVTAARKIR-------------NFPDTTEILLGTKATETPGILPFTGSYQL  64
         *.*..*:*  **:*:*:    ::  ..:.:* :            ..  :*    ::  * * : **   :

SpnEndA  IVNGNKTNLDAKVSSKPYADNKTKTVGKETVPTVANALLSKATRQYKNRKETGNGSTSWTPPGWHQVK--NLKGSYTH  154
SpyEndA             MLTRKKRKRVQEQKVPKRANALLTKSTRQYRNRYQTDNAYRNFKPAGWHRLH--HLKGSYDH  60
SpyEndB  VL----GDLDN-LQRPTFAHIQLKDQDE-----------------PNIKRKG---LKFNPPGWHNYKLTDANGKTTW  116
         ::    :** :    :  : * :                 *        : *:***::: : :*

SpnEndA  AVDRGHLLGYALIGGLDGFDASTSNPKNIAVQTAWANQAQAEYSTGQN-YYES----------------KVRKALDQN  215
SpyEndA  AVDRGHLLGYALVGGLKGFDASTGNPDNIATQLSWANQANKPYLTGQN-YYEGLVSTELECYT--------------  122
SpyEndB  LMDRGHLVGYQF-SGLN------DEPKNLVTMTKYLNTGFSDKNPLGMLYYENRLDSWLALHPNFWLDYKVTPVYHKN  189
         :*****:**  :  :**:    :*:*:.  :* * :  ***          ** ::::*: *

SpnEndA  KRVRYRVTLYYA---SNEDLVPSASQIEAKSSDGELEFNVLVPNVQKGLQLDYRTG-EVTVTQ--------------  274
SpyEndB  ELVPRQVVLQYVGIDENGDLLQIKLGSEKESVDNFGVTSVTLDNVSPLAELDYQTGMMLDSTQNEEDSNLETEEFEEAA  268
         *  :*.* *.  .* **:    * :* *.   .* : **.   :***:** :  **
```

(b)

```
EcoEndA  MYR---YLSIAAVVLSAAFSGPALAEG------INS--FSQAKAAAVKVHADAPGTFYCGCKINWQGKKGVV--DLQS  65
BsuEndA  MTKKAWFLPLVCVLLISGWLAPAASASAQTTLSLNDRLASSPSGTGSLLSLAAPAAPYADTDTYYEGAEGKTGDSLKS  78
         * :  :*.::.*:*  :..**  :.    :*.   :*  ::.:  **.:.*...  ::* .*.  .*:*

EcoEndA  CGYQVRKNENRASRVEW---------EHVVP-----AWQFGHQRQCWQDGGRKNCAKDPVYRKME----------SDM  119
BsuEndA  TLHRIISGHTMLSYSEVWNNALKETDEDPRNPNNVILLYTNESRSKNLNGGNVGDWNREHVWAKSHGDFGTSKGPGTDI  156
         ::: ....  *  *.      :    *       :  : :.*. .: :  *:  * .          :*:

EcoEndA  HNLQPSVGEVNGDRGNFMYSQWNGGEGQYGQCAMKVDFKEKAAEPPARARGAIARTYFYMR------DQY--------  183
BsuEndA  HHLRPADVQVNSARGNMFD--NGG-TEY-AKAPGNYYDGDSWEPRDDVKGDVARMLFYMAVRYEGDDGYPDLELNDK  230
         *:*:*:.:**. ***: **     :*  *      :...: **   .:* :** ***      * *

EcoEndA  --NLTLSRQQTQ-LFNAWNKMYPVTDWECERDERI-AKVQGNHNPYVQRACQARKS--  235
BsuEndA  TGNGSAPYHGKQSVLLEWNKQDPVDDRERKRNEIIYEKYQHNRNPFIDHPEWADEIWP  288
         * :. : .* :: *** ** * *:*;* * * *  *:**::: . * :
```

Fig. 5. Comparison of homologous endonucleases. (a) *S. pneumoniae* EndA compared to putative *S. pyogenes* nucleases SpyEndA and SpyEndB. (b) Comparison of *E. coli* and *B. subtilis* EndA nucleases, the latter encoded by *yurI* (= *endA*). Symbols as in Fig. 2.

and its product was characterized as an RNase with no DNase activity (Nakamura *et al.*, 1992). However, only a truncated, extracellular form of the enzyme was examined, and that form may have lost its DNase activity. In addition, the nucleotide sequence of *bsn* was only 73 % identical to the gene in the transformable strain 168 (shown in Fig. 5b).

Is S. pyogenes transformable?

In addition to the *cgl* operon, many homologues of *S. pneumoniae* late competence genes are under combox control in *S. pyogenes* (Håverstein & Morrison, 1999). This suggests that bacteria of this species may be transformable, albeit with a system different from *S. pneumoniae* for eliciting competence development. It was therefore of interest to look for the presence of *endA* in the *S. pyogenes* genomic sequence (SGSP database). A three-gene operon with the first two genes very closely homologous to those of *S. pneumoniae* (77 % and 50 % identity, respectively) is present. However, the *endA* homologue, although very similar to that of *S. pneumoniae* (61 %

identity; see Fig. 5a), is truncated and missing 94 residues from its N-terminus. Because the hydrophobic segment of the protein is absent, the protein could not be bound to the membrane so as to function in DNA uptake. But *S. pyogenes* contains yet another nuclease gene, which we designate *endB*, unlinked to the first, and also encoding a product similar to *S. pneumoniae* EndA (29% identity), albeit less so than the *S. pyogenes endA* product (Fig. 5a). The EndB protein, however, does contain a potential membrane-binding segment. It could support transformation in *S. pyogenes*.

Cell wall lysis and rebuilding

The first gene in the operon encoding EndA in *S. pneumoniae* is a *murA*-like gene that encodes UDP-*N*-acetylglucosamine 1-carboxyvinyltransferase, an enzyme required for cell wall synthesis. It is one of two *murA*-like genes in *S. pneumoniae*, and it may participate in cell wall synthesis and reconstruction during competence development. *B. subtilis* similarly contains two such genes, *murA* and *murZ*. Only one such gene is present in *E. coli* (Blattner *et al.*, 1997). The lytic fragility of cells of *S. pneumoniae* during competence and the ability of such cells to spontaneously form protoplasts have been reported (Lacks & Neuberger, 1975). The latter propensity was shown not to depend on the *cwl* (=*lytA*) gene that encodes the major pneumococcal autolysin. It could depend on a competence-induced gene, which may be one of those with unknown function. CoiA, a putative peptidase, and CoiD, a putative cell wall serine proteinase, are possible candidates. The purpose of such remodelling would be to install into the cell wall an appendage, perhaps formed of *cgl*-encoded proteins, that could allow attachment of DNA and its transport through the peptidoglycan layer of the cell wall and the cell membrane.

STRUCTURE AND FUNCTION OF THE UPTAKE MACHINERY

A speculative scenario of how the late competence gene products and other cell envelope components might interact to enable DNA uptake is depicted in Fig. 6. As a first step, CglA and CglB may act together to export CglC, D, E, F and G, with CilC processing the N-termini of CglC, D and F. The five downstream *cgl* products would form an appendage extending into the cell wall layer and anchored in the cell membrane. Prior action of *coi*-encoded peptidases and MurA might disrupt and rebuild the cell wall in the vicinity of the pilin-related appendage. The Cgl structure would act as a scaffold for other proteins involved in DNA uptake, such as CelA, CelB, CflA, CflB and EndA. The scaffold would essentially trap CflA and CflB, which are not directly bound to the membrane. This structure would allow access of donor DNA to CelA, which could bind it reversibly. Combined action of CelA and CflA might then nick the DNA in an ATP-driven reaction, which would

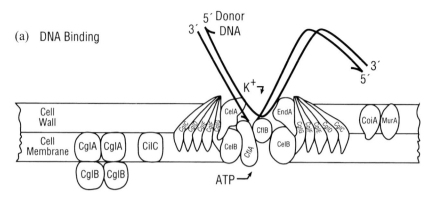

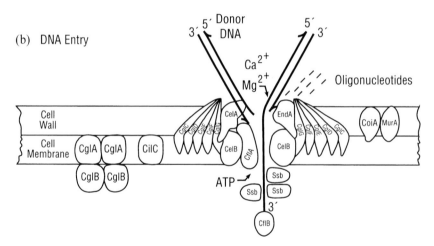

Fig. 6. Proposed mechanism for DNA uptake in *S. pneumoniae*. (a) DNA binding. All components except EndA, a membrane nuclease, and MurA, a cell wall synthetic enzyme, are induced during competence development. CoiA, a peptidase, and MurA reconstruct the cell wall for insertion of the DNA pore structure. CglA and CglB are responsible for export of CglC–G, prepilin-like components of the uptake structure. The latter are processed by CilC; their final arrangement in the cell envelope is unknown and arbitrarily drawn in the figure. CelA binds to donor DNA, and in conjunction with CflA makes a single strand break in it, possibly attaching the new 3′-end covalently to CflA or CflB. (b) DNA entry. CflB may act as a pilot protein bringing the single strand through a membrane pore formed by CelB. Entry of the donor strand may be facilitated by degradation of the complementary strand by EndA and helicase action of CflA. A competence-induced protein, Ssb, binds to the incoming donor strand. Potassium ions are needed for binding, and calcium and magnesium ions for entry.

leave the DNA irreversibly bound to either CflA or CflB and, hence, the cell. Under normal conditions, helicase activity of CflA would separate the donor strands, and EndA would degrade one strand, both of which actions facilitate entry of the other strand into the cell through a membrane pore formed by CelB. Presumably, CflA or CflB, attached to the incoming strand at its 3′-end, acts as a pilot protein to bring it through the transmembrane pore. One or more of these proteins may be able to pass through the transmembrane pore in either direction.

Once inside the cell, the entering strand is covered by SsbB protein. This may or may not assist its entry into the cell. Analysis of mutants defective in SsbB should answer this question. If EndA is defective, only a small amount of DNA is internalized by action of CflA. Similarly, if CflA is defective, only a small amount is internalized by action of EndA. When EndA is functional and CelB or CflA is defective, degradation of a donor strand could proceed, but the complementary strand would be unable to enter the cell, as found for mutants exhibiting degradation but no entry of DNA. Null mutants of CelA would not bind DNA at all.

ACKNOWLEDGEMENTS

I thank Donald Morrison and David Dubnau for generously providing information prior to publication. Genomic sequence data for *S. pneumoniae* was obtained from The Institute for Genomic Research (TIGR) website at http://www.tigr.org. Sequence data for *S. pyogenes* were obtained from the Streptococcal Genome Sequencing Project (SGSP) conducted at the University of Oklahoma by B. A. Roe, S. P. Linn, L. Song, X. Yuan, S. Clifton, M. McShan & J. Ferretti, funded by USPHS/NIH grant AI38406, and accessed at http://dna1.chem.ou.edu/strep.html. GenBank sequence data were obtained from the National Center for Biotechnology Information. This chapter was written at Brookhaven National Laboratory under the auspices of the US Department of Energy Office of Biological and Environmental Research.

REFERENCES

Albano, M., Breitling, R. & Dubnau, D. A. (1989). Nucleotide sequence and genetic organization of the *Bacillus subtilis comG* operon. *Journal of Bacteriology* **171**, 5386–5404.

Avery, O. T., MacLeod, C. M. & McCarty, M. (1944). Studies on the chemical nature of the substance inducing transformation of pneumococcal types. Induction of transformation by a desoxyribonucleic acid fraction isolated from pneumococcus type III. *Journal of Experimental Medicine* **89**, 137–158.

Barany, F. & Boeke, J. D. (1983). Genetic transformation of *Streptococcus pneumoniae* by DNA cloned into the single-stranded bacteriophage f1. *Journal of Bacteriology* **153**, 200–210.

Barany, F., Kahn, M. E. & Smith, H. O. (1983). Directional transport and integration of donor DNA in *Haemophilus influenzae* transformation. *Proceedings of the National Academy of Sciences, USA* **80**, 7274–7278.

Biswas, G. D., Sox, T., Blackman, E. & Sparling, P. F. (1977). Factors affecting genetic transformation of *Neisseria gonorrhoeae*. *Journal of Bacteriology* **129**, 983–992.

Blattner, F. R., Plunkett, G., III, Bloch, C. A. & 14 other authors (1997). The complete genome sequence of *Escherichia coli* K-12. *Science* **277**, 1453–1474.

Campbell, E. A., Choi, S. Y. & Masure, H. R. (1998). A competence regulon in *Streptococcus pneumoniae* revealed by genome analysis. *Molecular Microbiology* **27**, 929–939.

Cerritelli, M. E., Wall, J. S., Simon, M. N., Conway, J. F. & Steven, A. C. (1996). Stoichiometry and domainal organization of the long tail-fiber of bacteriophage T4: a hinged viral adhesin. *Journal of Molecular Biology* **260**, 767–780.

Cerritelli, S., Springhorn, S. S. & Lacks, S. A. (1989). DpnA, a methylase for single-strand DNA in the *Dpn*II restriction system, and its biological function. *Proceedings of the National Academy of Sciences, USA* **86**, 9223–9227.

Chang, D. C., Chassy, B. M., Saunders, J. A. & Sowers, A. E. (editors) (1992). *Guide to Electroporation and Electrofusion*. New York: Academic Press.

Chung, Y. S., Breidt, F. & Dubnau, D. (1998). Cell surface localization and processing of the ComG proteins required for DNA binding during transformation of *Bacillus subtilis*. *Molecular Microbiology* **29**, 905–913.

Citovsky, V. & Zambryski, P. (1993). Transport of nucleic acids through membrane channels: snaking through small holes. *Annual Review of Microbiology* **47**, 167–197.

Clavé, C. & Trombe, M.-C. (1989). Intracellular pH, calcium, and ATP involvement in DNA uptake by *Streptococcus pneumoniae* competent for genetic transformation. *FEMS Microbiology Letters* **65**, 113–118.

Clifton, S. W., McCarthy, D. & Roe, B. A. (1994). Sequence of the *rec-2* locus of *Haemophilus influenzae*: homologies to *comE*-ORF3 of *Bacillus subtilis* and *msbA* of *Escherichia coli*. *Gene* **146**, 95–100.

Cohen, A., Fisher, W. D., Curtiss, R., III & Adler, H. I. (1968). DNA isolated from *Escherichia coli* minicells mated with F+ cells. *Proceedings of the National Academy of Sciences, USA* **61**, 61–68.

Danner, D. B., Deich, R. A., Sisco, K. L. & Smith, H. O. (1980). An 11-base pair sequence determines the specificity of DNA uptake in *Haemophilus* transformation. *Gene* **11**, 311–318.

Dorocicz, I. R., Williams, P. M. & Redfield, R. J. (1993). The *Haemophilus influenzae* adenylate cyclase gene: cloning, sequence and essential role in competence. *Journal of Bacteriology* **175**, 7142–7149.

Dreiseikelmann, B. (1994). Translocation of DNA across bacterial membranes. *Microbiological Reviews* **58**, 293–316.

Dubnau, D. (1991). Genetic competence in *Bacillus subtilis*. *Microbiological Reviews* **55**, 395–424.

Dubnau, D. (1997). Binding and transport of transforming DNA by *Bacillus subtilis*: the role of type-IV pilin-like proteins – a review. *Gene* **192**, 191–198.

Dubnau, D. & Cirigliano, C. (1972). Fate of transforming DNA following uptake by competent *Bacillus subtilis*. III. Formation and properties of products isolated from transformed cells which are derived entirely from donor DNA. *Journal of Molecular Biology* **64**, 9–29.

Dürrenberger, F., Crameri, A., Hohn, B. & Koukolikova-Nicola, Z. (1989). Covalently bound VirD2 protein of *Agrobacterium tumefaciens* protects the T-DNA from exonucleolytic degradation. *Proceedings of the National Academy of Sciences, USA* **86**, 9154–9158.

Fleischmann, R. D., Adams, M. D., White, O. & 37 other authors (1995). Whole-genome random sequencing and assembly of *Haemophilus influenzae* Rd. *Science* **269**, 496–512.

Fox, M. S. (1957). Deoxyribonucleic acid incorporation by transformed bacteria. *Biochimica et Biophysica Acta* **26**, 83–85.

Fox, M. S. & Allen, M. K. (1964). On the mechanism of deoxyribonucleate integration in pneumococcal transformation. *Proceedings of the National Academy of Sciences, USA* **52**, 412–419.

Fujii, T., Naka, D., Toyoda, N. & Seto, H. (1987). LiCl treatment releases a nickase implicated in genetic transformation of *Streptococcus pneumoniae*. *Journal of Bacteriology* **169**, 4901–4906.

Ghei, O. K. & Lacks, S. A. (1967). Recovery of donor deoxyribonucleic acid marker activity from eclipse in pneumococcal transformation. *Journal of Bacteriology* **93**, 816–829.

Glaser-Wuttke, G., Keppner, J. & Rasched, I. (1989). Pore-forming properties of the adsorption protein of filamentous phage fd. *Biochimica et Biophysica Acta* **985**, 239–247.

Goodgal, S. H. & Herriott, R. M. (1961). Studies of transformation of *Haemophilus influenzae*. 1. Competence. *Journal of General Physiology* **44**, 1201–1227.

Goodman, S. D. & Scocca, J. J. (1988). Identification and arrangement of the DNA sequence recognized in specific transformation of *Neisseria gonorrhoeae*. *Proceedings of the National Academy of Sciences, USA* **85**, 6982–6986.

Gray, C. W., Brown, R. S. & Marvin, D. A. (1981). Adsorption complex of filamentous fd virus. *Journal of Molecular Biology* **146**, 621–627.

Hahn, J., Inamine, G., Kozlov, Y. & Dubnau, D. (1993). Characterization of *comE*, a late competence operon of *Bacillus subtilis* required for the binding and uptake of transforming DNA. *Molecular Microbiology* **10**, 99–111.

Hamoen, L. W., Van Werkhoven, A. F., Bijlsma, J. J. E., Dubnau, D. & Venema, G. (1998). The competence transcription factor of *Bacillus subtilis* recognizes short A/T-rich sequences arranged in a unique, flexible pattern along the DNA helix. *Genes & Development* **12**, 1539–1550.

Harlander, S. K. (1987). Transformation of *Streptococcus lactis* by electroporation. In *Streptococcal Genetics*, pp. 229–233. Edited by J. J. Ferretti & R. Curtiss, III. Washington, DC: American Society for Microbiology.

Håverstein, L. S. & Morrison, D. A. (1999). Quorum-sensing and peptide phero-mones in streptococcal competence for genetic transformation. In *Cell–cell Signalling in Bacteria*, pp. 9–26. Edited by G. M. Dunny & S. C. Winans. Washington, DC: American Society for Microbiology.

Håverstein, L. S., Coomaraswamy, G. & Morrison, D. A. (1995). An unmodified heptadecapeptide induces competence for genetic transformation in *Streptococcus pneumoniae*. *Proceedings of the National Academy of Sciences, USA* **92**, 11140–11144.

Hershey, A. D. & Chase, M. (1952). Independent functions of viral protein and nucleic acid in growth of bacteriophage. *Journal of General Physiology* **36**, 39–56.

Inamine, G. S. & Dubnau, D. (1995). ComEA, a *Bacillus subtilis* integral membrane protein required for genetic transformation, is needed for both DNA binding and transport. *Journal of Bacteriology* **177**, 3045–3051.

Kado, C. I. (1991). Molecular mechanism of crown gall tumorigenesis. *Critical Reviews in Plant Sciences* **10**, 1–32.

Kahn, M. E., Maul, G. & Goodgal, S. H. (1982). Possible mechanisms for donor DNA binding and transport in *Haemophilus*. *Proceedings of the National Academy of Sciences, USA* **80**, 6927–6931.

Karudapuram, S. & Barcak, G. J. (1997). The *Haemophilus influenzae dprABC* genes constitute a competence-inducible operon that requires the product of the *tfoX* (*sxy*) gene for transcriptional activation. *Journal of Bacteriology* **179**, 4815–4820.

Karudapuram, S., Zhao, X. & Barcak, G. J. (1995). DNA sequence and characterization of *Haemophilus influenzae dprA*$^+$, a gene required for chromosomal but not plasmid DNA transformation. *Journal of Bacteriology* **177**, 3235–3240.

Kunst, F., Ogasawaka, N., Mozer, I. & 148 other authors (1997). The complete genome sequence of the Gram-positive bacterium *Bacillus subtilis*. *Nature* **390**, 249–256.

Lacks, S. (1962). Molecular fate of DNA in genetic transformation of pneumococcus. *Journal of Molecular Biology* **5**, 119–131.

Lacks, S. (1977a). Binding and entry of DNA in pneumococcal transformation. In *Modern Trends in Bacterial Transformation and Transfection*, pp. 35–44. Edited by A. Portoles, R. Lopez & M. Espinosa. Amsterdam: Elsevier/North-Holland.

Lacks, S. A. (1977b). Binding and entry of DNA in bacterial transformation. In *Microbial Interactions*, pp. 179–232. Edited by J. L. Reissig. London: Chapman & Hall.

Lacks, S. (1979a). Steps in the process of DNA binding and entry in transformation. In *Transformation 1978*, pp. 27–41. Edited by S. W. Glover & L. O. Butler. Oxford: Cotswold Press.

Lacks, S. (1979b). Uptake of circular deoxyribonucleic acid and mechanism of deoxyribonucleic acid transport in genetic transformation of *Streptococcus pneumoniae*. *Journal of Bacteriology* **138**, 404–409.

Lacks, S. A. (1988). Mechanisms of genetic recombination in gram-positive bacteria. In *Genetic Recombination*, pp. 43–85. Edited by R. Kucherlapati & G. R. Smith. Washington, DC: American Society for Microbiology.

Lacks, S. & Greenberg, B. (1973). Competence for deoxyribonucleic acid uptake and deoxyribonuclease action external to cells in the genetic transformation of *Diplococcus pneumoniae*. *Journal of Bacteriology* **114**, 152–163.

Lacks, S. & Greenberg, B. (1976). Single-strand breakage on binding of DNA to cells in the genetic transformation of *Diplococcus pneumoniae*. *Journal of Molecular Biology* **101**, 255–275.

Lacks, S. & Neuberger, M. (1975). Membrane location of a deoxyribonuclease implicated in the genetic transformation of *Diplococcus pneumoniae*. *Journal of Bacteriology* **124**, 1321–1329.

Lacks, S. A. & Springhorn, S. S. (1984). Transfer of recombinant plasmids containing the gene for *Dpn*II DNA methylase into strains of *Streptococcus pneumoniae* that produce *Dpn*I or *Dpn*II restriction endonucleases. *Journal of Bacteriology* **158**, 905–909.

Lacks, S., Greenberg, B. & Carlson, K. (1967). Fate of donor DNA in pneumococcal transformation. *Journal of Molecular Biology* **29**, 327–347.

Lacks, S., Greenberg, B. & Neuberger, M. (1974). Role of a deoxyribonuclease in the genetic transformation of *Diplococcus pneumoniae*. *Proceedings of the National Academy of Sciences, USA* **71**, 2305–2309.

Lacks, S., Greenberg, B. & Neuberger, M. (1975). Identification of a deoxyribonuclease implicated in genetic transformation of *Diplococcus pneumoniae*. *Journal of Bacteriology* **123**, 222–232.

Larson, T. G. & Goodgal, S. H. (1991). Sequence and transcriptional regulation of *com101A*, a locus required for genetic transformation in *Haemophilus influenzae*. *Journal of Bacteriology* **173**, 4683–4691.

Lederberg, J. & Tatum, E. (1946). Gene recombination in *E. coli*. *Nature* **158**, 558.

Lerman, R. S. & Tolmach, L. J. (1957). Genetic transformation. I. Cellular incorporation of DNA accompanying transformation in pneumococcus. *Biochimica et Biophysica Acta* **28**, 68–82.

Levengood, S. K. & Webster, R. E. (1989). Nucleotide sequences of the *tolA* and *tolB* genes and localization of their products, components of a multistep translocation system in *Escherichia coli*. *Journal of Bacteriology* **171**, 6600–6609.

Lipinska, B., Krishna Rao, A. S. M., Bolten, B. M., Balakrishnan, R. & Goldberg, E. B. (1989). Cloning and identification of bacteriophage T4 gene 2 product gp2 and action of gp2 on infecting DNA in vivo. *Journal of Bacteriology* **171**, 488–497.

Londoño-Vallejo, J. A. & Dubnau, D. (1993). *comF*, a *Bacillus subtilis* late competence locus, encodes a protein similar to ATP-dependent RNA/DNA helicases. *Molecular Microbiology* **9**, 119–131.

Londoño-Vallejo, J. A. & Dubnau, D. (1994). Membrane association and role in DNA uptake of the *Bacillus subtilis* PriA analog ComF1. *Molecular Microbiology* **13**, 197–205.

Lopez, P., Espinosa, M., Stassi, D. L. & Lacks, S. A. (1982). Facilitation of plasmid transfer in *Streptococcus pneumoniae* by chromosomal homology. *Journal of Bacteriology* **150**, 692–701.

Lorenz, M. G. & Wackernagel, W. (1994). Bacterial gene transfer by natural genetic transformation in the environment. *Microbiological Reviews* **58**, 563–602.

Lunsford, R. D. (1998). Streptococcal transformation: essential features and applications of a natural gene exchange system. *Plasmid* **39**, 10–20.

Lunsford, R. D. & Roble, A. G. (1997). *comYA*, a gene similar to *comGA* of *Bacillus subtilis* is essential for competence-factor-dependent DNA transformation in *Streptococcus gordonii*. *Journal of Bacteriology* **179**, 3122–3126.

Lunsford, R. D., Nguyen, N. & London, J. (1996). DNA-binding activities in *Streptococcus gordonii*: identification of a receptor-nickase and a histonelike protein. *Current Microbiology* **32**, 95–100.

Mandel, M. & Higa, A. (1970). Calcium-dependent bacteriophage DNA infection. *Journal of Molecular Biology* **53**, 159–162.

Manoil, C. & Rosenbusch, J. P. (1982). Conjugation-deficient mutants of *E. coli* distinguish classes of functions of the outer membrane OmpA protein. *Molecular & General Genetics* **187**, 148–156.

Martin, B., Garcia, P., Castanie, M. P. & Claverys, J. P. (1995). The *recA* gene of *Streptococcus pneumoniae* is part of a competence-induced operon and controls lysogenic induction. *Molecular Microbiology* **15**, 367–379.

Matson, S. W., Nelson, W. C. & Morton, B. S. (1993). Characterization of the reaction product of the *oriT* nicking reaction catalyzed by *Escherichia coli* DNA helicase I. *Journal of Bacteriology* **175**, 2599–2606.

Mejean, V. & Claverys, J. P. (1993). DNA processing during entry in transformation of *Streptococcus pneumoniae*. *Journal of Biological Chemistry* **268**, 5594–5599.

Miao, R. & Guild, W. R. (1970). Competent *Diplococcus pneumoniae* accept both single- and double-stranded deoxyribonucleic acid. *Journal of Bacteriology* **101**, 361–364.

Morrison, D. A. (1978). Transformation in pneumococcus: protein content of eclipse complex. *Journal of Bacteriology* **136**, 548–557.

Morrison, D. A. (1997). Streptococcal competence for genetic transformation: regulation by peptide pheromones. *Microbial Drug Resistance* **3**, 27–37.

Morrison, D. A. & Baker, M. F. (1979). Competence for genetic transformation in pneumococcus depends on synthesis of a small set of proteins. *Nature* **282**, 215–217.

Morrison, D. A. & Guild, W. R. (1972). Transformation and deoxyribonucleic acid size: extent of degradation on entry varies with size of donor. *Journal of Bacteriology* **112**, 1157–1168.

Morrison, D. A. & Guild, W. R. (1973). Breakage prior to entry of donor DNA in *pneumococcus* transformation. *Biochimica et Biophysica Acta* **299**, 545–556.

Morrison, D. A., Lacks, S. A., Guild, W. R. & Hageman, J. M. (1983). Isolation and characterization of three new classes of transformation-deficient mutants of *Streptococcus pneumoniae* that are defective in DNA transport and genetic recombination. *Journal of Bacteriology* **156**, 281–290.

Nakamura, A., Koide, Y., Miyazaki, H., Kitamura, A., Masaki, H., Beppu, T. & Uozumi, T. (1992). Gene cloning and characterization of novel extracellular ribonuclease of *Bacillus subtilis*. *European Journal of Biochemistry* **209**, 121–127.

Notani, N. & Goodgal, S. H. (1966). On the nature of recombinants formed during transformation in *Hemophilus influenzae*. *Journal of General Physiology* **49**, 197–209.

Orren, D. K. & Sancar, A. (1989). The (A)BC excinuclease of *Escherichia coli* has only the UvrB and UvrC subunits in the incision complex. *Proceedings of the National Academy of Sciences, USA* **86**, 5237–5241.

Pakula, R. & Walczak, W. (1963). On the nature of competence of transformable streptococci. *Journal of General Microbiology* **31**, 125–133.

Palmen, R., Driessen, A. J. M. & Hellingwerf, K. J. (1994). Bioenergetic aspects of the translocation of macromolecules across bacterial membranes. *Biochimica et Biophysica Acta* **1183**, 417–451.

Pearce, B. J., Naughton, A. M., Campbell, E. A. & Masure, H. R. (1995). The *rec* locus, a competence-induced operon in *Streptococcus pneumoniae*. *Journal of Bacteriology* **177**, 86–93.

Pestova, E. V. & Morrison, D. A. (1998). Isolation and characterization of three *Streptococcus pneumoniae* transformation-specific loci by use of a lacZ reporter insertion vector. *Journal of Bacteriology* **180**, 2701–2710.

Piechowska, M. & Fox, M. S. (1971). Fate of transforming deoxyribonucleate in *Bacillus subtilis*. *Journal of Bacteriology* **108**, 680–689.

Pifer, M. L. (1986). Plasmid establishment in competent *Haemophilus influenzae* occurs by illegitimate transformation. *Journal of Bacteriology* **168**, 683–687.

Provvedi, R. & Dubnau, D. (1999). ComEA is a DNA receptor for transformation of competent *Bacillus subtilis*. *Molecular Microbiology* **31**, 271–280.

Puyet, A., Greenberg, B. & Lacks, S. A. (1990). Genetic and structural characterization of EndA, a membrane-bound nuclease required for transformation of *Streptococcus pneumoniae*. *Journal of Molecular Biology* **213**, 727–738.

Redfield, R. J. (1991). *sxy-1*, a *Haemophilus influenzae* mutation causing greatly enhanced spontaneous competence. *Journal of Bacteriology* **173**, 5612–5618.

Rees, C. E. D. & Wilkins, B. M. (1990). Protein transfer into the recipient cell during bacterial conjugation: studies with F and RP4. *Molecular Microbiology* **4**, 1199–1205.

Rosenthal, A. L. & Lacks, S. A. (1977). Nuclease detection in SDS-polyacrylamide gel electrophoresis. *Analytical Biochemistry* **80**, 76–90.

Rosenthal, A. L. & Lacks, S. A. (1980). Complex structure of the membrane nuclease of *Streptococcus pneumoniae* revealed by two-dimensional electrophoresis. *Journal of Molecular Biology* **141**, 133–146.

Sabelnikov, A. G. (1994). Nucleic acid transfer through cell membranes: towards the underlying mechanisms. *Progress in Biophysics and Molecular Biology* **62**, 119–152.

Sabelnikov, A. G., Greenberg, B. & Lacks, S. A. (1995). An extended −10 promoter alone directs transcription of the *Dpn*II operon of *Streptococcus pneumoniae*. *Journal of Molecular Biology* **250**, 144–155.

Sastry, P. A., Pasloske, B. L., Paranchych, W., Pearlstone, J. R. & Smillie, L. B. (1985). Nucleotide sequence and transcriptional initiation site of two *Pseudomonas aeruginosa* pilin genes. *Journal of Bacteriology* **164**, 571–577.

Saunders, C. W. & Guild, W. R. (1981). Monomer plasmid DNA transforms *Streptococcus pneumoniae*. *Molecular & General Genetics* **181**, 57–62.

Seto, H. & Tomasz, A. (1974). Early stages in DNA binding and uptake during genetic transformation of pneumococci. *Proceedings of the National Academy of Sciences, USA* **71**, 1493–1498.

Seto, H. & Tomasz, A. (1976). Calcium-requiring step in the uptake of deoxyribonucleic acid molecules through the surface of competent pneumococci. *Journal of Bacteriology* **126**, 1113–1118.

Shirasu, K. & Kado, C. I. (1993). Membrane location of the Ti plasmid VirB proteins involved in the biosynthesis of a pilin-like conjugative structure on *Agrobacterium tumefaciens*. *FEMS Microbiology Letters* **111**, 287–294.

van Sinderen, D., Kiewiet, R. & Venema, G. (1995a). Differential expression of two closely related deoxyribonuclease genes, *nucA* and *nucB*, in *Bacillus subtilis*. *Molecular Microbiology* **15**, 213–223.

van Sinderen, D., Luttinger, A., Kong, L., Dubnau, D., Venema, G. & Hamoen, L. (1995b). *comK* encodes the competence transcription factor, the key regulatory protein for competence development in *Bacillus subtilis*. *Molecular Microbiology* **15**, 455–462.

Smith, H., Wiersma, K., Bron, S. & Venema, G. (1983). Transformation in *Bacillus subtilis*: purification and partial characterization of a membrane-bound DNA-binding protein. *Journal of Bacteriology* **156**, 101–108.

Solomon, J. M. & Grossman, A. D. (1996). Who's competent and when: regulation of natural genetic competence in bacteria. *Trends in Genetics* **12**, 150–155.

Strom, M. S., Nunn, D. N. & Lory, S. (1993). A single bifunctional enzyme, PilD, catalyzes cleavage and N-methylation of proteins belonging to the type IV pilin family. *Proceedings of the National Academy of Sciences, USA* **90**, 2404–2408.

Sugawara, E. & Nikaido, H. (1992). Pore-forming activity of OmpA protein of *Escherichia coli*. *Journal of Biological Chemistry* **267**, 2507–2511.

Sun, T. P. & Webster, R. E. (1986). *fii*, a bacterial locus required for filamentous phage infection and its relation to colicin-tolerant *tolA* and *tolB*. *Journal of Bacteriology* **165**, 107–115.

Tarahovsky, Y. S., Khusainov, A. A., Deev, A. A. & Kim, Y. V. (1991). Membrane fusion during infection of *Escherichia coli* cells by phage T4. *FEBS Letters* **289**, 18–22.

Thompson, J. D., Higgins, D. G. & Gibson, T. J. (1994). CLUSTAL W: improving the sensitivity of progressive multiple sequence alignment through sequence weighting, position specific gap penalties and weight matrix choice. *Nucleic Acids Research* **22**, 4673–4680.

Tomasz, A. (1965). Control of the competent state in *Pneumococcus* by a hormone-like cell product: an example for a new type of regulatory mechanism in bacteria. *Nature* **208**, 155–159.

Tomasz, A. & Hotchkiss, R. D. (1964). Regulation of the transformability of pneumococcal cultures by macromolecular cell products. *Proceedings of the National Academy of Sciences, USA* **51**, 480–487.

Tomasz, A. & Mosser, J. L. (1966). On the nature of the pneumococcal activator substance. *Proceedings of the National Academy of Sciences, USA* **55**, 58–66.

Tomb, J.-F., El-Hajj, H. & Smith, H. O. (1991). Nucleotide sequence of a cluster of genes involved in the transformation of *Haemophilus influenzae* Rd. *Gene* **104**, 1–10.

Trautner, T., Pawlek, B., Bron, S. & Anagnostopoulos, C. (1974). Restriction and modification in *Bacillus subtilis*. Biological aspects. *Molecular & General Genetics* **131**, 181–191.

Wilkins, B. M. (1995). Gene transfer by bacterial conjugation: diversity of systems and functional specializations. In *Population Genetics of Bacteria*, pp. 59–88. Edited by S. Baumberg, J. P. W. Young, E. M. H. Wellington & J. R. Saunders. Cambridge: Cambridge University Press.

Zulty, J. J. & Barcak, G. J. (1995). Identification of a DNA transformation gene required for *com101A+* expression and supertransformer phenotype in *Haemophilus influenzae*. *Proceedings of the National Academy of Sciences, USA* **92**, 3616–3620.

ESCHERICHIA COLI SIGNAL RECOGNITION PARTICLE – A HISTORICAL PERSPECTIVE

QUIDO A. VALENT[1], EDWARD N. S. O'GORMAN[2], PIER SCOTTI[1], JOEN LUIRINK[1] AND STEPHEN HIGH[2]

[1]*Department of Microbiology, Institute of Molecular Biological Sciences, Biocentrum Amsterdam, De Boelelaan 1087, 1081 HV Amsterdam, The Netherlands*
[2]*School of Biological Sciences, University of Manchester, 2.205 Stopford Building, Oxford Road, Manchester M13 9PT, UK*

THE EUKARYOTIC SIGNAL RECOGNITION PARTICLE

In contrast to prokaryotes such as *Escherichia coli*, eukaryotic cells have a number of different, membrane-enclosed, subcellular compartments. Many of the newly synthesized proteins destined for these compartments are made in the cytosol and must then be 'targeted' to the appropriate organelle. This targeting is achieved by incorporating a molecular postcode, or targeting signal, during the synthesis of the protein and it is these signals that target proteins to the correct subcellular destination.

The targeting of newly synthesized secretory and membrane proteins to the endoplasmic reticulum (ER) of eukaryotic cells has long been known to involve a signal sequence comprised from a stretch of hydrophobic amino acid residues. These signal sequences are first recognized by a cytosolic ribonucleoprotein complex called the signal recognition particle or SRP (see Fig. 1; Rapoport *et al.*, 1996; Brodsky, 1998). Mammalian SRP was first identified over 20 years ago and was later found to be composed of six different proteins assembled on to an RNA scaffold. Further studies revealed that only one of these proteins, the 54 kDa subunit of SRP (SRP54), actually recognizes and binds the hydrophobic ER signal sequences, thereby initiating the chain of events that results in membrane targeting (Lütcke, 1995; Rapoport *et al.*, 1996).

The complex of SRP, and the ribosome-bound nascent chain to which it binds, is specifically targeted to the ER membrane because it contains the SRP receptor, a dimer of the SRP receptor α and β subunits. SRP54 and the SRP receptor α and β subunits are all GTP-binding proteins, and the SRP-dependent targeting process is controlled by GTP binding and hydrolysis (see Fig. 1 and Rapiejko & Gilmore, 1997). Upon its release from SRP at the ER membrane, the signal sequence interacts with

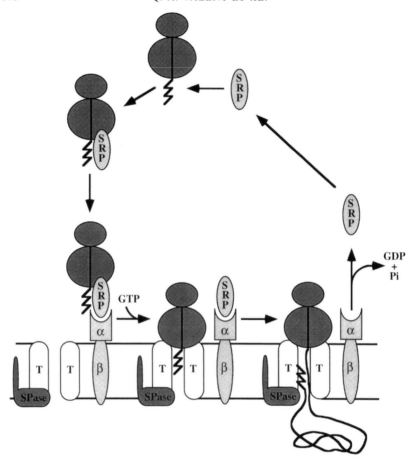

Fig. 1. SRP-dependent protein targeting to the eukaryotic ER. A hydrophobic signal sequence is recognized by the SRP as it emerges from the ribosome. SRP retards the translation of the nascent polypeptide and targets the ribosome/nascent chain complex to the α subunit of the SRP receptor at the ER membrane. GTP binding to both the SRP receptor α subunit and the 54 kDa subunit of SRP releases the ribosome/nascent chain complex from the SRP, thus relieving the translational block. The signal sequence is recognized by the Sec61 complex and the ribosome makes a tight seal with the translocon (T). Ongoing translation drives the translocation of the polypeptide through the aqueous channel of the translocon and into the ER lumen. Signal peptidase (SPase) cleaves the targeting signal from the mature region of the preprotein. GTP hydrolysis by both the SRP54 subunit and the SRP receptor α subunit separates these components and recycles them for another round of targeting.

components of the ER membrane translocation machinery (see Fig. 1; High *et al.*, 1991; Jungnickel & Rapoport, 1995), resulting in the integration of a membrane protein into the ER membrane, or the transport of a secretory protein across it.

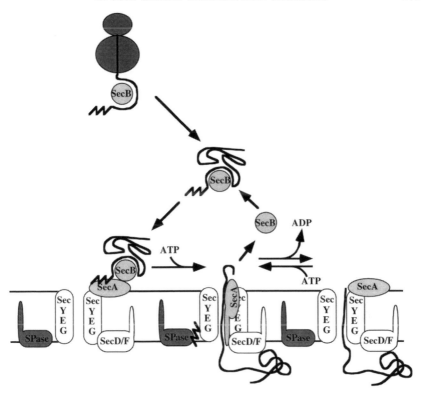

Fig. 2. SecB-dependent protein targeting to the *E. coli* inner membrane. SecB binds to the mature region of proteins destined for secretion and targets them to the inner membrane translocon (the SecYEG) and associated SecA. The SecA has affinity for both SecB and the preprotein and ATP binding by SecA initiates protein translocation and releases SecB. Cycles of ATP binding and hydrolysis regulate the membrane insertion and de-insertion of SecA which functions to push the preprotein through the translocon. SecD/F appears to stabilize the membrane-inserted state of SecA.

E. COLI SRP?

From the earliest comparisons of eukaryotic and prokaryotic signal sequences, it was clear that they bore striking similarities and that any differences were subtle (von Heijne, 1985). The respective signal peptidases that subsequently cleave off these signals have also proved to be closely related (Dalbey & von Heijne, 1992; Dalbey *et al.*, 1997). Despite the similarities of eukaryotic and prokaryotic signal sequences, substantial studies of protein secretion in *E. coli* initially failed to identify its SRP. Thus, at one time there was a widely held view that the molecular details of protein targeting to the *E. coli* inner membrane were very different from protein targeting to the ER of a mammalian cell (see Fig. 2 and Wickner *et al.*, 1991).

SRP54

P48

SRα

FtsY

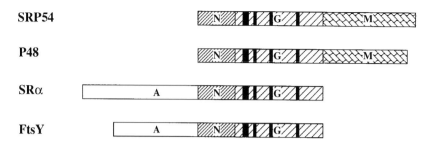

Fig. 3. An alignment of the amino acid sequences of the homologous proteins SRP54, P48, the SRP receptor α subunit and FtsY. The G-domains are well-conserved throughout all four proteins – the elements of the consensus regions for GTP-binding proteins are indicated in black. The C-terminal methionine-rich 'M-domain' is only present in SRP54-like proteins such as SRP54 and P48.

The first indication that *E. coli* may actually have a signal recognition particle of some kind was when similarities between the *E. coli* 4.5S RNA and the well-characterized 7S RNA of mammalian SRP were identified (Poritz *et al.*, 1988). The central role of the mammalian SRP54 kDa subunit during signal sequence recognition had resulted in a concerted effort to obtain its cDNA. This was achieved by the laboratories of Bernhard Dobberstein and Peter Walter (Bernstein *et al.*, 1989; Römisch *et al.*, 1989) shortly after the relationship between *E. coli* 4.5S and mammalian 7S RNA had been highlighted. The amino acid sequence of the mammalian SRP54 subunit was remarkably informative, providing insights into both the molecular basis for the GTP requirement of the targeting process, and a potential mechanism for the specific recognition of the relatively diverse, hydrophobic, ER signal sequences.

Equally striking were the similarities between the amino acid sequences of SRP54 and those of proteins/open reading frames already lodged in databases. A significant region of mammalian SRP54 was found to be similar to the α subunit of the SRP receptor, something that had not been expected on the basis of the contemporary experimental information. Even more exciting was the finding that two *E. coli* proteins, P48 (or Ffh) and FtsY, were homologous to SRP54 and SRα, respectively (Bernstein *et al.*, 1989; Römisch *et al.*, 1989; Fig. 3).

The similarity between SRP54 and P48, and SRα and FtsY, led both laboratories to suggest that an SRP-dependent protein targeting pathway may exist in prokaryotes (Bernstein *et al.*, 1989; Römisch *et al.*, 1989). Initially this proposal was not well received by many researchers working in the field of *E. coli* protein secretion. The lack of convincing experimental evidence and the failure to detect any of the putative SRP components by a variety of genetic screens were seen as major weaknesses of the *E. coli* SRP hypothesis (Bassford *et al.*, 1991; Beckwith, 1991; Brown, 1991). In the remainder of this article we will review the large amount of evidence that has now been accumulated to show that *E. coli* SRP does play a crucial role in protein targeting to the *E. coli* inner membrane. For an in-depth analysis of eukaryotic SRPs, the reader should consult one of the recent reviews about this subject (Lütcke, 1995; Walter & Johnson, 1994).

A BONA FIDE SRP-DEPENDENT TARGETING PATHWAY OPERATES IN *E. COLI*

Shortly after their initial identification, it was shown that *E. coli* P48 and 4.5S RNA can form a ribonucleoprotein complex reminiscent of mammalian SRP (Poritz *et al.*, 1990; Ribes *et al.*, 1990). Furthermore, the secretory protein precursor pre-β-lactamase was found to accumulate upon the depletion of the 4.5S RNA, providing the first direct evidence that *E. coli* SRP was involved in protein secretion. Two years later the depletion of P48, the protein component of *E. coli* SRP, was also shown to cause the accumulation of several secretory protein precursors including pre-β-lactamase (Phillips & Silhavy, 1992).

Meanwhile, whilst working in Bernhard Dobberstein's laboratory at the European Molecular Biology Laboratory, we had been adapting the cross-linking technology developed for investigating mammalian protein targeting and applied it to the study of the putative *E. coli* SRP. The results of these experiments seemed clear cut, and we found an obvious functional relationship between mammalian SRP54 and *E. coli* P48 (Luirink *et al.*, 1992). In particular, we found that *E. coli* SRP (i.e. the P48/4.5S RNA complex) bound specifically to a functional signal sequence as it emerged from the ribosome. We concluded that the *E. coli* cytosol contained a true signal recognition particle and suggested that its role was akin to that of mammalian SRP (Luirink *et al.*, 1992).

WHICH PROTEINS USE *E. COLI* SRP DURING THEIR BIOSYNTHESIS?

Although not all of the physiological substrates of the *E. coli* SRP have been accurately defined, several studies have shown that it is the efficient insertion of inner membrane proteins that is most frequently dependent upon a functional SRP pathway (MacFarlane & Müller, 1995; de Gier *et al.*, 1996; Ulbrandt *et al.*, 1997). The role of *E. coli* SRP during protein secretion is more variable (Phillips & Silhavy, 1992), and the dependency of any particular precursor upon the SRP-dependent pathway seems to correlate with the hydrophobicity of its signal sequence (see below).

SIGNAL SEQUENCE RECOGNITION

The relative diversity of eukaryotic and prokaryotic signal sequences was a key factor in driving scientific investigations into the underlying molecular mechanism that allowed these signals to be recognized and distinguished from other stretches of amino acids. The sequencing of the mammalian SRP54 subunit, and the identification of the *E. coli* homologue P48, provided a substantial amount of information about this process. Our detailed understanding has recently been dramatically enhanced by the publication of high-resolution crystal structures of key domains of P48-like proteins.

The hydrophobicity of the signal sequences responsible for the targeting of secretory and integral membrane proteins is a crucial factor influencing substrate selection by both prokaryotic and eukaryotic SRPs (Ng *et al.*, 1996; Hatsuzawa *et al.*, 1997; High *et al.*, 1997; Valent *et al.*, 1995, 1997). These signals show a wide variation in length and amino acid composition whilst maintaining the common trait of hydrophobicity (von Heijne, 1985).

The methionine-rich C-terminal M-domains of *E. coli* P48 and mammalian SRP54 are well-conserved and distinguish them from the partially homologous membrane receptors SRα and FtsY, which lack this region (see Fig. 3; Bernstein *et al.*, 1989; Römisch *et al.*, 1989). The amino acid sequence of the P48 and SRP54 M-domains, combined with the properties of the methionine residues that are abundant in both, led Bernstein and colleagues to propose that the M-domain might form a flexible, amphipathic, groove that could act as a hydrophobic binding pocket for such diverse signal sequences (Bernstein *et al.*, 1989).

Subsequent experiments designed to test this hypothesis showed that it is indeed the M-domain that binds to the hydrophobic signal sequences and that this region also binds to the RNA component which characterizes SRPs as ribonucleoprotein complexes (High & Dobberstein, 1991; Lütcke *et al.*, 1992; Römisch *et al.*, 1990; Zopf *et al.*, 1990, 1993). Almost 10 years after obtaining the original cDNAs, the high-resolution crystal structure of the M-domain of the *Thermus aquaticus* P48 (Ffh) shows that this region does indeed form a hydrophobic groove lined with the side-chains of flexible amino acid residues (Keenan *et al.*, 1998). This putative signal sequence binding groove is of sufficient size and hydrophobicity to accommodate signal sequences of different sizes and amino acid sequences. This study also identified an arginine-rich α helix as the region of the M-domain to which the negatively charged SRP RNA binds (Keenan *et al.*, 1998).

SRP AND THE RIBOSOME

In eukaryotes, SRP is intimately associated with incomplete or 'nascent' polypeptides that are still being synthesized and are therefore still attached to the ribosome. After binding to the signal sequence as it emerges from the ribosomes, one of the first functions performed by mammalian SRP is to slow down the rate of protein translation. This increases the time window during which the nascent polypeptide can interact productively with the translocation sites present in the ER membrane (see Walter & Johnson, 1994; Brodsky, 1998). Eukaryotic SRPs are significantly more complex than their prokaryotic counterparts (Lütcke, 1995) and *E. coli* SRP is unable to retard translation when added to a eukaryotic translation system (Powers & Walter, 1997). It therefore remains to be established whether *E. coli* SRP retards the translation rate of its substrates, or whether such a mechanism is not required in *E. coli* (de Gier *et al.*, 1997).

FTSY – THE RECEPTOR FOR *E. COLI* SRP

Since the SRP receptor α subunit was known to act as an ER membrane receptor for SRP-mediated protein targeting, its similarity to the *E. coli* FtsY protein led to the obvious suggestion that FtsY carried out a related function in *E. coli* (Bernstein *et al.*, 1989; Römisch *et al.*, 1989). The depletion of FtsY and the perturbation of P48 expression were both shown to result in an accumulation of the same subset of secretory protein precursors, indicating that both components participate in the same targeting pathway (Luirink *et al.*, 1994). The idea that FtsY functions as a receptor for *E. coli* SRP was further supported by the observation that a significant fraction of FtsY is localized at the cytoplasmic membrane (Luirink *et al.*, 1994). Furthermore, it was shown that FtsY can interact with *E. coli* SRP *in vitro* (Miller *et al.*, 1994; Kusters *et al.*, 1995).

In eukaryotic cells, the SRP receptor α subunit is bound to the ER membrane via a specific interaction with its β subunit, an integral membrane protein that also contains a GTP-binding site (Miller *et al.*, 1995). However, it has recently been shown that whilst an intact GTP-binding motif is required for SRP receptor β to function correctly, its transmembrane domain is dispensable (Ogg *et al.*, 1998). Thus, a stable ER membrane localization of the SRP receptor α subunit, via its interaction with the β subunit, is not a prerequisite for SRP-dependent protein targeting (Ogg *et al.*, 1998). This means that the role of the SRP receptor complex may be to act as a regulatory switch that only needs to associate transiently with the ER membrane (Ogg *et al.*, 1998).

In this light, the fact that no *E. coli* equivalent of the mammalian SRP receptor β subunit has been identified probably reflects the relative simplicity of the SRP targeting pathway in *E. coli*. Nevertheless, exactly how the hydrophilic FtsY protein accomplishes its partial membrane localization is presently unknown (de Leeuw *et al.*, 1997; Zelazny *et al.*, 1997). Prior to contacting the membrane-embedded translocon, the SRP–ribosome/nascent chain complex is targeted to the *E. coli* inner membrane and the SRP is released from the nascent protein (Valent *et al.*, 1998). Although a significant amount of FtsY is associated with the inner membrane, the soluble fraction of FtsY may also play a functional role. Hence it can bind to *E. coli* SRP–ribosome/nascent chain complexes and assist their delivery to the *E. coli* inner membrane translocon (Valent *et al.*, 1998). The overall function of FtsY may simply be to increase the effective concentration of SRP substrates near the *E. coli* inner membrane translocation complex (Powers & Walter, 1997).

ROLE OF GTP IN SRP-MEDIATED PROTEIN TARGETING

Like their mammalian counterparts, both P48 and FtsY contain a GTP-binding (G-) domain (Bernstein *et al.*, 1989; Römisch *et al.*, 1989; Fig. 3). Both of these G-domains contain the consensus sequence for a GTP-binding site consisting of four short motifs at conserved positions (see Fig. 3). Furthermore, the structures of the P48 and FtsY G-domains are almost identical (Freymann *et al.*, 1997; Montoya *et al.*, 1997). GTPases have been linked to various and diverse cellular functions, and frequently act as molecular switches whose 'on' and 'off' states are triggered by the binding and hydrolysis of GTP (see Bourne *et al.*, 1990). The irreversible nature of GTP hydrolysis makes these cycles unidirectional. Both the mammalian and *E. coli* SRP-dependent protein targeting pathways require GTP binding for the release of the nascent chain from SRP at the membrane (Connolly & Gilmore, 1989, 1993; Connolly *et al.*, 1991; High *et al.*, 1991) and GTP hydrolysis for dissociation of the SRP–SRP receptor complex (Kusters *et al.*, 1995; Miller *et al.*, 1993; Rapiejko & Gilmore, 1994, 1997; Valent *et al.*, 1998).

Since it is 10 years older than the *E. coli* SRP field, the detailed molecular mechanisms of the mammalian SRP-dependent targeting pathway have been studied in more detail than the equivalent process in *E. coli*. Nevertheless, the precise details of the mammalian SRP pathway are still contested and readers should consult Bacher *et al.* (1996) and Rapiejko & Gilmore (1997) if they wish to view the two predominant models (see also Neuhof *et al.*, 1998; Raden & Gilmore, 1998). The model for SRP-dependent targeting in mammals that appears to most closely mirror the events occurring in *E. coli* is as follows. (i) SRP cycles on and off translating ribosomes 'sampling' all of the nascent chains for the presence of a signal sequence (Ogg & Walter, 1995). (ii) The binding of the signal sequence by the 54 kDa subunit of SRP initiates the formation of a targeting complex; the SRP54 GTP-binding site is probably empty at this stage (see Fig. 1). The high-resolution structures derived for the P48 and FtsY G-domains show that the active-site side-chains present in the nucleotide-free forms of the proteins are effectively sequestered, suggesting that the proteins would be perfectly stable in their nucleotide-free forms (Freymann *et al.*, 1997; Montoya *et al.*, 1997). This fits well with the suggestion that the GTP-binding sites of both SRP54 and SRα are empty prior to docking at the ER membrane as indicated in Fig. 1 (see Rapiejko & Gilmore, 1997). (iii) The targeting complex (SRP, ribosome and nascent polypeptide chain) arrives at the ER membrane and interacts with its cognate receptor, the SRP receptor complex (α and β subunits). Conformational changes in SRP54 and the SRP receptor stimulate GTP binding to SRP54 and the SRP receptor α subunit (see Rapiejko & Gilmore, 1997). The binding of GTP to SRP54 and the SRP receptor α subunit is likely to require significant structural rearrangements and these may well form the interaction of SRP and its receptor (see Freymann et al., 1997; Montoya et al., 1997). This process also requires that the SRP receptor β subunit has a functional GTP-binding site and it may be that the SRP receptor β subunit ensures that the nascent chain is only released from SRP when an active ER translocon is available to accept it (see Ogg et al., 1998). The binding of GTP results in a reduction in the affinity of SRP54 for the signal sequence which is therefore released. GTP hydrolysis is not required at this point since non-hydrolysable analogues of GTP will also allow this step to occur (Connolly & Gilmore, 1986; High et al., 1991). (iv) After the signal sequence is released from SRP, it interacts directly with the ER membrane translocation machinery. The SRP–SRP receptor complex is released from any transient association with the ER translocon and GTP hydrolysis by the SRP receptor α subunit and SRP54 allows SRP to dissociate from the SRP receptor and be recycled back into the cytosol (see Rapiejko & Gilmore, 1997 and references therein).

The sequence of events occurring during SRP-dependent protein targeting in E. coli appears to be very similar to that outlined above (see Fig. 4). (i) E. coli SRP binds to signal sequences as they emerge from the ribosome (Luirink et al., 1992; Valent et al., 1995, 1997, 1998). (ii) Although soluble FtsY can interact with the SRP–ribosome/nascent chain complex, the release of SRP from the ribosome/nascent chain complex requires the presence of E. coli inner membranes. This ensures that any premature release of SRP from the nascent polypeptide in the cytosol is prevented and that the ribosome-bound polypeptide is delivered to the E. coli inner membrane translocon (Valent et al., 1998). Quite how the FtsY-dependent release of E. coli SRP is specifically triggered by arrival at the inner membrane remains to be established. (iii) Upon binding to each other, P48 and FtsY act reciprocally to stimulate GTP hydrolysis by each other and hence E. coli SRP and FtsY regulate each other's activity (Powers & Walter, 1995). As observed in the mammalian system, GTP binding, but not GTP hydrolysis, is a prerequisite for the release of E. coli SRP from the signal sequence (Valent et al., 1998). (iv) After the targeting reaction has been completed the hydrolysis of GTP enables the components of the E. coli SRP-dependent pathway to be recycled (Powers & Walter, 1995; Fig. 4).

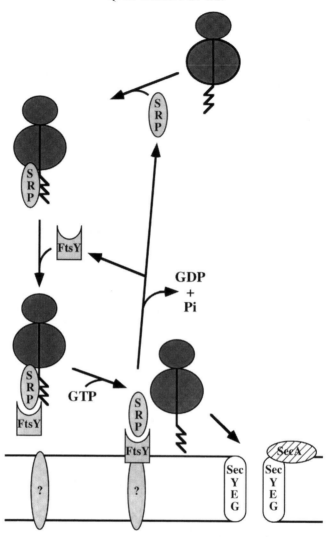

Fig. 4. Current model for SRP-mediated protein targeting to the *E. coli* inner membrane. Particularly hydrophobic signal sequences are recognized and bound by SRP as soon as they emerge from the ribosome. The ribosome/nascent chain/SRP complex can interact with FtsY in the cytosol. The growing ribosome/nascent chain is released from the SRP after the docking of FtsY at an unidentified site in the membrane. This process occurs during or after the binding of GTP to both FtsY and SRP. The released ribosome/nascent chain inserts into the SecYEG translocon, where SecA may facilitate the translocation reaction. GTP hydrolysis at both SRP and FtsY serves to dissociate and recycle the targeting components.

EVERYTHING NEEDS AN SRP?

A number of completed genome sequencing projects have shown that all organisms contain at least sequence homologues of *E. coli* P48 and FtsY. This is best illustrated by the fact that *Mycoplasma genitalium*, which has the smallest known genome of any free-living organism, has open reading frames that encode both P48 (Ffh)- and FtsY-like proteins (Fraser *et al.*, 1995). These components are likely to be functional homologues of their *E. coli* counterparts (see Macao *et al.*, 1997 and references therein).

ORGANELLAR SRP

The first hint that eukaryotic organelles may contain distinct SRPs was the discovery of an SRP54 homologue present within the stroma of pea chloroplasts. In fact the sequence of the cpSRP54 protein was much more similar to *E. coli* P48 than mammalian SRP54 (Franklin & Hoffmann, 1993). This similarity of P48 and cpSRP54 fits well with the idea that chloroplasts have evolved from what were originally endosymbiotic bacteria that lived inside a host cell (see deDuve, 1991). The cpSRP54 protein is encoded by a nuclear gene and is synthesized as a precursor with an N-terminal transit peptide which results in its import into the chloroplast stroma. Like other SRPs, cpSRP54 shows a similar preference for particularly hydrophobic signal sequences, in this case present on proteins destined for integration into the thylakoid membrane (High *et al.*, 1997). Stromal cpSRP54 has been shown to directly participate in the post-translational integration of a polytopic membrane protein at the thylakoid (Li *et al.*, 1995), in this case acting as a subunit of a novel signal recognition particle formed with a 43 kDa protein, cpSRP43, which seems to substitute for the usual SRP RNA (see Schuenemann *et al.*, 1998). Thus, the role of stromal SRP appears to mirror that of *E. coli* SRP with both components promoting efficient membrane integration into their respective target membranes. Whilst one assumes that a stromal equivalent of FtsY must also exist, any such component remains to be identified.

TRIGGER FACTOR AND NEWLY SYNTHESIZED PROTEINS

In addition to the signal-sequence-specific interaction that we observed between nascent polypeptides and *E. coli* SRP, we also detected a 55 kDa component of the *E. coli* cytosol that interacted weakly with several newly synthesized nascent chains including those lacking a functional signal sequence (Luirink *et al.*, 1992; Valent *et al.*, 1995). We were able to identify this component as the so-called 'trigger factor' (Valent *et al.*, 1995). Trigger factor has an intrinsic affinity for the ribosome (Stoller *et al.*, 1995) and interacts specifically with ribosome-bound nascent chains. It most probably functions as a molecular chaperone to promote the co-translational folding of many newly synthesized proteins including those destined for the *E. coli* cytosol (Hesterkamp *et al.*, 1996; Valent *et al.*, 1995).

AFTER SRP WHERE NEXT?

After the release of the mammalian signal sequence from SRP at the ER membrane, the ribosome-bound nascent polypeptide interacts with the ER translocation machinery (High & Laird, 1997; Rapoport *et al.*, 1996). The principal component of the ER translocon is the heterotrimeric Sec61 complex and oligomers of this complex form a transmembrane pore (Beckmann *et al.*, 1997; Hanein *et al.*, 1996). This aqueous channel (Johnson, 1997) extends from the ribosome and through the ER membrane, providing the site where proteins are inserted into and translocated across the lipid bilayer (Beckmann *et al.*, 1997; Hanein *et al.*, 1996; Johnson, 1997).

In *E. coli*, the classical Sec-dependent targeting pathway (see Fig. 2) has long been known to utilize SecB to deliver full-length precursor proteins to SecA and the *E. coli* inner membrane translocon complex SecYEG. It is striking that the SecYEG complex bears a significant resemblance to that of the mammalian Sec61 heterotrimer (Rapoport *et al.*, 1996). We recently showed that *E. coli* SRP also presents nascent, ribosome-bound, precursor proteins to a membrane-embedded translocon that is composed of the same core, namely SecYEG (Valent *et al.*, 1998). Although we used ribosome-bound chains for this study, we still observed an interaction between SecA, the molecular motor of post-translational translocation (Schatz & Dobberstein, 1996), and the nascent precursor proteins (Valent *et al.*, 1998). This suggests that SecA may be involved in the translocation of all polypeptides across the *E. coli* inner membrane, regardless of the targeting mechanism that presents the precursors to the translocon (cf. Figs 2 and 4).

WHY DOES *E. COLI* HAVE MULTIPLE TARGETING PATHWAYS?

As outlined above, the case for the existence of an SRP-dependent targeting pathway in *E. coli* is now well-established, and its importance is underlined by the fact that both P48 and FtsY are essential for cell survival (Phillips & Silhavy, 1992; Luirink *et al.*, 1994). What remains less clear is precisely why the SRP is necessary given the existence of the alternative 'Sec' pathway. In fact, the use of multiple routes for targeting proteins has been widely observed for destinations such as the ER (Ng *et al.*, 1996), the *E. coli* inner membrane (de Gier *et al.*, 1997) and the thylakoid membrane (Cline & Henry, 1996). In most cases particular precursors favour one or other of the available pathways, although in some cases proteins can use more than one route, suggesting some level of redundancy (e.g. Ng *et al.*, 1996).

The SRP-dependent targeting route in *E. coli* may well favour the co-translational translocation of proteins in exactly the same way as eukaryotic SRP does (cf. Rapoport *et al.*, 1996; Rapiejko & Gilmore, 1997). The co-translational integration of hydrophobic membrane proteins may be particularly important in ensuring that these proteins are made efficiently. In the case of multiple-spanning membrane proteins this would ensure that the bulk of the hydrophobic transmembrane domains are synthesized *in situ* at the translocon, thereby preventing the opportunity for these regions to become aggregated or misfolded. Whilst this is a nice model, it is clearly too simplistic. SRP can clearly function in promoting the post-translational targeting of full-length precursors to the thylakoid membrane (Schuenemann *et al.*, 1998), and it will be no surprise if it can also carry out this role in *E. coli*.

CONCLUSION

Over the past decade tremendous advances in our understanding of the structure and function of *E. coli* SRP have been made. These studies have themselves raised many issues that remain to be addressed, for example: what is the role of the *E. coli* ribosome during SRP-dependent membrane targeting and subsequent membrane insertion/translocation; can *E. coli* SRP function post-translationally to target completed proteins; what ensures that a signal sequence is only released from *E. coli* SRP when an inner membrane translocon is available to accept it; and what role does SecA play during the translocation of the ribosome-bound polypeptides that are targeted via *E. coli* SRP? Clearly there will be plenty to do for at least the next 10 years.

ACKNOWLEDGEMENTS

Our interest in the *E. coli* signal recognition particle was initially sparked by a happy and productive period spent working in Bernhard Dobberstein's group at the European Molecular Biology Laboratory and we thank him (S. H. & J. L.). Over the last 10 years, the authors' contributions to the work described above have been made possible by funding from the following sources: BBSRC Advanced Research Fellowship (S. H.), EMBO Fellowship (S. H.), NWO (J. L., Q. V.) and the EU (J. L., S. H.).

REFERENCES

Bacher, G., Lütcke, H., Jungnickel, B., Rapoport, T. A. & Dobberstein, B. (1996). Regulation by the ribosome of the GTPase of the signal-recognition particle during protein targeting. *Nature* **381**, 248–251.

Bassford, P., Beckwith, J., Ito, K. & 7 other authors (1991). The primary pathway of protein export in *E. coli. Cell* **65**, 367–368.

Beckmann, R., Bubeck, D., Grassucci, R., Penczek, P., Verschoor, A., Blobel, G. & Frank, J. (1997). Alignment of conduits for the nascent polypeptide chain in the Ribosome-Sec61 complex. *Science* **278**, 2123–2126.

Beckwith, J. (1991). "Sequence gazing?" *Science* **241**, 1161–1162.

Bernstein, H. D., Poritz, M. A., Strub, K., Hoben, P. J., Brenner, S. & Walter, P. (1989). Model for signal sequence recognition from amino-acid sequence of 54K subunit of signal recognition particle. *Nature* **340**, 482–486.

Bourne, H. R., Sanders, D. A. & McCormick, F. (1990). The GTPase superfamily: a conserved switch for diverse cell functions. *Nature* **348**, 125–132.

Brodsky, J. L. (1998). Translocation of proteins across the endoplasmic reticulum membrane. *International Review of Cytology* **178**, 277–327.

Brown, S. (1991). 4.5S RNA: does form predict function? *New Biologist* **3**, 430–438.

Cline, K. & Henry, R. (1996). Import and routing of nucleus-encoded chloroplast proteins. *Annual Review of Cell and Developmental Biology* **12**, 1–26.

Connolly, T. & Gilmore, R. (1986). Formation of a functional ribosome-membrane junction during translocation requires the participation of a GTP-binding protein. *Journal of Cell Biology* **103**, 2253–2261.

Connolly, T. & Gilmore, R. (1989). The signal recognition particle receptor mediates the GTP-dependent displacement of SRP from the signal sequence of the nascent polypeptide. *Cell* **57**, 599–610.

Connolly, T. & Gilmore, R. (1993). GTP hydrolysis by complexes of the signal recognition particle and the signal recognition particle receptor. *Journal of Cell Biology* **123**, 799–807.

Connolly, T., Rapiejko, P. J. & Gilmore, R. (1991). Requirement of GTP hydrolysis for dissociation of the signal recognition particle from its receptor. *Science* **252**, 1171–1173.

Dalbey, R. E. & von Heijne, G. (1992). Signal peptidases in prokaryotes and eukaryotes – a new protease family. *Trends in Biochemical Sciences* **17**, 474–478.

Dalbey, R. E., Lively, M. O., Bron, S. & van Dijl, J. M. (1997). The chemistry and enzymology of the type I signal peptidases. *Protein Science* **6**, 1129–1138.

deDuve, C. (1991). *Blueprint for a Cell: the Nature and Origin of Life*. Burlington: Carolina Biological Supply Company.

Franklin, A. E. & Hoffmann, N. E. (1993). Characterization of a chloroplast homologue of the 54-kDa subunit of the signal recognition particle. *Journal of Biological Chemistry* **268**, 22175–22180.

Fraser, C. M., Gocayne, J. D., White, O. & 26 other authors (1995). The minimal gene complement of *Mycoplasma genitalium. Science* **270**, 397–403.

Freymann, D. M., Keenan, R. J., Stroud, R. M. & Walter, P. (1997). Structure of the conserved GTPase domain of the signal recognition particle. *Nature* **385**, 361–364.

de Gier, J. W. L., Mansournia, P., Valent, Q. A., Phillips, G. J., Luirink, J. & von Heijne, G. (1996). Assembly of a cytoplasmic membrane protein in *Escherichia coli* is dependent on the signal recognition particle. *FEBS Letters* **399**, 307–309.

de Gier, J. W. L., Valent, Q. A., von Heijne, G. & Luirink, J. (1997). The *E. coli* SRP: preferences of a targeting factor. *FEBS Letters* **408**, 1–4.

Hanein, D., Matlack, K. E. S., Jungnickel, B., Plath, K., Kalies, K.-U., Miller, K. R., Rapoport, T. A. & Akey, C. W. (1996). Oligomeric rings of the Sec61p complex induced by ligands required for protein translocation. *Cell* **87**, 721–732.

Hatsuzawa, K., Tagaya, M. & Mizushima, S. (1997). Hydrophobic region of signal peptides is a determinant for SRP recognition and protein translocation across the ER membrane. *Journal of Biochemistry* **121**, 270–277.

von Heijne, G. (1985). Signal sequences. The limits of variation. *Journal of Molecular Biology* **184**, 99–105.

Hesterkamp, T., Hauser, S., Lütcke, H. & Bukau, B. (1996). *Escherichia coli* trigger factor is a prolyl isomerase that associates with nascent polypeptide chains. *Proceedings of the National Academy of Sciences, USA* **93**, 4437–4441.

High, S. & Dobberstein, B. (1991). The signal sequence interacts with the methionine-rich domain of the signal recognition particle. *Journal of Cell Biology* **113**, 229–233.

High, S. & Laird, V. (1997). Membrane protein biosynthesis – all sewn up? *Trends in Cell Biology* **7**, 206–210.

High, S., Flint, N. & Dobberstein, B. (1991). Requirements for the membrane insertion of signal-anchor type proteins. *Journal of Cell Biology* **113**, 25–34.

High, S., Henry, R., Mould, R. M., Valent, Q. A., Meacock, S., Cline, K., Gray, J. C. & Luirink, J. (1997). Chloroplast SRP54 interacts with a specific subset of thylakoid precursor proteins. *Journal of Biological Chemistry* **272**, 11622–11628.

Johnson, A. E. (1997). Protein translocation at the ER membrane: a complex process becomes more so. *Trends in Cell Biology* **7**, 90–95.

Jungnickel, B. & Rapoport, T. A. (1995). A posttargeting signal sequence recognition event in the endoplasmic reticulum membrane. *Cell* **82**, 261–270.

Keenan, R. J., Freymann, D. M., Walter, P. & Stroud, R. M. (1998). Crystal structure of the signal sequence binding subunit of the signal recognition particle. *Cell* **94**, 181–191.

Kusters, R., Lentzen, G., Eppens, E., van Geel, A., van der Weijden, C. C., Wintermeyer, W. & Luirink, J. (1995). The functioning of the SRP receptor FtsY in protein-targeting in *E. coli* correlated with its ability to bind and hydrolyse GTP. *FEBS Letters* **372**, 253–258.

de Leeuw, E., Poland, D., Mol, O., Sinning, I., ten Hagen-Jongman, C. M., Oudega, B. & Luirink, J. (1997). Membrane association of FtsY, the *E. coli* SRP receptor. *FEBS Letters* **416**, 225–229.

Li, X., Henry, R., Yuan, J., Cline, K. & Hoffman, N. E. (1995). A chloroplast homologue of the signal recognition particle subunit SRP54 is involved in the posttranslational integration of a protein into thylakoid membranes. *Proceedings of the National Academy of Sciences, USA* **92**, 3789–3793.

Luirink, J., High, S., Wood, H., Giner, A., Tollervey, D. & Dobberstein, B. (1992). Signal sequence recognition by an *Escherichia coli* ribonucleoprotein complex. *Nature* **359**, 741–743.

Luirink, J., ten Hagen-Jongman, C. M., van der Weijden, C. C., Oudega, B., High, S., Dobberstein, B. & Kusters, R. (1994). An alternative protein targeting pathway in *Escherichia coli*: studies on the role of FtsY. *EMBO Journal* **13**, 2289–2296.

Lütcke, H. (1995). Signal recognition particle (SRP), a ubiquitous initiator of protein translocation. *European Journal of Biochemistry* **228**, 531–550.

Lütcke, H., High, S., Römisch, K., Ashford, A. J. & Dobberstein, B. (1992). The methionine-rich domain of the 54 kDa subunit of signal recognition particle is sufficient for the interaction with signal sequences. *EMBO Journal* **11**, 1543–1551.

Macao, B., Luirink, J. & Samuelsson, T. (1997). Ffh and FtsY in a *Mycoplasma mycoides* signal-recognition particle pathway: SRP RNA and M domain of Ffh are not required for stimulation of GTPase activity *in vitro*. *Molecular Microbiology* **24**, 523–534.

MacFarlane, J. & Müller, M. (1995). The functional integration of a polytopic membrane protein of *Escherichia coli* is dependent on the bacterial signal-recognition particle. *European Journal of Biochemistry* **233**, 766–771.

Miller, J. D., Wilhelm, H., Gierasch, L., Gilmore, R. & Walter, P. (1993). GTP binding and hydrolysis by the signal recognition particle during initiation of protein translocation. *Nature* **366**, 351–354.

Miller, J. D., Bernstein, H. D. & Walter, P. (1994). Interaction of *E. coli* Ffh/4.5S ribonucleoprotein and FtsY mimics that of mammalian signal recognition particle and its receptor. *Nature* **367**, 657–659.

Miller, J. D., Tajima, S., Lauffer, L. & Walter, P. (1995). The beta subunit of the signal recognition particle receptor is a transmembrane GTPase that anchors the alpha subunit, a peripheral membrane GTPase, to the endoplasmic reticulum membrane. *Journal of Cell Biology* **128**, 273–282.

Montoya, G., Svensson, C., Luirink, J. & Sinning, I. (1997). Crystal structure of the NG domain from the signal-recognition particle receptor FtsY. *Nature* **385**, 365–368.

Neuhof, A., Rolls, M. M., Jungnickel, B., Kalies, K. U. & Rapoport, T. A. (1998). Binding of signal recognition particle gives ribosome/nascent chain complexes a competitive advantage in endoplasmic reticulum membrane interaction. *Molecular Biology of the Cell* **9**, 103–115.

Ng, D. T. W., Brown, J. D. & Walter, P. (1996). Signal sequences specify the targeting route to the endoplasmic reticulum membrane. *Journal of Cell Biology* **134**, 269–278.

Ogg, S. C. & Walter, P. (1995). SRP samples nascent chains for the presence of signal sequences by interacting with ribosomes at a discrete step during translation elongation. *Cell* **81**, 1075–1084.

Ogg, S. C., Barz, W. P. & Walter, P. (1998). A functional GTPase domain, but not its transmembrane domain, is required for function of the SRP receptor β-subunit. *Journal of Cell Biology* **142**, 341–354.

Phillips, G. J. & Silhavy, T. J. (1992). The *E. coli ffh* gene is necessary for viability and efficient protein export. *Nature* **359**, 744–746.

Poritz, M. A., Strub, K. & Walter, P. (1988). Human SRP RNA and *E. coli* 4.5S RNA contain a highly homologous structural domain. *Cell* **55**, 4–6.

Poritz, M. A., Bernstein, H. D., Strub, K., Zopf, D., Wilhelm, H. & Walter, P. (1990). An *E. coli* ribonucleoprotein containing 4.5S RNA resembles mammalian signal recognition particle. *Science* **250**, 1111–1117.

Powers, T. & Walter, P. (1995). Reciprocal stimulation of GTP hydrolysis by two directly interacting GTPases. *Science* **269**, 1422–1424.

Powers, T. & Walter, P. (1997). Co-translational protein targeting catalyzed by the *Escherichia coli* signal recognition particle and its receptor. *EMBO Journal* **16**, 4880–4886.

Raden, D. & Gilmore, R. (1998). Signal recognition particle-dependent targeting of ribosomes to the rough endoplasmic reticulum in the absence and presence of the nascent polypeptide-associated complex. *Molecular Biology of the Cell* **9**, 117–130.

Rapiejko, P. J. & Gilmore, R. (1994). Signal sequence recognition and targeting of ribosomes to the endoplasmic reticulum by the signal recognition particle do not require GTP. *Molecular Biology of the Cell* **5**, 887–897.

Rapiejko, P. J. & Gilmore, R. (1997). Empty site forms of the SRP54 and SRα GTPases mediate targeting of ribosome nascent chain complexes to the endoplasmic reticulum. *Cell* **89**, 703–713.

Rapoport, T. A., Jungnickel, B. & Kutay, U. (1996). Protein transport across the eukaryotic endoplasmic reticulum and bacterial inner membranes. *Annual Review of Biochemistry* **65**, 271–303.

Ribes, V., Römisch, K., Giner, A., Dobberstein, B. & Tollervey, D. (1990). *E. coli* 4.5S RNA is part of a ribonucleoprotein particle that has properties related to signal recognition particle. *Cell* **63**, 591–600.

Römisch, K., Webb, J., Herz, J., Prehn, S., Frank, R., Vingron, M. & Dobberstein, B. (1989). Homology of 54K protein of signal-recognition particle, docking protein and two *E. coli* proteins with putative GTP-binding domains. *Nature* **340**, 478–482.

Römisch, K., Webb, J., Lingelbach, K., Gausepohl, H. & Dobberstein, B. (1990). The 54-kD protein of signal recognition particle contains a methionine-rich RNA binding domain. *Journal of Cell Biology* **111**, 1793–1802.

Schatz, G. & Dobberstein, B. (1996). Common principles of protein translocation across membranes. *Science* **271**, 1519–1526.

Schuenemann, D., Gupta, S., Persello-Cartieaux, F., Klimyuk, V. I., Jones, J. D. G., Nussaume, L. & Hoffman, N. E. (1998). A novel signal recognition particle targets light-harvesting proteins to the thylakoid membranes. *Proceedings of the National Academy of Sciences, USA* **95**, 10312–10316.

Stoller, G., Rücknagel, K. P., Nierhaus, K. H., Schmid, F. X., Fischer, G. & Rahfeld, J.-U. (1995). A ribosome associated peptidyl-prolyl *cis/trans* isomerase identified as the trigger factor. *EMBO Journal* **14**, 4939–4948.

Ulbrandt, N. D., Newitt, J. A. & Bernstein, H. D. (1997). The *E. coli* signal recognition particle is required for the insertion of a subset of inner membrane proteins. *Cell* **88**, 187–196.

Valent, Q. A., Kendall, D. A., High, S., Kusters, R., Oudega, B. & Luirink, J. (1995). Early events in preprotein recognition in *E. coli*: interaction of SRP and trigger factor with nascent polypeptides. *EMBO Journal* **14**, 5494–5505.

Valent, Q. A., de Gier, J. W.-L., von Heijne, G., Kendall, D. A., ten Hagen-Jongman, C. M., Oudega, B. & Luirink, J. (1997). Nascent membrane and presecretory proteins synthesised in *Escherichia coli* associate with signal recognition particle and trigger factor. *Molecular Microbiology* **25**, 53–64.

Valent, Q. A., Scotti, P. A., High, S., de Gier, J.-W. L., von Heijne, G., Lentzen, G., Wintermeyer, W., Oudega, B. & Luirink, J. (1998). The *Escherichia coli* SRP and SecB targeting pathways converge at the translocon. *EMBO Journal* **17**, 2504–2512.

Walter, P. & Johnson, A. E. (1994). Signal sequence recognition and protein targeting to the endoplasmic reticulum membrane. *Annual Review of Cell Biology* **10**, 87–119.

Wickner, W., Driessen, A. J. M. & Hartl, F.-U. (1991). The enzymology of protein translocation across the *Escherichia coli* plasma membrane. *Annual Review of Biochemistry* **60**, 101–124.

Zelazny, A., Seluanov, A., Cooper, A. & Bibi, E. (1997). The NG domain of the prokaryotic signal recognition particle receptor, FtsY, is fully functional when fused to an unrelated integral membrane polypeptide. *Proceedings of the National Academy of Sciences, USA* **94**, 6025–6029.

Zopf, D., Bernstein, H. D., Johnson, A. E. & Walter, P. (1990). The methionine-rich domain of the 54 kd protein subunit of the signal recognition particle contains an RNA binding site and can be crosslinked to a signal sequence. *EMBO Journal* **9**, 4511–4517.

Zopf, D., Bernstein, H. D. & Walter, P. (1993). GTPase domain of the 54-kD subunit of the mammalian signal recognition particle is required for protein translocation but not for signal sequence binding. *Journal of Cell Biology* **120**, 1113–1121.

PROTEIN TRANSLOCATION INTO THE ENDOPLASMIC RETICULUM

BARRY P. YOUNG, JUDY K. BROWNSWORD AND COLIN J. STIRLING

School of Biological Sciences, University of Manchester, Oxford Road, Manchester M13 9PT, UK

INTRODUCTION

The eukaryotic cell is characterized by the presence of membrane-bound compartments, or organelles, that perform a variety of highly specialized functions. The structure and function of each organelle is largely defined by its unique complement of constituent proteins. Thus many proteins which begin their synthesis in the cytoplasm must be accurately targeted to specific compartments, usually by virtue of sorting signals located within the polypeptide chain itself. The delivery of a protein from the cytosol to an organelle involves its translocation across at least one lipid bilayer. Clearly the translocation of large, hydrophilic polypeptides across any lipid bilayer represents a major feat, made all the more difficult by the need to maintain the integrity of the membrane so as to prevent the catastrophic mixing of cytosolic and organellar contents. The endoplasmic reticulum (ER) plays a major role in intracellular protein trafficking. Proteins translocated into the ER can either remain resident there, or may be transported to the Golgi apparatus, the lysosomal network or the cell surface. Transport from the ER occurs by vesicular budding and fusion and does not require cargo proteins to undergo any further membrane translocation events. Protein translocation across the ER membrane is therefore fundamental to organelle biogenesis and protein secretion in eukaryotes. The aim of this review is to highlight some of the most important recent findings that have revolutionized our understanding of this translocation mechanism.

PROTEIN TARGETING TO THE ER

Proteins entering the secretory pathway contain an ER targeting signal, commonly referred to as the signal sequence. This signal sequence is usually located towards the N-terminus and consists of a continuous stretch of hydrophobic residues (6–20) flanked by one or more basic residues to the N-terminal side of the hydrophobic core (von Heijne, 1990). In most cases the signal sequence is cleaved during translocation of the nascent chain and the remainder of the polypeptide is translocated into the lumen of the ER.

Integral membrane proteins are targeted to the ER by either a cleavable signal sequence or a signal-anchor domain that becomes stably integrated in the bilayer (for review see High & Dobberstein, 1992).

Recent studies have demonstrated that proteins can be targeted to the ER either co-translationally or post-translationally (see High & Stirling, 1993). Signal recognition particle (SRP) and its cognate membrane receptor, SRP receptor (SR), mediate the co-translational route in which nascent secretory precursors are targeted to the ER membrane, thus enabling translocation to be coupled to translational elongation (for reviews see Valent and others this volume; Walter & Johnson, 1994). Neither SRP nor SRP receptor is involved in the translocation reaction but rather they deliver the nascent chain/ ribosome complex to the Sec61 protein complex that then mediates the translocation process (Fig. 1).

The post-translational route is independent of SRP/SR but requires cytosolic Hsp70s and Ydj1, which maintain precursors in a translocation-competent conformation (Deshaies et al., 1988; Chirico et al., 1988; Cyr et al., 1992; Caplan et al., 1992; Cyr & Douglas, 1994), plus a further membrane protein complex comprising Sec62p, Sec63p, Sec71p and Sec72p (Deshaies & Schekman, 1987; Rothblatt et al., 1989; Deshaies et al., 1991; Panzner et al., 1995). Precursors are first targeted to this so-called Sec63 complex and then delivered, in an ATP-dependent manner, to the Sec61 complex which again mediates the translocation reaction (Fig. 1). The role of the Sec63 complex in precursor targeting is thus analogous to that of SRP receptor, but the Sec63 protein plays a further role in that it recruits a lumenal Hsp70, Kar2p, to the inner surface of the translocation site. This Hsp70 interacts directly with the translocating polypeptide and is believed to drive the post-translational translocation reaction, which also requires ATP, by sequential rounds of substrate binding and release that effectively 'pull' the pre-protein into the ER lumen (Sanders et al., 1992; Lyman & Schekman, 1995, 1997; Wilkinson et al., 1997a; McClellan et al., 1998). A second lumenal Hsp70, Lhs1p, is also required for efficient post-translational translocation but the precise nature of its role has yet to be determined (Craven et al., 1996, 1997). The hydrophobicity of the signal sequence dictates which of the two targeting pathways a specific precursor will follow, with more hydrophobic sequences being substrates for the SRP pathway (Ng et al., 1996).

COMPONENTS OF THE ER TRANSLOCON

Protein translocation is thought to occur via a protein-conducting channel in the membrane often referred to as the 'translocon' or 'translocase'. The first component of this machinery to be identified was found through genetic studies in the budding yeast Saccharomyces cerevisiae. This led to the identification of the Sec61 protein (Sec61p), a multispanning integral membrane protein of the ER required for the translocation of both secretory

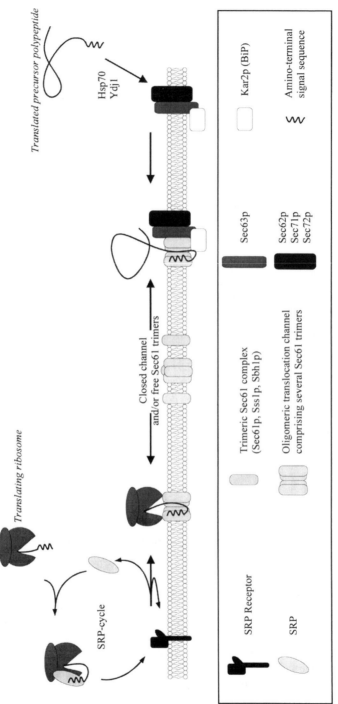

Fig. 1. Protein translocation across the ER membrane in yeast. Nascent polypeptides are targeted to the ER co-translationally by the concerted action of SRP and SRP receptor. The targeted ribosome/mRNA/nascent chain complex then associates with the translocon which comprises three to four copies of the Sec61 complex. Post-translationally targeted preproteins interact initially with components of the Sec63 complex before being delivered to the Sec61-containing translocon. It is not yet certain whether the translocon assembles *de novo* from free Sec61 trimers only in the presence of precursor, or whether it exists in a tightly closed state prior to the arrival of the targeted precursor.

and membrane proteins (Deshaies & Schekman, 1987; Stirling *et al.*, 1992; Wilkinson *et al.*, 1996). Cross-linking experiments have shown that a translocating polypeptide is in intimate contact with Sec61p during its membrane translocation, suggesting that Sec61p corresponds to a component of the translocon per se (Musch *et al.*, 1992; Sanders *et al.*, 1992). A mammalian homologue of Sec61p (Sec61α) sharing 56% identity with the yeast protein has been identified which can also be cross-linked to translocating chains (High *et al.*, 1991, 1993a, b; Gorlich *et al.*, 1992b; High & Stirling, 1993) and is in tight association with membrane-bound ribosomes during translocation (Gorlich *et al.*, 1992b; Kalies *et al.*, 1994). Sec61α has been isolated from dog pancreas microsomes as a complex with two other integral membrane proteins, Sec61β and Sec61γ, both belonging to the C-terminal anchor class (Gorlich & Rapoport, 1993; Hartmann *et al.*, 1994). A similar Sec61 complex is found in yeast where Sec61p interacts directly with Ss s1p (homologous to Sec61γ) (Esnault *et al.*, 1993, 1994; Wilkinson *et al.*, 1997b), and a Sec61β homologue, Sbh1p (Panzner *et al.*, 1995). Significantly, mammalian Sss1p/Sec61γ has been shown to function in budding yeast, as has the Sec61p/Sec61α homologue from *Yarrowia lipolytica*, emphasizing the extent of functional conservation of individual components of the Sec61 complex amongst eukaryotes (Hartmann *et al.*, 1994; Broughton *et al.*, 1997). The tripartite nature of the Sec61 complex also draws comparison with the molecular nature of the SecY/E/G translocase of *Escherichia coli* (Brundage *et al.*, 1990; Douville *et al.*, 1994). Bacterial SecY appears to be related to Sec61p/Sec61α (Stirling, 1993; Stirling *et al.*, 1992), whilst SecE shares a lesser degree of sequence identity with Sss1p/Sec61γ (Hartmann *et al.*, 1994). Interestingly, bacterial SecG and yeast Sbh1p show no convincing sequence similarity but, perhaps significantly, neither of these components is essential for viability in their respective organisms (Finke *et al.*, 1996; Nishiyama *et al.*, 1994).

Multiple roles for the Sec61 complex

The genetic data and cross-linking evidence discussed above strongly suggest a direct role for the Sec61 complex in protein translocation. This has been confirmed by the reconstitution of protein translocation using purified components. Co-translational translocation of proteins into mammalian-derived proteoliposomes requires only SRP, SR and the Sec61 complex, although some precursors display an additional requirement for the presence of the TRAM protein (Gorlich *et al.*, 1992a; Gorlich & Rapoport, 1993; Oliver *et al.*, 1995; Voigt *et al.*, 1996). The role of TRAM is uncertain but it has been suggested that it may be involved in regulating the activity of the translocon (Hegde *et al.*, 1998c). Interestingly, no TRAM homologue is encoded by the yeast genome, indicating that specialized components may have evolved in some systems.

The Sec61 complex co-purifies with ribosomes from detergent-solubilized membranes and these ribosomes can be removed by treatment with puromycin and high salt (Gorlich et al., 1992b; Gorlich & Rapoport, 1993; Kalies et al., 1994). The same treatment had been previously demonstrated to remove ribosomes from rough microsomal membranes (Adelman et al., 1973). Puromycin stimulates the premature release of nascent polypeptides from the translating ribosome and thus implicates the nascent chain itself in the ribosome/Sec61p interaction. This is consistent with a direct interaction between the Sec61 complex and the ribosome/nascent chain complex thereby providing a molecular basis for coupling between translation and translocation.

In addition to its interaction with ribosome the Sec61 protein also interacts specifically with signal sequence prior to the initiation of translocation. This allows the translocon to proof-read the signal sequence on a targeted precursor thus enhancing the accuracy of the overall targeting process (Jungnickel & Rapoport, 1995; Plath et al., 1998). It seems highly likely that lipids also play a role in discriminating signal sequences, especially given that there is a strong correlation between the hydrophobicity of a peptide sequence and both its ability to interact with lipid and to function as a signal sequence (Hoyt & Gierasch, 1991). Moreover, photocross-linking studies have shown that lipids are in close contact with signal sequence during translocation, leading to the proposal that the protein-conducting channel is open laterally towards the lipid bilayer during an early stage of protein insertion (Martoglio et al., 1995).

In addition to co-purifying with ribosomes the yeast Sec61 complex (Sec61p, Sss1p and Sbh1p) can also be co-purified with the Sec63 complex (Sec62p, Sec63p, Sec71p and Sec72p; see Fig. 1). This combined heptameric complex together with Kar2p is sufficient to reconstitute the post-translational translocation reaction into proteoliposomes (Panzner et al., 1995; Matlack et al., 1997).

NATURE OF THE TRANSLOCATION CHANNEL

Until recently, it was assumed that the translocation channel would be so narrow as to accommodate only unfolded polypeptide chains. This assumption seemed consistent with the co-translational translocation of nascent polypeptides, and with the perceived role of cytosolic Hsp70s in preventing precursor folding prior to translocation by the post-translational route. However, a variety of recent data indicate that the channel is very much larger than expected. Firstly, electrophysiological techniques identified a high-conductance aqueous channel that was opened when partially synthesized proteins were released with puromycin under conditions where the ribosome remained membrane-bound. Removal of the ribosome by increasing salt concentration led to closure of this channel observed as a loss of conductance

(Simon & Blobel, 1991). Fluorophore quenching experiments also revealed the presence of an aqueous channel with a diameter of between 40 and 60 Å formed in the presence of a ribosome/nascent chain complex (Hamman et al., 1997). This would be the largest hole identified to date in any semi-permeable biological membrane. The translocation of a polypeptide through such a large channel raises a number of interesting questions, the most obvious being how could such a large pore be regulated so as to remain impermeable even to ions? The regulation of the channel will be discussed in the following section, but first another unexpected possibility is that this very large channel might provide a protected environment within which the translocating polypeptide might begin to fold. To put this in perspective, biophysical measurements indicate that a variety of polypeptides behave as spheres in aqueous solution with diameters smaller than that of the ER translocation channel. For example, ribonuclease (14 kDa) and ovalbumin (45 kDa) have effective diameters in aqueous solution of 40 Å and 56 Å, respectively (Tanford, 1961). It would therefore appear that the channel is large enough to accommodate very substantial degrees of protein folding. Some degree of folding during translocation might be desirable as opposed to the simple extrusion of an unfolded polypeptide into the milieu of the ER lumen where non-productive protein–protein interactions might be more likely. Moreover, the formation of secondary structure may be essential prior to the translocating chain becoming exposed to the modifying enzymes located within the lumenal compartment. This might favour the formation of disulphide bonds only between appropriate cysteine residues, or the addition of asparagine-linked core oligosaccharides only to those glycosylation consensus sequences (Asn-x-Ser/Thr) located on the exposed surfaces of the protein.

STRUCTURE OF THE TRANSLOCON AND GATING OF THE CHANNEL

Current data suggest that the translocon is a dynamic structure that assembles and opens only in the presence of a suitable substrate. Recent images of the purified Sec61 complex suggest that it forms an oligomeric ring structure comprising two to four Sec61 trimers. The overall diameter of the ring is some 85–95 Å, with the internal pore having a mean diameter of 20–35 Å (Hanein et al., 1996; Beckmann et al., 1997). Beckmann et al. (1997) have produced startling images of the yeast Sec61-ring complexed with the ribosome (Fig. 2). The central pore of the Sec61 oligomer is seen to align with the site on the large ribosomal subunit that is believed to correspond to the exit channel for the nascent polypeptide. The translocon channel can therefore be viewed as an extension of the ribosome with the translating polypeptide having nowhere to go but through the transmembrane channel.

The apparent 'hole' in the Sec61 doughnut structure observed by cryo-electron microscopy is smaller (20–35 Å) than the ion-conductance channel discussed above (40–60 Å). This might be due to the particular cryofixation

(a)   (b)

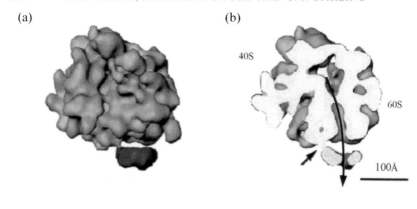

(c)

(i)   (ii)   (iii)

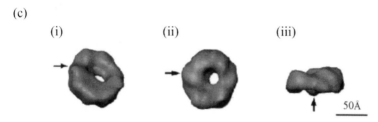

Fig. 2. Three-dimensional reconstruction of the ribosome/Sec61 complex. (a) Side view with the ribosome shown in light shading and the Sec61 oligomer in darker shading at the bottom of the image. (b) The same as in (a) but cut along a plane that cross-sections the Sec61 oligomer and the ribosome tunnel through which the nascent polypeptide is believed to pass (indicated by a long arrow). The point of attachment between the ribosome and Sec61 oligomer is indicated by a short arrow. (c) Close-up view of the Sec61 oligomer. (i) Surface facing the ribosome; (ii) surface facing away from the ribosome; (iii) side view. The arrows indicate the point of ribosomal attachment seen in (a) and (b). Reprinted with kind permission from Beckmann, R., Bubeck, D., Grassucci, R., Penczek, P., Verschoor, A., Blobel, G. & Frank, J. (1997). Alignment of conduits for the nascent polypeptide chain in the ribosome-Sec61 complex. *Science* **278**, 2123–2126. Copyright 1997 American Association for the Advancement of Science.

or staining procedures used. Alternatively, the complexes imaged by cryo-electron microscopy may represent the closed state of a dynamic channel. Interestingly, Hanein *et al.* (1996) observed a similar Sec61 doughnut structure in both yeast and mammals and most strikingly they found that this ring is induced by the presence of either ribosome or the Sec63 complex. This observation suggests that individual Sec61 trimers (comprising Sec61p, Sss1p and Sbh1p) are recruited to form a functional oligomeric translocon only in the presence of a suitable ligand. It is clear that ribosome alone can hold a channel in its open state after the release of the nascent chain by puromycin (as discussed above), but current evidence indicates that the precursor is required to promote channel opening. Johnson and co-workers have examined the accessibility of fluorescently tagged nascent chains to fluorophore-quenching ions from either the cytosolic or lumenal face of microsomal membranes. Their data confirm that the translocation channel can be sealed at the cytosolic face (presumably at the ribosome/Sec61 junction) (Crowley *et al.*, 1993, 1994), but furthermore that the aqueous channel remains sealed to the ER lumen until the nascent polypeptide reaches a length of ≈ 70 amino acids (Crowley *et al.*, 1994), which is highly suggestive of a role for the signal sequence in initiating channel opening (Fig. 3). The authors have proposed that the lumenal seal might provide a safety mechanism that would maintain the permeability barrier of the ER membrane by preventing channel opening until the cytosolic face of the channel is sealed by a properly engaged ribosome. The lumenal seal depends upon the presence of active BiP (Crowley *et al.*, 1994; Hamman *et al.*, 1998). In yeast, BiP/Kar2p has been shown to be recruited to the yeast post-translational translocon by Sec63p, but it is not known how it might be recruited in the co-translational reaction. Moreover, it remains to be established what factors might replace the ribosome in sealing the cytosolic face of the channel in the post-translational reaction. Obvious candidates might include the cytosolic domains of the Sec63 complex, or even the cytosolic Hsp70s associated with the targeted precursor, but any such role remains speculative.

MEMBRANE PROTEIN INSERTION

The ER represents the major site of membrane biogenesis in eukaryotes. Many membrane proteins are first inserted into the ER bilayer and then transported to their final destination by membrane-bound transport vesicles. Most membrane protein precursors appear to be targeted to the ER by SRP and are thus inserted into the membrane co-translationally. Thus the initial targeting of proteins to the ER is indistinguishable from that of SRP-dependent secretory proteins. Moreover, the insertion process clearly involves the same trimeric Sec61 complex as that involved in secretory protein translocation (Stirling *et al.*, 1992; Gorlich & Rapoport, 1993; Oliver *et al.*, 1995; Laird & High, 1997). However, the two processes can be distinguished by the fact that during membrane protein insertion, the polypeptide chain must, at some stage, pass laterally from the translocation channel into the lipid phase to allow the integration of transmembrane domains. In principle, one might imagine that this process is rather straightforward, requiring only the partitioning of a hydrophobic domain into the lipid bilayer driven by obvious thermodynamic forces. However, cross-linking studies suggest that the mechanism is more complex, requiring the nascent protein to interact with several proteinaceous components prior to its eventual release into the lipid (Do *et al.*, 1996; Laird & High, 1997). More recently, Hegde *et al.* (1998a) have reported that the human prion protein can exist in several forms including a secreted species and at least two topologically distinct integral membrane forms. The membrane insertion of

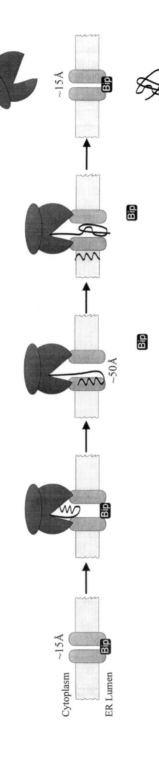

Fig. 3. Dynamic gating of the translocation channel at both the cytosolic and lumenal faces. In this model BiP seals the lumenal face of a closed translocation channel prior to initiation of translocation. The ribosome binds to the cytosolic face of the translocon where it forms a tight seal. The lumenal seal remains intact until the precursor polypeptide is some 70 residues long, whereupon a 40–60 Å channel opens to the lumen. The location/role of BiP during this stage is not known. Release of the ribosome upon completion of translation leaves a 9–15 Å aqueous pore that is open to the cytoplasm but that is tightly sealed at the lumenal face in a manner that is dependent upon the presence of BiP.

PrP into reconstituted membranes requires the Sec61 complex and TRAM, but also requires unidentified ER components termed _Translocation Accessory Factor(s)_ (TrAFs) (Hegde _et al._, 1998b). The extent to which these TrAFs might be generally involved in membrane protein assembly is currently unknown.

Orientation of membrane domain insertion

Some membrane proteins possess a cleavable signal sequence followed by a hydrophobic _stop-transfer_ domain that integrates into the bilayer thus terminating the translocation of C-terminal sequences (High & Laird, 1997). Alternatively, targeting may be via a non-cleavable _signal-anchor_ domain that constitutes the first transmembrane sequence in the mature protein. A signal sequence is presumed to initiate translocation by inserting into the membrane in a loop conformation, thus exposing the C-terminal signal cleavage site to signal peptidase activity within the lumen (see Wilkinson _et al._, 1997a). However, signal-anchor domains come in two guises: those that insert with their N-terminus in the cytosol, and those with their N-terminus in the lumen (see High & Laird, 1997). Exhaustive studies on the _cis_-acting factors which dictate the orientation of a signal-anchor domain have demonstrated the significance of positively charged residues which tend to be clustered on the cytosolic side of the membrane (Sipos & von Heijne, 1993). This so-called _positive inside rule_ appears valid in both prokaryotes and in eukaryotes. In bacteria it has been suggested that it may simply mirror the charge differential across the plasma membrane (i.e. negative inside). However, there is no known charge differential across the eukaryotic ER membrane, therefore any _trans_-acting components required to decode the polarity of membrane protein insertion remain to be defined. Intriguingly, the translocase in the bacterial plasma membrane is clearly related to the Sec61 complex, suggesting that the regulation of these events may be similar throughout biology (Stirling, 1993).

Multispanning membrane protein integration

The insertion of many multispanning integral membrane proteins can be conveniently explained in terms of the signal-anchor/stop-transfer hypothesis (Wessels & Spiess, 1988). In this model, the orientation of the first transmembrane sequence determines the relative orientation of all subsequent transmembrane sequences which insert sequentially, and alternately, as either signal-anchor or stop-transfer domains. During this process each stop-transfer domain terminates translocation, thus requiring each subsequent signal-anchor domain to reinitiate insertion. Although this model predicts the independent insertion of each transmembrane domain, there are several examples where specific transmembrane domains are required to interact closely with downstream transmembrane sequences in order to integrate stably in the bilayer (Sengstag *et al.*, 1990; Wilkinson *et al.*, 1996). Indeed, in some cases it is suggested that closely spaced transmembrane domains insert simultaneously into the bilayer as a helical hairpin (Gafvelin & von Heijne, 1994; Gafvelin *et al.*, 1997).

An unexpected role for the ribosome?

Data discussed above indicate that the translocation pore can be gated at either its cytosolic or lumenal face. Evidence suggests that the lumenal gate remains closed until a nascent polypeptide is ≈ 70 residues long (Crowley *et al.*, 1994). At this point the lumenal gate opens, but the integrity of the channel is maintained by the cytosolic seal. However, the question of gating becomes more complex when one considers the co-translational integration of a multispanning membrane protein. The initial insertion of a type II membrane domain (i.e. C-terminus in the lumen) and the subsequent translocation of its lumenal domain might be very similar to secretory protein translocation. However, for multispanning proteins, translocation is terminated by stop-transfer sequences, and cytosolic domains must presumably be translated directly into the cytosol. The next signal-anchor domain then reinitiates the translocation process. Thus any gating mechanism must be versatile enough to maintain the integrity of the channel as it switches from signal-anchor to stop-transfer integration modes. Perhaps the most striking observation of recent years has been the finding that both the lumenal and cytosolic gates appear to be regulated from within the ribosome (Liao *et al.*, 1997). This has led to the suggestion that integral membrane domains are first recognized, not by the Sec61 complex, but within the translating ribosome. It has been further proposed that this recognition event may trigger a long-range conformational change that signals a switch in the translocon between translocation mode and membrane insertion model (Liao *et al.*, 1997). The close association between the ribosome and Sec61p during the translocation process might mediate such a signal (Gorlich *et al.*, 1992b). The nature of this recognition event, and the associated switch, are currently unknown.

TWO-WAY TRAFFIC ACROSS THE ER MEMBRANE

When proteins are translocated into the ER they may be subject to a variety of covalent modifications including signal-peptide cleavage, glycosylation, disulphide bond formation, etc. These modifications often occur during the chaperone-assisted folding of the precursor as it attains its functional conformation. A failure in any stage of protein folding or modification may lead to aggregation or to degradation of the malfolded polypeptide. Recent studies have indicated that this degradation takes place not in the ER, as expected, but rather in the cytosol, where it requires the activity of the cytosolic ubiquitin-proteasome pathway (Hampton *et al.*, 1996; Hiller *et al.*, 1996; Hilt & Wolf, 1996; McCracken & Brodsky, 1996; Werner *et al.*, 1996; Wiertz *et al.*, 1996; Hazes & Read, 1997). Of course this would require that malfolded polypeptides are translocated from the ER lumen to the cytosol by a process that has been variously termed dislocation, retrotranslocation or retrograde transport. Current evidence implicates components of the Sec61 complex in the dislocation process. For example, mammalian membrane proteins destined for degradation have been shown to be transiently associated with the Sec61β component of the Sec61 complex (Wiertz *et al.*, 1996; Bebok *et al.*, 1998). In addition, a number of mutant alleles of yeast *SEC61* have been shown to be defective in the degradation of malfolded polypeptides both *in vivo* and *in vitro* (Pilon *et al.*, 1997; Plemper *et al.*, 1997). Studies in yeast have also implicated ER-lumenal chaperones, including calnexin and Kar2p (BiP), in the <u>ER</u>-<u>A</u>ssociated <u>D</u>egradation (ERAD) of malfolded proteins (McCracken & Brodsky, 1996; Plemper *et al.*, 1997). The roles played by these chaperones in the ERAD pathway remain uncertain. Mammalian calnexin has been shown to participate in a 'quality-control' pathway which involves its binding to incompletely folded polypeptides that are tagged by virtue of differential processing of sugar residues on *N*-linked oligosaccharide side chains (for a recent review see Trombetta & Helenius, 1998). Any direct role for calnexin in dislocation remains speculative but it is intriguing to note that the rate of ERAD in yeast is determined by the structure of the oligosaccharides present on a glycoprotein substrate, suggesting that a specific lectin might be involved in the targeting of proteins for degradation (Jakob *et al.*, 1998). The involvement of Kar2p/BiP might simply reflect this Hsp70's general role as a molecular chaperone involved in the binding and ER-retention of malfolded polypeptide. However, there are also genetic data implicating *SEC63* in the ERAD pathway leading to the seductive hypothesis that the dislocation reaction may correspond to a reversal of the post-translational translocation mechanism known to require Kar2p, Sec63p and Sec61p. Several genetic screens have been designed to identify components of the ERAD pathway and have led to the identification of genes encoding novel ER membrane proteins such as Der1p, Der3p/Hrd1p or Hrd3p (Bordallo *et al.*, 1998; Hampton *et al.*, 1996; Knop *et al.*, 1996), and it has been suggested that these novel factors might function in switching the translocon into dislocation mode.

IN CONCLUSION

Considerable advances have been made both in identifying components of the ER translocon and in understanding the role that each plays in the translocation reaction. This has been achieved using a combination of approaches ranging from yeast genetics to biochemical reconstitution. Nonetheless several key questions remain. For example, it is not yet known whether the post-translational targeting route requires any cytosolic targeting factor(s) analogous to SRP. Nor is it clear how the translocation channel can be sealed at the cytosolic face during the post-translational translocation reaction. Moreover, the very existence of a Sec62/Sec63-dependent post-translational pathway in mammals has yet to be established. In the co-translational reaction the role of the ribosome becomes ever more complex. A molecular description of its interaction with the translocon and its apparent role in modulating the integration of multispanning membrane proteins will be vital to our understanding of this most crucial aspect of membrane biogenesis. Finally, the dimensions of the translocation channel raise new and unexpected possibilities, most notably the concept that translocating polypeptides might fold extensively within a protected environment prior to release into the ER lumen. Another surprise has been the finding that the translocation channel also functions in reverse to facilitate the degradation of misfolded proteins. If the present rate of progress in this area continues then we might expect a few more surprises in the near future.

REFERENCES

Adelman, M. R., Sabatini, D. D. & Blobel, G. (1973). Ribosome-membrane interaction. Nondestructive disassembly of rat liver rough microsomes into ribosomal and membranous components. *Journal of Cell Biology* **56**, 206–229.

Bebok, Z., Mazzochi, C., King, S. A., Hong, J. S. & Sorscher, E. J. (1998). The mechanism underlying cystic fibrosis transmembrane conductance regulator transport from the endoplasmic reticulum to the proteasome includes Sec61 beta and a cytosolic, deglycosylated intermediary. *Journal of Biological Chemistry* **273**, 29873–29878.

Beckmann, R., Bubeck, D., Grassucci, R., Penczek, P., Verschoor, A., Blobel, G. & Frank, J. (1997). Alignment of conduits for the nascent polypeptide chain in the ribosome-Sec61 complex. *Science* **278**, 2123–2126.

Bordallo, J., Plemper, R. K., Finger, A. & Wolf, D. H. (1998). Der3p/Hrd1p is required for endoplasmic reticulum-associated degradation of misfolded lumenal and integral membrane proteins. *Molecular Biology of the Cell* **9**, 209–222.

Broughton, J., Swennen, D., Wilkinson, B. M., Joyet, P., Gaillardin, C. & Stirling, C. J. (1997). Cloning of SEC61 homologues from *Schizosaccharomyces pombe* and *Yarrowia lipolytica* reveals the extent of functional conservation within this core component of the ER translocation machinery. *Journal of Cell Science* **110**, 2715–2727.

Brundage, L., Hendrick, J. P., Schiebel, E., Driessen, A. J. & Wickner, W. (1990). The purified E. coli integral membrane protein SecY/E is sufficient for reconstitution of SecA-dependent precursor protein translocation. *Cell* **62**, 649–657.

Caplan, A. J., Cyr, D. M. & Douglas, M. G. (1992). Ydj1p facilitates polypeptide translocation across different intracellular membranes by a conserved mechanism. *Cell* **71**, 1143–1155.

Chirico, W. J., Waters, M. G. & Blobel, G. (1988). 70K heat shock related proteins stimulate protein translocation into microsomes. *Nature* **332**, 805–810.

Craven, R. A., Egerton, M. & Stirling, C. J. (1996). A novel Hsp70 of the yeast ER lumen is required for the efficient translocation of a number of protein precursors. *EMBO Journal* **15**, 2640–2650.

Craven, R. A., Tyson, J. R. & Stirling, C. J. (1997). A novel subfamily of Hsp70s in the endoplasmic reticulum. *Trends in Cell Biology* **7**, 277–282.

Crowley, K. S., Reinhart, G. D. & Johnson, A. E. (1993). The signal sequence moves through a ribosomal tunnel into a noncytoplasmic aqueous environment at the ER membrane early in translocation. *Cell* **73**, 1101–1115.

Crowley, K. S., Liao, S., Worrell, V. E., Reinhart, G. D. & Johnson, A. E. (1994). Secretory proteins move through the endoplasmic reticulum membrane via an aqueous, gated pore. *Cell* **78**, 461–471.

Cyr, D. M. & Douglas, M. G. (1994). Differential regulation of Hsp70 subfamilies by the eukaryotic DnaJ homologue YDJ1. *Journal of Biological Chemistry* **269**, 9798–9804.

Cyr, D. M., Lu, X. Y. & Douglas, M. G. (1992). Regulation of Hsp70 function by a eukaryotic DnaJ homolog. *Journal of Biological Chemistry* **267**, 20927–20931.

Deshaies, R. J. & Schekman, R. (1987). A yeast mutant defective at an early stage in import of secretory protein precursors into the endoplasmic reticulum. *Journal of Cell Biology* **105**, 633–645.

Deshaies, R. J., Koch, B. D., Werner-Washburne, M., Craig, E. A. & Schekman, R. (1988). A subfamily of stress proteins facilitates translocation of secretory and mitochondrial precursor polypeptides. *Nature* **332**, 800–805.

Deshaies, R. J., Sanders, S. L., Feldheim, D. A. & Schekman, R. (1991). Assembly of yeast Sec proteins involved in translocation into the endoplasmic reticulum into a membrane-bound multisubunit complex. *Nature* **349**, 806–808.

Do, H., Falcone, D., Lin, J., Andrews, D. W. & Johnson, A. E. (1996). The cotranslational integration of membrane proteins into the phospholipid bilayer is a multistep process. *Cell* **85**, 369–378.

Douville, K., Leonard, M., Brundage, L., Nishiyama, K., Tokuda, H., Mizushima, S. & Wickner, W. (1994). Band 1 subunit of *Escherichia coli* preprotein translocase and integral membrane export factor P12 are the same protein. *Journal of Biological Chemistry* **269**, 18705–18707.

Esnault, Y., Blondel, M. O., Deshaies, R. J., Scheckman, R. & Kepes, F. (1993). The yeast *SSS1* gene is essential for secretory protein translocation and encodes a conserved protein of the endoplasmic reticulum. *EMBO Journal* **12**, 4083–4093.

Esnault, Y., Feldheim, D., Blondel, M. O., Schekman, R. & Kepes, F. (1994). *SSS1* encodes a stabilizing component of the Sec61 subcomplex of the yeast protein translocation apparatus. *Journal of Biological Chemistry* **269**, 27478–27485.

Finke, K., Plath, K., Panzner, S., Prehn, S., Rapoport, T. A., Hartmann, E. & Sommer, T. (1996). A second trimeric complex containing homologs of the Sec61p complex functions in protein transport across the ER membrane of *S. cerevisiae*. *EMBO Journal* **15**, 1482–1494.

Gafvelin, G. & von Heijne, G. (1994). Topological "frustration" in multispanning E. coli inner membrane proteins. *Cell* **77**, 401–412.

Gafvelin, G., Sakaguchi, M., Andersson, H. & von Heijne, G. (1997). Topological rules for membrane protein assembly in eukaryotic cells. *Journal of Biological Chemistry* **272**, 6119–6127.

Gorlich, D. & Rapoport, T. A. (1993). Protein translocation into proteoliposomes reconstituted from purified components of the endoplasmic reticulum membrane. *Cell* **75**, 615–630.

Gorlich, D., Hartmann, E., Prehn, S. & Rapoport, T. A. (1992a). A protein of the endoplasmic reticulum involved early in polypeptide translocation. *Nature* **357**, 47–52.

Gorlich, D., Prehn, S., Hartmann, E., Kalies, K. U. & Rapoport, T. A. (1992b).

A mammalian homolog of SEC61p and SECYp is associated with ribosomes and nascent polypeptides during translocation. *Cell* **71**, 489–503.

Hamman, B. D., Chen, J. C., Johnson, E. E. & Johnson, A. E. (1997). The aqueous pore through the translocon has a diameter of 40–60 angstrom during co-translational protein translocation at the ER membrane. *Cell* **89**, 535–544.

Hamman, B. D., Hendershot, L. M. & Johnson, A. E. (1998). BiP maintains the permeability barrier of the ER membrane by sealing the lumenal end of the translocon pore before and early in translocation. *Cell* **92**, 747–758.

Hampton, R. Y., Gardner, R. G. & Rine, J. (1996). Role of 26S proteasome and HRD genes in the degradation of 3-hydroxy-3-methylglutaryl-CoA reductase, an integral endoplasmic reticulum membrane protein. *Molecular Biology of the Cell* **7**, 2029–2044.

Hanein, D., Matlack, K. E., Jungnickel, B., Plath, K., Kalies, K. U., Miller, K. R., Rapoport, T. A. & Akey, C. W. (1996). Oligomeric rings of the Sec61p complex induced by ligands required for protein translocation. *Cell* **87**, 721–732.

Hartmann, E., Sommer, T., Prehn, S., Gorlich, D., Jentsch, S. & Rapoport, T. A. (1994). Evolutionary conservation of components of the protein translocation complex. *Nature* **367**, 654–657.

Hazes, B. & Read, R. J. (1997). Accumulating evidence suggests that several AB-toxins subvert the endoplasmic reticulum-associated protein degradation pathway to enter target cells. *Biochemistry* **36**, 11051–11054.

Hegde, R. S., Mastrianni, J. A., Scott, M. R., Defea, K. A., Tremblay, P., Torchia, M., DeArmond, S. J., Prusiner, S. B. & Lingappa, V. R. (1998a). A transmembrane form of the prion protein in neurodegenerative disease. *Science* **279**, 827–834.

Hegde, R. S., Voigt, S. & Lingappa, V. R. (1998b). Regulation of protein topology by *trans*-acting factors at the endoplasmic reticulum. *Molecular Cell* **2**, 85–91.

Hegde, R. S., Voigt, S., Rapoport, T. A. & Lingappa, V. R. (1998c). TRAM regulates the exposure of nascent secretory proteins to the cytosol during translocation into the endoplasmic reticulum. *Cell* **92**, 621–631.

von Heijne, G. (1990). The signal peptide. *Journal of Membrane Biology* **115**, 195–201.

High, S. & Dobberstein, B. (1992). Mechanisms that determine the transmembrane disposition of proteins. *Current Opinion in Cell Biology* **4**, 581–586.

High, S. & Laird, V. (1997). Membrane protein biosynthesis – all sewn up? *Trends in Cell Biology* **7**, 206–210.

High, S. & Stirling, C. J. (1993). Protein translocation across membranes: common themes in divergent organisms. *Trends in Cell Biology* **3**, 335–339.

High, S., Gorlich, D., Wiedmann, M., Rapoport, T. A. & Dobberstein, B. (1991). The identification of proteins in the proximity of signal-anchor sequences during their targeting to and insertion into the membrane of the ER. *Journal of Cell Biology* **113**, 35–44.

High, S., Andersen, S. S., Gorlich, D., Hartmann, E., Prehn, S., Rapoport, T. A. & Dobberstein, B. (1993a). Sec61p is adjacent to nascent type I and type II signal-anchor proteins during their membrane insertion. *Journal of Cell Biology* **121**, 743–750.

High, S., Martoglio, B., Gorlich, D. & 7 other authors (1993b). Site-specific photocross-linking reveals that Sec61p and TRAM contact different regions of a membrane-inserted signal sequence. *Journal of Biological Chemistry* **268**, 26745–26751.

Hiller, M. M., Finger, A., Schweiger, M. & Wolf, D. H. (1996). ER degradation of a misfolded luminal protein by the cytosolic ubiquitin-proteasome pathway. *Science* **273**, 1725–1728.

Hilt, W. & Wolf, D. H. (1996). Proteasomes: destruction as a programme. *Trends in Biochemical Sciences* **21**, 96–102.

Hoyt, D. W. & Gierasch, L. M. (1991). Hydrophobic content and lipid interactions of wild-type and mutant OmpA signal peptides correlate with their in vivo function. *Biochemistry* **30**, 10155–10163.

Jakob, C. A., Burda, P., Roth, J. & Aebi, M. (1998). Degradation of misfolded endoplasmic reticulum glycoproteins in *Saccharomyces cerevisiae* is determined by a specific oligosaccharide structure. *Journal of Cell Biology* **142**, 1223–1233.

Jungnickel, B. & Rapoport, T. A. (1995). A post-targeting signal sequence recognition event in the endoplasmic reticulum membrane. *Cell* **82**, 261–270.

Kalies, K. U., Gorlich, D. & Rapoport, T. A. (1994). Binding of ribosomes to the rough endoplasmic reticulum mediated by the Sec61p-complex. *Journal of Cell Biology* **126**, 925–934.

Knop, M., Finger, A., Braun, T., Hellmuth, K. & Wolf, D. H. (1996). Der1, a novel protein specifically required for endoplasmic reticulum degradation in yeast. *EMBO Journal* **15**, 753–763.

Laird, V. & High, S. (1997). Discrete cross-linking products identified during membrane protein biosynthesis. *Journal of Biological Chemistry* **272**, 1983–1989.

Liao, S. R., Lin, J. L., Do, H. & Johnson, A. E. (1997). Both lumenal and cytosolic gating of the aqueous ER translocon pore are regulated from inside the ribosome during membrane protein integration. *Cell* **90**, 31–41.

Lyman, S. K. & Schekman, R. (1995). Interaction between BiP and Sec63p is required for the completion of protein translocation into the ER of *Saccharomyces cerevisiae*. *Journal of Cell Biology* **131**, 1163–1171.

Lyman, S. K. & Schekman, R. (1997). Binding of secretory precursor polypeptides to a translocon sub-complex is regulated by BiP. *Cell* **88**, 85–96.

McClellan, A. J., Endres, J. B., Vogel, J. P., Palazzi, D., Rose, M. D. & Brodsky, J. L. (1998). Specific molecular chaperone interactions and an ATP-dependent conformational change are required during posttranslational protein translocation into the yeast ER. *Molecular Biology of the Cell* **9**, 3533–3545.

McCracken, A. A. & Brodsky, J. L. (1996). Assembly of ER-associated protein degradation in vitro: dependence on cytosol, calnexin, and ATP. *Journal of Cell Biology* **132**, 291–298.

Martoglio, B., Hofmann, M. W., Brunner, J. & Dobberstein, B. (1995). The protein-conducting channel in the membrane of the endoplasmic reticulum is open laterally toward the lipid bilayer. *Cell* **81**, 207–214.

Matlack, K. S., Plath, K., Misselwitz, B. & Rapoport, T. A. (1997). Protein transport by purified yeast Sec complex and Kar2p without membranes. *Science* **277**, 938–941.

Musch, A., Wiedmann, M. & Rapoport, T. A. (1992). Yeast Sec proteins interact with polypeptides traversing the endoplasmic reticulum membrane. *Cell* **69**, 343–353.

Ng, D. T., Brown, J. D. & Walter, P. (1996). Signal sequences specify the targeting route to the endoplasmic reticulum membrane. *Journal of Cell Biology* **134**, 269–278.

Nishiyama, K., Hanada, M. & Tokuda, H. (1994). Disruption of the gene encoding p12 (SecG) reveals the direct involvement and important function of SecG in the protein translocation of *Escherichia coli* at low temperature. *EMBO Journal* **13**, 3272–3277.

Oliver, J., Jungnickel, B., Gorlich, D., Rapoport, T. & High, S. (1995). The Sec61 complex is essential for the insertion of proteins into the membrane of the endoplasmic reticulum. *FEBS Letters* **362**, 126–130.

Panzner, S., Dreier, L., Hartmann, E., Kostka, S. & Rapoport, T. (1995). Post-translational protein transport in yeast reconstituted with a purified complex of Sec proteins and Kar2p. *Cell* **81**, 561–570.

Pilon, M., Schekman, R. & Romisch, K. (1997). Sec61p mediates export of a misfolded secretory protein from the endoplasmic reticulum to the cytosol for degradation. *EMBO Journal* **16**, 4540–4548.

Plath, K., Mothes, W., Wilkinson, B. M., Stirling, C. J. & Rapoport, T. A. (1998). Signal sequence recognition in post-translational protein translocation across the yeast ER membrane. *Cell* **94**, 795–807.

Plemper, R. K., Bohmler, S., Bordallo, J., Sommer, T. & Wolf, D. H. (1997). Mutant analysis links the translocon and BiP to retrograde protein transport for ER degradation. *Nature* **388**, 891–895.

Rothblatt, J. A., Deshaies, R. J., Sanders, S. L., Daum, G. & Schekman, R. (1989). Multiple genes are required for proper insertion of secretory proteins into the endoplasmic reticulum in yeast. *Journal of Cell Biology* **109**, 2541–2552.

Sanders, S. L., Whitfield, K. M., Vogel, J. P., Rose, M. D. & Schekman, R. W. (1992). Sec61p and BiP directly facilitate polypeptide translocation into the ER. *Cell* **69**, 353–365.

Sengstag, C., Stirling, C. J., Schekman, R. & Rine, J. (1990). Genetic and biochemical evaluation of eucaryotic membrane protein topology: multiple transmembrane domains of *Saccharomyces cerevisiae* 3-hydroxy-3-methylglutaryl coenzyme A reductase. *Molecular and Cellular Biology* **10**, 672–680.

Simon, S. M. & Blobel, G. (1991). A protein-conducting channel in the endoplasmic reticulum. *Cell* **65**, 371–380.

Sipos, L. & von Heijne, G. (1993). Predicting the topology of eukaryotic membrane proteins. *European Journal of Biochemistry* **213**, 1333–1340.

Stirling, C. J. (1993). Similarities between *S. cerevisiae* Sec61p and *E. coli* SecY suggest a common origin for protein translocases of the eukaryotic ER and the bacterial plasma membrane. *Protein Synthesis and Targeting in Yeast (NATO ASI series)* **H71**, 293–306.

Stirling, C. J., Rothblatt, J., Hosobuchi, M., Deshaies, R. & Schekman, R. (1992). Protein translocation mutants defective in the insertion of integral membrane proteins into the endoplasmic reticulum. *Molecular Biology of the Cell* **3**, 129–142.

Tanford, C. (1961). *Physical Chemistry of Macromolecules*. New York: John Wiley.

Trombetta, E. S. & Helenius, A. (1998). Lectins as chaperones in glycoprotein folding. *Current Opinion in Structural Biology* **8**, 587–592.

Voigt, S., Jungnickel, B., Hartmann, E. & Rapoport, T. A. (1996). Signal sequence-dependent function of the TRAM protein during early phases of protein transport across the endoplasmic reticulum membrane. *Journal of Cell Biology* **134**, 25–35.

Walter, P. & Johnson, A. E. (1994). Signal sequence recognition and protein targeting to the endoplasmic reticulum membrane. *Annual Review of Cell Biology* **10**, 87–119.

Werner, E. D., Brodsky, J. L. & McCracken, A. A. (1996). Proteasome-dependent endoplasmic reticulum-associated protein degradation: an unconventional route to a familiar fate. *Proceedings of the National Academy of Sciences, USA* **93**, 13797–13801.

Wessels, H. P. & Spiess, M. (1988). Insertion of a multispanning membrane protein occurs sequentially and requires only one signal sequence. *Cell* **55**, 61–70.

Wiertz, E. J. H. J., Tortorella, D., Bogyo, M., Yu, J., Mothes, W., Jones, T. R., Rapoport, T. A. & Ploegh, H. L. (1996). Sec61-mediated transfer of a membrane protein from the endoplasmic reticulum to the proteasome for destruction. *Nature* **384**, 432–438.

Wilkinson, B. M., Critchley, A. J. & Stirling, C. J. (1996). Determination of the transmembrane topology of yeast Sec61p, an essential component of the endoplasmic reticulum translocation complex. *Journal of Biological Chemistry* **271**, 25590–25597.

Wilkinson, B., Regnacq, M. & Stirling, C. J. (1997a). Protein translocation across the membrane of the endoplasmic reticulum. *Journal of Membrane Biology* **155**, 189–197.

Wilkinson, B. M., Esnault, Y., Craven, R. A., Skiba, F., Fieschi, J., Kepes, F. & Stirling, C. J. (1997b). Molecular architecture of the ER translocase probed by chemical crosslinking of Sss1p to complementary fragments of Sec61p. *EMBO Journal* **16**, 4549–4559.

PEROXISOME BIOGENESIS

EDUARDO LOPEZ-HUERTAS AND ALISON BAKER

Centre for Plant Sciences, Leeds Institute for Plant Biotechnology and Agriculture, University of Leeds, Leeds LS2 9JT, UK

INTRODUCTION

Peroxisomes were first isolated by Christian de Duve in the early 1960s, who proposed the name 'peroxisome' to designate a special group of cytoplasmic particles distinct from mitochondria, lysosomes and microsomes which contained hydrogen-peroxide-producing oxidases and catalase (de Duve, 1996). In the years following the discovery of peroxisomes, these cellular organelles were found in almost all eukaryotic cells. Although peroxisome is the accepted name for this subcellular organelle (the ancient name microbody is still used in the literature), peroxisomes are given different names according to their specific metabolic functions, which may change according to developmental and physiological circumstances. More than 80 different enzymic activities have been described so far in peroxisomes (Tables 1 and 2).

In higher plants four types of peroxisome are recognized as follows (Huang *et al.*, 1983). Glyoxysomes are specialized peroxisomes which contain the enzymes for the glyoxylate cycle. In oil seeds they are responsible for the conversion of stored fatty acids into precursors for gluconeogenesis. They have also been described in senescent tissues where they are involved in the recycling of degraded membrane lipids. Leaf-type peroxisomes in photosynthetic tissues contain the enzymes for the photorespiration. Root nodule peroxisomes contain the enzyme uricase and are involved in nitrogen assimilation. Finally, unspecialized peroxisomes are found in other tissues but their function is unknown.

Fungal peroxisomes, which are induced by growth on fatty acids and two-carbon compounds and contain enzymes of the glyoxylate cycle, are also glyoxysomes. In contrast mammals lack the two unique glyoxylate cycle enzymes isocitrate lyase and malate synthase and consequently cannot carry out gluconeogenesis from acetyl-CoA. Therefore mammalian peroxisomes are not glyoxysomes. In trypanosomes many of the glycolytic enzymes are sequestered within a peroxisome-type organelle, the glycosome. Despite this wide range of functions all these organelles have catalase and flavin oxidases, a similar morphology and appear to share a common biogenetic mechanism. Because of this 'peroxisome' is used as a generic term for all these related but functionally different organelles. Examples of peroxisomes from different sources are shown in Fig. 1.

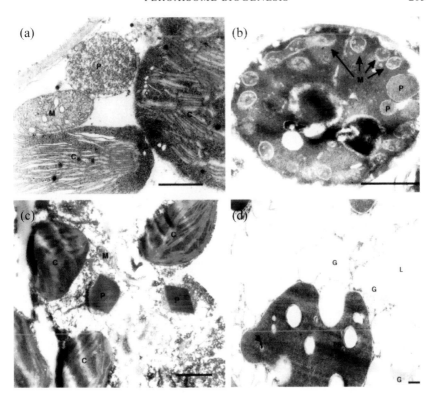

Fig. 1. Peroxisomes from different sources. (a) Pea leaves; (b) *S. cerevisiae* grown on oleate; (c) tobacco leaves; (d) sunflower cotyledons. P, Peroxisome; M, mitochondria; C, chloroplast; PB, protein body; L, lipid body; G, glyoxysome. Bars, 1 µm.

In plants, peroxisome types can be interconverted by the synthesis and import of new enzyme activities (Fig. 2). During germination and early post-germinative growth, glyoxysomes are found in the cotyledons and/or endosperm of oil-storing plant species. Upon greening, the levels of glyoxylate cycle enzymes begin to drop and the photorespiration enzymes increase to become major components of the peroxisomes (Titus & Becker, 1985; Nishimura *et al.*, 1986). In senescence of cotyledons and leaves, a reverse transition from leaf-type peroxisomes to glyoxysomes has also been described (Gut & Matile, 1988; de Bellis *et al.*, 1990; de Bellis & Nishimura, 1991). Similarly, the enzyme content of yeast peroxisomes can vary and strongly depends on the carbon and nitrogen source they use (van den Bosch *et al.*, 1992).

An important feature of peroxisomal metabolism is that these organelles are very sensitive to external signals. Hypolipidaemic drugs and plasticizers induce proliferation of the peroxisomal population in the liver of rodents (van den Bosch *et al.*, 1992). In methylotrophic yeasts peroxisomes are

Table 1. *Functions of peroxisomes in animal cells*

Lipid biosynthesis (cholesterol, dolichol, ether-phospholipids, bile acids)
β-Oxidation of fatty acids
Glyoxylate metabolism
β-Oxidation of dicarboxylic acids
Metabolism of xenobiotics
Oxidation of polyamines
Phytanic and pipecolic acid degradation
Metabolism of amino acids
Metabolism of purines
Metabolism of reactive oxygen species

Table 2. *Functions of peroxisomes in plant and yeast cells*

Photo-respiration
β-Oxidation of fatty acids
Glyoxylate cycle
Nitrogen assimilation
Metabolism of ureides
Metabolism of polyamines
Metabolism of reactive oxygen species
Metabolism of methanol (yeasts, fungi)
Oxalate synthesis (fungi)
Metabolism of amines (yeasts)
Metabolism of alcans (yeasts)

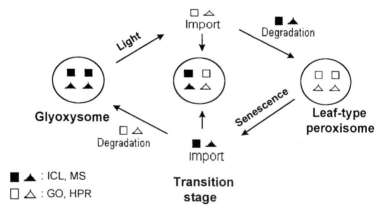

Fig. 2. Glyoxysome to leaf-type peroxisome conversion. During early stages of germination glyoxylate cycle enzymes convert stored lipids into precursors for gluconeogenesis. Upon greening, glyoxysomes import the enzymes for photorespiration and the enzymes for the glyoxylate cycle begin to be degraded, which leads to a transition stage organelle where enzymes from both metabolic pathways coexist. Senescence reverts the process. ICL, Isocitrate lyase; MS, malate synthase; GO, glycolate oxidase; HPR, hydroxypyruvate reductase.

massively induced upon growth in media containing methanol, fatty acids or
D-alanine (Goodman *et al.*, 1990). In plants, peroxisomes are induced in
response to different stress situations (herbicides, xenobiotics, senescence)
and a role in the cell response to stress has been proposed for these organelles
(Palma *et al.*, 1991; del Río *et al.*, 1992, 1998).

THE PEROXISOMAL MEMBRANE

Peroxisomes contain a single boundary membrane, 6–7 nm thick, with a
typical trilaminar appearance and a phospholipid-protein ratio of 140–
200 nmol mg^{-1}. Phosphatidylcholine and phosphatidylethanolamine are
the main phospholipids in the peroxisomal membrane of rat liver and castor
bean endosperm (Lazarow & Fujiki, 1985, and references therein), whilst in
cotton cotyledon glyoxysome membranes phosphatidylcholine is the most
abundant phospholipid but phosphatidylinositol, phosphatidylglycerol and
phosphatidylethanolamine are present in significant amounts as are nonpolar lipids (mostly triacylglycerol) which make up 36–62 % by weight of the
glyoxysome membrane at various developmental stages (Chapman & Trelease, 1991). Peroxisomal membranes contain no cholesterol, which allows
the use of digitonin to selectively permeabilize the plasma membrane without
affecting the peroxisomal membrane (Wendland & Subramani, 1993).

The peroxisomal membrane is involved in many of the functions described
for peroxisomes and has many roles of its own. Fujiki *et al.* (1982) described
the elimination of peripheral and adhered matrix proteins from the peroxisomal membrane, leaving only the integral membrane proteins and the
lipids, by a series of washes with Na_2CO_3 at pH 11.5. This technique allowed
the study of the integral membrane proteins (IMP) of the different types of
peroxisomes. Being such a versatile organelle, it is unsurprising that the IMP
profiles reported for the different types of peroxisomes vary considerably
(van den Bosch *et al.*, 1992 and references therein; Luster *et al.*, 1988;
Chapman & Trelease, 1992; Corpas *et al.*, 1994; Lopez-Huertas *et al.*, 1995).
Most of the functions of these IMPs are still unknown and this is an area of
active research at present.

Although peroxisomes are sealed compartments, several metabolites have
to cross the peroxisomal membrane. The permeability properties of the
peroxisomal membrane are still a matter of debate. Isolated peroxisomes are
freely permeable to small substrates like sucrose. Because of this unusual
property, they equilibrate at a particularly high density in sucrose gradients
(de Duve, 1996). The mammalian peroxisomal membrane has been shown to
possess pores by electrophysiological and solute permeability measurements.
The diameter of these peroxisomal pores was reported to be between 1.5 and
3 nm (Lemmens *et al.*, 1989). Isolated rat liver peroxisomes were reported to
be permeable to molecules like glucose, sucrose, urea, methanol, uric acid
and the co-factors of the β-oxidation, i.e. NAD, CoA, ATP and carnitine

(Van Veldhoven *et al.*, 1987), although the possibility of physical damage to these fragile organelles during the isolation procedure cannot be excluded.

Plant and yeast peroxisomal membranes are reported to be less permeable to substrates like co-factors of the β-oxidation or glyoxylate cycle (Donaldson *et al.*, 1981; Van Roermund *et al.*, 1995; van den Bosch *et al.*, 1992). Recently it has been demonstrated that highly purified membranes of leaf peroxisomes and glyoxysomes contain a pore-forming protein which represents a small and specific diffusion channel. This porin is anion selective with a binding site for dicarboxylic anions (Reumann *et al.*, 1995, 1997). In leaf-type peroxisomes the channel has a diameter of 0.6 nm and possesses a high affinity for photorespiratory intermediates (Reumann *et al.*, 1998). Perhaps such a difference in the pore size could explain the observed differences in the permeability of plant and mammalian peroxisomes.

In *Saccharomyces cerevisiae*, *in vivo* evidence suggests that the transport of NAD(H) and acetyl-CoA across the membrane is carried out by specific shuttles (Van Roermund *et al.*, 1995). Specific transporters for fatty acids, which belong to the ABC transporter superfamily, have also been identified (Hettema *et al.*, 1996). The integral peroxisomal membrane protein PMP47 from *Candida boidinii* has some sequence similarity to transporter proteins of the mitochondrial inner membrane. Porin-like channels with similar properties to the mitochondrial outer membrane porins have been described in the membrane of yeast peroxisomes and an integral membrane protein of 31 kDa has been proposed to be responsible for the permeability of yeast peroxisomes (Sulter *et al.*, 1993). Mitochondrial porins form general diffusion channels where substrates (up to 600 Da) can freely cross the outer membrane. However, the matrix of yeast peroxisomes has been reported to be acidic *in vivo* with a pH of approximately 6.0 (Waterham *et al.*, 1990; Nicolay *et al.*, 1987), and ATPase activity has also been described in peroxisomal membranes (van den Bosch *et al.*, 1992). Collectively these results indicate some kind of regulation of the membrane permeability and argue against the existence of non-specific diffusion pores *in vivo*. It is clear that the membrane regulates peroxisomal metabolic pathways by being a selective barrier that contains transporters, pores or channels for the transport of metabolites and proteins.

PEROXISOMAL PROTEIN TARGETING

Unlike mitochondria or chloroplasts, peroxisomes contain no DNA and therefore all peroxisomal proteins are nuclear-encoded and have to either cross the peroxisomal membrane in the case of matrix proteins, or be inserted in the membrane in the case of membrane proteins.

Although recent years have seen a considerable progress in the understanding of peroxisome biogenesis, comparatively little is known about the proteins involved in the import machinery, the events that occur before and

1. Matrix proteins

PTS-1

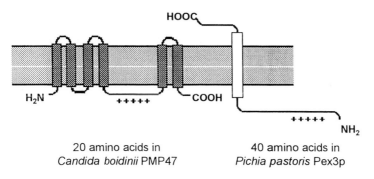

H₂N ▨▨▨▨▨▨▨▨▨▨▨▨▨▨▨▨▨▨▨ **SKL-COOH**

PTS-1 variants: **S** or neutral, small amino acid
K or charged amino acid
L or hydrophobic amino acid

PTS-2

H₂N-(R/K)(L/N/I/Q)X₅(Q/H)(L/A) ▨▨▨▨▨COOH

2. Membrane proteins

Fig. 3. Peroxisome targeting sequences.

20 amino acids in
Candida boidinii PMP47

40 amino acids in
Pichia pastoris Pex3p

after the import process and how peroxisomal membranes are synthesized. The need to understand the peroxisomal protein import process and biogenesis in detail is enhanced by the existence of a number of devastating human disorders in which peroxisome biogenesis is impaired (Subramani, 1997). In recent years a number of proteins involved in peroxisome protein import have been identified, mainly through genetic studies using various yeasts (see 'Components of the peroxisome import machinery' section). A unified nomenclature for peroxisome biogenesis factors has recently been adopted (Distel *et al.*, 1996) in which proteins involved in peroxisome biogenesis are designated 'peroxins' (Pex1p-PexNp) and the corresponding genes are named *PEX1–PEXN* (Table 3).

It is generally established that peroxisomal proteins are synthesized in the cytosol and are post-translationally imported into the peroxisomes (Lazarow & Fujiki, 1985). Peroxisomal proteins contain targeting signals (PTSs) in their sequence that have to be recognized by the import machinery. These signals are different for matrix and membrane proteins (Fig. 3).

Table 3. *Some mammalian and fungal peroxins*[a]

Gene designation		Description/proposed function of the protein
Fungi	Mammals	
ScPEX1, PpPEX1	*HsPEX1*	117–127 kDa AAA ATPase; cytosolic and vesicle-associated; interacts with Pex6p; defective in CG1 of the PBDs
PpPEX2, PaPEX2	*RnPEX2, HsPEX2*	35–52 kDa C3H4 zinc-binding integral peroxisome membrane protein; defective in CG10 of the PBDs
PpPEX3, ScPEX3, HpPEX3	*HsPEX3*	51–53 kDa integral membrane protein involved in peroxisome biogenesis; may traffic via the ER
PpPEX4, ScPEX4		21–24 kDa peroxisome-associated ubiquitin conjugating enzyme
PpPEX5, ScPEX5, HpPEX5, YlPEX5	*HsPEX5*	64–69 kDa PTS1 receptor; defective in CG2 of the PBDs
PpPEX6, ScPEX6, YlPEX6	*RnPEX6, HsPEX6*	112–127 kDa AAA ATPase; cytosolic and vesicle-associated; interacts with Pex1p; defective in CG4 of the PBDs
ScPEX7, KlPEX7	*MmPEX7, HsPEX7*	37–42 kDa PTS2 receptor; defective in CG11/RCDP patients
PpPEX8, HpPEX8		71–81 kDa peroxisome-associated protein; has PTS1
YlPEX9		42 kDa integral peroxisome membrane protein with no known orthologues in other organisms
PpPEX10, HpPEX10	*HsPEX10*	34–48 kDa C3H4 zinc-binding integral membrane protein; defective in CG7 of the PBDs
ScPEX11, CbPEX11	*RnPEX11*	27–32 kDa membrane protein involved in peroxisome proliferation; the Rat protein binds coatomer via a KXKXX C-terminal motif
PpPEX12, ScPEX12	*HsPEX12*	40–48 kDa zinc-binding integral membrane protein; defective in CG3 of the PBDs
PpPEX13, ScPEX13	*HsPEX13*	40–43 kDa SH3-domain-containing integral membrane protein; docking protein for Pex5p
HpPEX14, ScPEX14	*HsPEX14*	39 kDa peroxisome membrane protein; binds Pex5p and Pex7p; probable point of convergence for PTS1 and 2 import pathways
ScPEX15		Integral peroxisome membrane protein; probably traffics via ER
YlPEX16		Matrix-located peripheral membrane protein
ScPEX17		Peripheral membrane protein required for PTS1 and PTS2 import; interacts with Pex5p, Pex7p and Pex13p via Pex14p
ScPEX19	*HsPEX19*	Farnesylated protein required for import of matrix proteins; interacts with Pex3p
YlPEX20		Predominantly cytosolic protein required for import of thiolase
DJP1		Cytosolic DnaJ homologue required for import of some matrix proteins

[a] Genes are given the abbreviation *PEX* for peroxisome expressed and preceded by the initials of the species, e.g. Sc = *S. cerevisiae*, Hs = *Homo sapiens*, etc. Where not referenced in the text, references are cited in Olsen (1998), with the exception of *DJP1* (Hettema *et al.*, 1998); *ScPEX17* (Hühse *et al.*, 1998); *ScPEX19* and *HsPEX19* (Götte *et al.*, 1998); *YlPEX20* (Titorenko *et al.*, 1998).

Targeting of matrix proteins

Most matrix proteins are imported via PTS1-type signal, which consists of a tripeptide located at the carboxyl-terminal position of the protein (Fig. 3). The first PTS1 (Ser-Lys-Leu) was discovered in firefly luciferase by Subramani and co-workers (Gould et al., 1987, 1989). Although several peroxisomal proteins contain SKL as the carboxy-terminal tripeptide, this tripeptide is in fact variable and a general consensus sequence consists of a neutral and small amino acid in the first position, a positively charged amino acid in the second position, and a hydrophobic amino acid in the third position (Olsen, 1998).

A small subset of matrix proteins like thiolases, plant malate dehydrogenase and citrate synthase are imported via PTS2 signal (R/K)(L/V/I)X5(Q/H)(L/A) located near or at the amino-terminal position of the protein (Fig. 3) (de Hoop & Ab, 1992; Swinkels et al., 1991). Some PTS2-containing proteins are proteolytically processed after import although this event does not seem to be related to the import process. Interestingly, the same protein can be imported by different pathways in different organisms. Unlike their plant counterparts, yeast citrate synthase and malate dehydrogenase are targeted by PTS1 signals.

There are a number of peroxisomal matrix proteins that are imported without having an obvious PTS1 or PTS2 signal in their sequence or that are imported when the PTS signal is removed (Kragler et al., 1993; Elgersma et al., 1995; Behari & Baker, 1993; Horng et al., 1995). Examples of this are an internal domain of acyl-CoA oxidase from *Candida tropicalis* which is sufficient to target a passenger protein to peroxisomes *in vitro* (Small et al., 1988), and *S. cerevisiae* catalase A which has also been reported to contain internal targeting information (probably between residues 104 and 126) (Kamiryo et al., 1989). Glycolate oxidase is a PTS1-containing protein (Volokita, 1991) but amino- and carboxyl-terminal truncations of this protein are still imported into peroxisomes *in vitro* (Horng et al., 1995), indicating that some sort of internal signal mediates the targeting. All these examples illustrate the existence of internal signal for peroxisome protein import but neither the actual sequences nor the receptors for these signals have been described so far. As proteins can be imported into peroxisomes as oligomers (see 'Import of folded proteins' section) it is possible that some of the 'internal targeting signals' which have been identified could in fact represent oligomerization domains.

Peroxisomal targeting signals PTS1 and PTS2 are recognized in the cytosol by receptors which bind to the newly synthesized proteins and take them to the peroxisomal membrane. The receptor for PTS1-containing proteins is Pex5p, which binds with high affinity (K_d = 460–500 nM) (McCollum et al., 1993; Terlecky et al., 1995) to the tripeptide signal. The receptor for PTS2-containing proteins is Pex7p, which also binds the targeting sequence in the amino-terminal part of the protein (Marzioch *et al.*, 1994; Rehling *et al.*, 1996) (see '*PEX7*' section). Both PTS1 (Keller *et al.*, 1991) and PTS2 signals and their corresponding receptors, which are encoded by *PEX5* and *PEX7* genes, respectively, are conserved from yeasts to plants and mammals.

Targeting of peroxisomal membrane proteins

There is much less information available on the targeting of membrane proteins due to the fact that very few peroxisomal membrane proteins (PMPs) have been characterized so far. PMPs also contain signals (named mPTSs) which have to be recognized for their insertion in the membrane (Fig. 3). mPTSs have been described in yeasts from two integral membrane proteins: PMP47 from *C. boidinii* (Dyer *et al.*, 1996; McCammon *et al.*, 1994) and Pex3p from *Pichia pastoris* and *S. cerevisiae* (Wiemer *et al.*, 1996; Erdmann *et al.*, 1997). PMP47 from *C. boidinii* is thought to have six transmembrane domains, and the mPTS is proposed to be located in the hydrophilic loop between transmembrane domains 4 and 5 which faces the peroxisomal matrix (Fig. 3). Twenty amino acids which span this region are both necessary and sufficient for targeting a passenger protein to the peroxisomal membrane. The mPTS in *P. pastoris* Pex3p is 40 amino acids long and is also located facing the peroxisomal matrix in a hydrophilic domain which is close to the amino-terminal part of the protein. The 40-amino-acid mPTS of *P. pastoris* and *S. cerevisiae* Pex3p is sufficient to insert a protein in the peroxisomal membrane. Although there is no clear consensus targeting sequence for PMPs, a common feature among these two mPTSs is that they contain five positively charged residues in their sequences. The charged PMP47 sequence (KIKKR) when fused to a passenger protein causes it to be inserted in the peroxisomal membrane, and the same effect is seen when the amino acids next to the charged sequence were fused to the same protein. This indicates that both sequences contain some kind of sorting information (Dyer *et al.*, 1996; Subramani, 1998).

Peroxisome targeting signals for membrane proteins are remarkably different from the PTS1 and 2 signals reported for matrix proteins and there seems to be no overlap in their import pathways. This is emphasized by three lines of evidence. Firstly, immunoelectron microscopy experiments with antibodies raised against PTS1 label the peroxisomal matrix but not the membrane (Keller *et al.*, 1991). Secondly, mutations which impair import of matrix proteins generally do not affect membrane protein insertion (see 'Components of the peroxisome import machinery' section). Finally, a recombinant hybrid PTS1-containing protein (protein A glycolate oxidase) competes for the membrane binding of authentic glycolate oxidase (a PTS1-targeted protein) and also for the binding of the PTS2-containing protein malate dehydrogenase to plant peroxisomes, but it does not compete for the binding of the membrane protein PMP22 (Pool *et al.*, 1998).

Although it is generally established that peroxisomal proteins are synthesized in the cytoplasm before being imported into the peroxisomes, an increasing number of reports are being published recently where the involvement of the endoplasmic reticulum (ER) in the import of some PMPs is described (see 'The endoplasmic reticulum and peroxisome biogenesis: back to the future?' section). All these reports taken together suggest that at least some PMPs are targeted to the peroxisomes via the ER, which argues in favour of the existence of at least two different import pathways for peroxisomal membrane proteins and two different types of mPTS. One of them (mPTS1) could target membrane proteins to the peroxisomes directly from the cytoplasm, and the other one (mPTS2) might contain some information to direct the protein from the ER to the peroxisome (Subramani, 1998). Since it has been proposed that the ER is the source of lipids for the peroxisomal membrane (Fahimi *et al.*, 1993), perhaps the transfer of lipids between these two compartments occurs together with the import of these mPTS2-containing proteins.

Conclusive proof of receptors in the cytoplasm or docking sites in the membrane for the import of peroxisomal membrane proteins is lacking. These receptors are very likely to exist though, as the treatment of peroxisomes with proteases abolishes both binding and membrane insertion of rat PMP22 in an *in vitro* import assay (Diestelkotter & Just, 1993). Using the same system, the eukaryotic molecular chaperone TRiC (equivalent of *Escherichia coli* GroEL) has been reported to cross-link and co-immunoprecipitate with PMP22. The co-ordinated activity of TRiC together with Hsp70/Hsp40 was proposed to mediate PMP22 insertion in the membrane (Pause *et al.*, 1997).

IMPORT OF FOLDED PROTEINS

One interesting aspect of peroxisomal protein import is the ability of peroxisomes to import folded substrates and oligomeric proteins (Fig. 4). A rather striking experiment was carried out by Walton *et al.* (1995), which demonstrated that human fibroblast peroxisomes are able to import 9 nm gold particles coupled to a peptide containing a PTS1 sequence. Other experiments have shown that peroxisomes can import folded proteins stabilized by internal cross-links (Walton *et al.*, 1995). Also the import of a dihydrofolate reductase–glycerate kinase fusion protein by glycosomes is not inhibited by aminopterin, a folate analogue which stabilizes the folded conformation of the dihydrofolate reductase moiety (Hausler *et al.*, 1996).

Peroxisomes can also import oligomeric proteins such as *S. cerevisiae* malate dehydrogenase, which is imported as a dimer (Elgersma *et al.*, 1996a). Piggy-backing experiments have also demonstrated that expressed proteins whose monomers are devoid of targeting signals are co-imported into peroxisomes with wild-type expressed proteins containing PTS (Glover

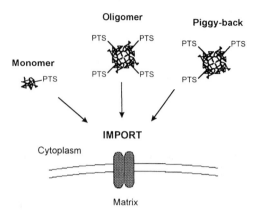

**Protein unfolding is a prerequisite for
protein import in mitochondria and chloroplasts**

Fig. 4. Peroxisomes can import folded proteins, oligomeric proteins, and also co-import subunits devoid of PTS by piggy-backing, whereas protein unfolding is a prerequisite for protein import into mitochondria and chloroplasts.

et al., 1994; McNew & Goodman, 1994; Lee *et al.*, 1997). This also indicates that the receptors do not necessarily have to recognize each monomer's PTS for the protein to be imported, which enhances the possibility of cytosolic polypeptides oligomerizing with peroxisomal matrix polypeptides to gain access to the peroxisome (Rachubinski & Subramani, 1995). Molecular chaperones could be among the cytosolic components which might be co-imported with matrix proteins although there is no proof of this event. The

PTS1 and PTS2 receptors Pex5p and Pex7p have also been detected in the peroxisome matrix by some groups (see '*PEX5*' and '*PEX7*' sections). This raises the possibility that the receptor may pick up its cargo in the cytoplasm and not only bring it to the peroxisome but also accompany it across the membrane (van der Klei *et al.*, 1995). This model would imply a mechanism for re-export of the receptor (van der Klei *et al.*, 1998).

Protein unfolding is therefore not a prerequisite for the import of proteins into peroxisomes as it is for other subcellular organelles like chloroplasts, mitochondria or ER (Fig. 4). This makes the peroxisomal import machinery rather unusual, although other examples of the transmembrane translocation of folded proteins are beginning to emerge (see 'Paradigms for the import of folded proteins' section). The fact that peroxisomes are able to import oligomeric proteins does not imply that all peroxisomal proteins are imported as oligomers. The eight subunits of alcohol oxidase are imported as monomers and assembly takes place within the peroxisome (Waterham *et al.*, 1997).

ENERGY REQUIREMENTS FOR PROTEIN IMPORT

The energy requirements for peroxisomal protein import were elucidated by using *in vitro* import systems (Imanaka *et al.*, 1987; Behari & Baker, 1993; Wendland & Subramani, 1993; Rapp *et al.*, 1993). Although these assays allow the manipulation of the conditions under which peroxisomes import proteins, they are difficult to work with because peroxisomes are extremely fragile organelles. *In vitro* import assays currently used are based on the isolation of peroxisomes from homogenates and the import of added substrate proteins under defined conditions, or they are based on plasma membrane permeabilized cells which contain intact peroxisomes that import substrate added from outside the cell.

Two stages can be distinguished for the import of peroxisomal proteins. The first stage represents the binding of the protein with the peroxisomal membrane. This process does not require ATP hydrolysis and is tempera-ture-insensitive. The second stage is the translocation across the membrane, which is ATP-dependent and temperature-dependent, as is the translocation through all membranes (Behari & Baker, 1993). GTP hydrolysis has also been described to support the protein import of glycolate oxidase in pumpkin peroxisomes (Brickner & Olsen, 1998) and, in fact, GTP-binding proteins have been described in peroxisomal membranes from rat (Verheyden *et al.*, 1992). It is not clear though at which step of the import process GTP hydrolysis is required and whether this requirement is specific to GTP, as the utilization of GTP by ATP-hydrolysing proteins (like for example mitochon-drial ATPases) is well-documented (Whitney & Bellion, 1991) and GTP may be used to resynthesize ATP in the presence of the ubiquitous enzyme nucleotide diphosphokinase.

COMPONENTS OF THE PEROXISOME IMPORT MACHINERY

Much of our current knowledge of the peroxisome protein import machinery stems from the identification and characterization of mutants in peroxisome biogenesis. The phenotypes of the mutants and the isolation of the corresponding genes has led to the identification of about 20 peroxisome expressed genes (*PEX* genes) which encode peroxins, proteins which play an essential role in protein import into peroxisomes and biogenesis of the organelle (Table 3).

Genetic screens exploiting various aspects of peroxisomal metabolism have been developed for the yeasts *S. cerevisiae*, *Hansenula polymorpha*, *P. pastoris* and *Yarrowia lipolytica* as well as mammalian Chinese hamster ovary (CHO) cells. Peroxisomes are the only site of β-oxidation in *S. cerevisiae* and so are required for growth on fatty acids as a sole carbon source. Erdmann *et al.* (1989) selected oleate non-utilizing (*onu*) mutants which are unable to grow on the 16-carbon monounsaturated fatty acid oleic acid (Fig. 5a). The mutants fell into two classes which could be distinguished by subcellular fractionation and electron microscopy; *fox* mutants lacked single enzyme activities required for fatty acid oxidation, but had otherwise normal peroxisomes, whereas *pas* mutants (now reclassified as *pex* mutants) were defective in peroxisome assembly and mislocalized peroxisomal proteins to the cytosol (Erdmann & Kunau, 1992). *pas* mutants are unable to carry out significant β-oxidation, even though in many cases all the necessary enzymes are present in the cytosol, presumably because these enzymes require to be in close physical proximity to sustain a reasonable metabolic flux.

Conceptually similar screens were used to isolate peroxisome-defective mutants in *H. polymorpha* (Cregg *et al.*, 1989), *P. pastoris* (Gould *et al.*, 1992) and *Y. lipolytica* (Nuttley *et al.*, 1993). *Hansenula* requires functional peroxisomes for growth on methanol. *mut* mutants analogous to the *onu* mutants of *S. cerevisiae* were isolated and further characterized by subcellular fractionation and electron microscopy. *Pichia* mutants were isolated which were defective for growth on methanol and fatty acids (both substrates require independent peroxisomal metabolic pathways) but functional for growth on ethanol (which requires a functional glyoxylate cycle). By analysing a combination of growth requirements, the proportion of mutants which were defective in peroxisome biogenesis/assembly relative to other mutants which were simply unable to use a particular carbon source was increased.

Although highly successful, these screens are rather laborious. Two groups developed positive selections based on hydrogen peroxide toxicity (Fig. 5b) (van der Leij *et al.*, 1992; Zhang *et al.*, 1993). The acyl-CoA oxidase step of peroxisomal β-oxidation generates hydrogen peroxide, which is detoxified by peroxisomal catalase. When catalase is inactivated by mutation, or the inhibitor 3-aminotriazole, the hydrogen peroxide produced kills the cells when fatty acids are present. Peroxisome assembly mutants survive as they cannot sustain significant rates of β-oxidation.

(a)   onu mutants fail to grow on fatty acids as a sole carbon source

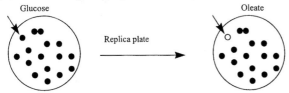

(b)   Hydrogen peroxide produced by β-oxidation is lethal in the absence of catalase

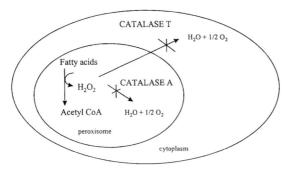

(c)   Phleomycin resistance is dependent upon the cytosolic localisation of Bleomycin R

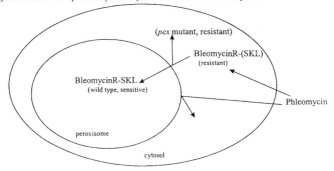

Fig. 5. Screens and selections for peroxisome-deficient mutants in *S. cerevisiae*. (a) Screen for oleate non-utilizing (*onu*) strains. Mutagenized cells are plated onto glucose-containing media and the resulting colonies are replica-plated onto plates which contain the monounsaturated fatty acid oleate as the sole carbon source. Mutants which fail to grow on oleate (*onu*) are taken for further analysis by subcellular fractionation and electron microscopy to determine whether the *onu* phenotype is caused by a peroxisome deficiency. (b) Selection of peroxisome-deficient mutants via hydrogen peroxide lethality. In wild-type cells peroxisomal β-oxidation produces hydrogen peroxide, which is detoxified by peroxisomal and cytosolic catalase. When catalase activity is blocked, for example in strains which are defective in catalase or in the presence of the catalase inhibitor 3-aminotriazole, β-oxidation is lethal and therefore the cells die in the presence of fatty acids. However, cells which are defective in β-oxidation or peroxisome assembly can survive. (c) Selection for compartment-dependent resistance to the antibiotic phleomycin. Yeast cells are sensitive to phleomycin, but are resistant if an enzyme which detoxifies phleomycin (bleomycin R) is expressed in the cytosol. If the bleomycin-resistance protein is efficiently transported to peroxisomes by means of a PTS1 signal (SKL) added to its carboxy-terminus, phleomycin is not detoxified and the cells are sensitive. Cells expressing bleomycin R–SKL were mutagenized and those which were now resistant to phleomycin were selected. A high proportion were defective in the transport of bleomycin–SKL into the peroxisome.

A further positive selection was developed based on the location-dependent phenotype of a selectable marker (Elgersma *et al.*, 1993; Fig. 5c). *S. cerevisiae* cells expressing a gene for bleomycin resistance are resistant to the related antibiotic phleomycin. The bleomycin-resistance gene was modified to end with the PTS1 signal SKL. Bleomycin–SKL was imported into peroxisomes and was unable to confer resistance to phleomycin. Cells expressing bleomycin–SKL were mutagenized and those cells which had acquired resistance were checked for the cytosolic location of bleomycin–SKL and other peroxisomal enzymes. This selection is especially efficient. Up to 25 % of phleomycin-resistant mutants were unable to grow on oleate (*onu*) and all *onu* mutants tested were also *pex* mutants (Elgersma *et al.*, 1993).

Detailed characterization of the mutants derived from the various screens shows that they fall into a number of classes. For example, the *pex13*, *pex14* and *pex17* mutants mislocalize all matrix proteins to the cytosol whereas the *pex5* and *pex7* mutants specifically mislocalize PTS1 and PTS2 proteins, respectively. Other mutants can be discriminated by the gross morphology of the peroxisome. For example, *pex6* mutants have abnormally small peroxisomes with severely reduced matrix content, whilst *pex11* mutants have abnormally large peroxisomes.

Most of these mutants have membrane remnants ('ghosts') which contain peroxisomal membrane proteins, suggesting that peroxisomal membrane proteins are inserted by a pathway that is distinct from matrix proteins (see 'Targeting of peroxisomal membrane proteins' section). The only mutant which does not appear to have ghosts is *pex3*, suggesting that it may be required for membrane formation or membrane protein insertion (see 'The endoplasmic reticulum and peroxisome biogenesis: back to the future?' section).

As all these yeast species are amenable to genetic manipulation it has been relatively straightforward to clone the corresponding *PEX* genes by complementation (Fig. 6). In some cases the gene sequence has given clues about the biochemical function of the protein, although in most cases these proteins are not homologous to any other proteins of known function. The availability of gene sequences has allowed the localization of gene products and in some cases functional studies to be carried out. The knowledge of yeast *PEX* genes has allowed the identification, primarily through homology searches of EST databases, of mammalian homologues (Table 3). In several cases these mammalian *PEX* genes have subsequently been shown to be responsible for many of the peroxisome biogenesis disorders.

Knowledge of the mechanism of peroxisome biogenesis and protein import is still fragmentary, but characterization of *PEX* genes and their corresponding mutants is yielding important insights, and the functions of some peroxins have now been deduced.

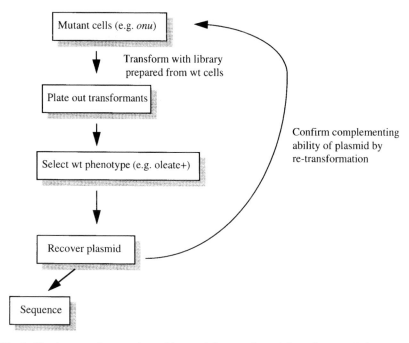

Fig. 6. Cloning genes for peroxisome biogenesis by complementation of a mutant phenotype.

PEX5

The phenotype of the *pex5* mutant is a mislocalization of PTS1-containing proteins to the cytosol. PTS2 proteins are correctly localized and membrane proteins are correctly inserted. This phenotype suggested that *PEX5* encoded a component specific for the PTS1 import pathway (see 'Targeting of matrix proteins' section). Pex5p is a protein of 64–70 kDa with seven tetratricopeptide repeats within the carboxy-terminus of the protein. The protein has been shown to bind to a peptide corresponding to the PTS1 signal via the TPR domain (Terlecky *et al.*, 1995) and to interact with various PTS1-containing proteins in the yeast two-hybrid system. The subcellular localization of Pex5p has been a subject of much debate. It has variously been reported to be entirely peroxisomal (McCollum *et al.*, 1993; Fransen *et al.*, 1995) or predominantly cytosolic with a small proportion associated with peroxisomes (Dodt *et al.*, 1995; Gould *et al.*, 1996). Some studies have suggested that some Pex5p is intraperoxisomal (van der Klei *et al.*, 1995). Current models favour the idea that Pex5p is a cycling receptor, binding PTS1 proteins in the cytosol and delivering them to the peroxisome (Dodt & Gould, 1996) or perhaps even being co-imported into the peroxisome with the cargo protein (see the 'Import of folded proteins' section).

In yeasts there appears to be only a single *PEX5* gene and a single Pex5p. However, in mammals two variants of Pex5p are produced from a single gene via alternative splicing (Otera *et al.*, 1998). The short form of Pex5p (also known as PTS1RS) is functionally equivalent to Pex5p from yeasts and its loss through mutation results in the loss of the PTS1 import pathway. The long form of Pex5p (also known as PTS1RL) results from a 37-amino-acid insert between amino acids 215 and 216 of the PTS1RS. A mutation in PTS1RL results in loss of both PTS1 *and* PTS2 import in CHO cells, and only the PTS1 import pathway is restored by transfection with a cDNA for PTS1RS. PTS2 import in CHO cells, as in yeast cells, requires Pex7p (see below). Why it also requires PTS1RL is not known.

PEX7

The phenotype of *pex7* cells is the opposite of *pex5* cells, namely a specific defect in the import of PTS2 proteins. Pex7p is a 42 kDa protein of the WD40 family. Pex7p has been shown to bind thiolase in a PTS2-specific manner in the yeast two-hybrid system and by co-immune precipitation (Rehling *et al.*, 1996). Pex7p contains a novel peroxisomal targeting signal within its amino-terminal 56 amino acids (Zhang & Lazarow, 1996). Like Pex5p the location of Pex7p has been the subject of debate. Marzioch *et al.* (1994) reported that it was predominantly cytosolic with a small amount associated with peroxisomes and postulate that, like Pex5p, Pex7p acts as a cycling receptor for PTS2 proteins. Zhang & Lazarow (1995, 1996) reported that Pex7p was predominantly intraperoxisomal. Part of the reason for the controversy over the localization was that in order to detect Pex7p both groups relied on overexpressed epitope-tagged proteins which in principle could affect the natural location of the protein. Native Pex7p has been studied in *P. pastoris* and is predominantly cytosolic, suggesting that the cycling receptor model may be correct (Elgersma *et al.*, 1998a). *S. cerevisiae* Pex5p and Pex7p interact in the two-hybrid system although the study of their respective null mutants indicates that in contrast to the situation in CHO cells, Pex5p is required only for import of PTS1 proteins and Pex7p is required only for the import of PTS2 proteins. A human orthologue of Pex7p has recently been identified by using the yeast *PEX7* sequences to search EST databases. HsPex7p has been shown to be mutated in patients with rhizomelic chondrodysplasia punctata (RCDP) (Purdue *et al.*, 1997). Cells from RCDP patients fail to import PTS2 proteins (Motley *et al.*, 1994) and are complemented by HsPex7p (Braverman *et al.*, 1997; Motley *et al.*, 1997; Purdue *et al.*, 1997).

PEX13

Pex13p is an integral peroxisome membrane protein known from yeasts and humans, with a cytosolically oriented src homology 3 (SH3) domain.

Δ*PEX13* mutants fail to import both PTS1 and PTS2 proteins but do insert at least some membrane proteins into membrane remnants. Pex13p interacts with Pex5p via the SH3 domain and so is a strong candidate for a docking protein for the PTS1 receptor (Elgersma *et al.*, 1996b; Erdmann & Blobel, 1996; Gould *et al.*, 1996).

PEX14

Pex14p is tightly associated with the cytosolic face of the peroxisomal membrane but is probably not an integral membrane protein, at least in yeasts. Δ*PEX14* mutants fail to import PTS1 and PTS2 proteins. ScPex14p has been shown to interact with Pex5p, Pex7p, Pex13p, Pex17p and itself in a two-hybrid assay, and with Pex5p, Pex7p, the PTS2 targeted protein thiolase and itself in co-immunoprecipitation experiments (Albertini *et al.*, 1997; Brocard *et al.*, 1997). It is therefore likely to be the point of convergence of the PTS1 and PTS2 import pathways. Overproduction of HpPex14p leads to a peroxisome-deficient phenotype and the accumulation of many small, clustered vesicles which contain Pex14p (Komori *et al.*, 1997). This may suggest that overproduction of Pex14p results in sequestration of other limiting components of the import machinery. HsPex14p was identified biochemically (Fransen *et al.*, 1998). Antibodies raised against a rat liver peroxisomal membrane protein fraction specifically recognized a 57 kDa integral membrane protein and inhibited the import in a permeabilized cell system of human serum albumin conjugated to a PTS1 peptide. This antibody specifically recognized recombinant HsPex14p. In ligand blots HsPex14p interacted with Pex5p and weakly with Pex13p, confirming the results from the two-hybrid experiments with the yeast homologues.

PEX17

ScPex17p is a small (199 amino acids) peripheral peroxisomal membrane protein. Δ*PEX17* mutants are defective in the import of PTS1- and PTS2-containing proteins but correctly localize membrane proteins to peroxisome 'ghosts' (Huhse *et al.*, 1998). Using the yeast two-hybrid system Pex17p was shown to interact with Pex14p and Pex5p. The interaction with Pex5p is dependent on Pex14p, but not Pex13p, suggesting that the interaction between Pex17p and Pex5p may be indirect and mediated by Pex14p. A complex comprising Pex17p, Pex14p and both PTS1 and 2 receptors can be detected by immunoprecipitation. Thus Pex13p, Pex14p and Pex17p are probably membrane-associated core components of an 'import complex' (Fig. 7).

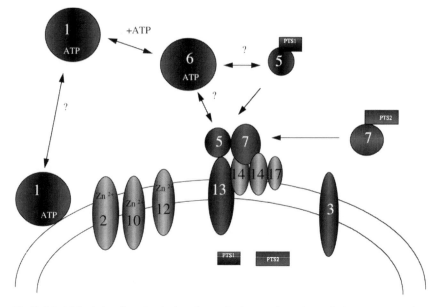

Fig. 7. Model depicting the role of selected peroxins in peroxisomal protein import. Interactions shown have been elucidated by use of the yeast two-hybrid system and/or direct binding studies. The evidence for association between Pex5p and Pex6p is indirect, based on the observation that the PTS1 receptor is less stable in cells deficient in Pex6p.

PEX2, PEX10 and PEX12

These three peroxins are all integral peroxisome membrane proteins and belong to the C_3HC_4 class of zinc-binding proteins (Tsukamoto *et al.*, 1991; Kalish *et al.*, 1995, 1996; Chang *et al.*, 1997; Okumoto *et al.*, 1998a, b). Mutations in any of these genes cause severe import defects. *PEX2, 10* and *12* are mutated in patients with peroxisome biogenesis disorders of groups 10, 7 and 3, respectively.

PEX1 and PEX6

Pex1p and Pex6p are proteins which belong to the AAA superfamily of ATPases. Members of this family share a conserved module of approximately 230 amino acids which includes an ATP-binding site, and perform diverse functions in many cellular processes such as DNA replication, protein complex assembly, protein degradation and membrane fusion (Patel & Latterich, 1998). The mechanism of AAA proteins is unknown but it is tempting to suggest that they may catalyse ATP-dependent rearrangement of protein complexes. Yeast mutants in *pex1* and *pex6* have small peroxisomes with much reduced (but not absent) matrix content. Recently, Pex1p and Pex6p of *P. pastoris* were shown to interact in an ATP-dependent manner and to be located on small membrane vesicles distinct from peroxisomes (Faber *et al.*, 1998). Human Pex1p is mutated in complementation group 1 of the peroxisome biogenesis disorders (Portsteffen *et al.*, 1997; Reuber *et al.*, 1997). Patients in complementation group 1 and complementation group 4 (in which HsPex6p is defective) have reduced levels of the PTS1 receptor (Yahraus *et al.*, 1996), suggesting that Pex1p and Pex6p may directly or indirectly stabilize the receptor.

PEX11

Pex11p is involved in regulation of constitutive division and proliferation of peroxisomes. Yeast mutants lacking *PEX11* produce one or two large organelles instead of several smaller ones in response to proliferators (Erdmann & Blobel, 1995; Sakai *et al.*, 1995), suggesting that peroxisome division does not occur efficiently in the absence of this protein. Over-expression of Pex11p results in abnormal hyperproliferation of peroxisomes. There are two isoforms of Pex11p, monomer and dimer, which apparently can interconvert according to the redox state of the cell. In proliferating peroxisomes, the protein exists as a monomer of about 30 kDa but in mature peroxisomes the inherent oxidative metabolism of these organelles seems to cause the protein to dimerize by a disulphide bond which stops the peroxisome proliferation (Marshall *et al.*, 1996).

While yeast genetics has been of enormous importance in identifying many key components of the peroxisome import pathway, and has led directly to the identification of many genes mutated in the peroxisome biogenesis disorders (Subramani, 1997), a combination of genetic, cell biological and biochemical approaches will be necessary to finally resolve the mechanism(s) of peroxisomal protein import. The various genetic screens have probably not reached saturation, and some further components, particularly chaperones, which have not been identified genetically have been implicated in peroxisomal protein import.

Chaperones and peroxisomal protein import

Peroxisomal protein import has been studied in semi-permeabilized mammalian cell systems and shown to be dependent upon cytosol (Wendland & Subramani, 1993; Rapp *et al.*, 1993). A role for the involvement of chaperones of the Hsp70 family has been suggested (Walton *et al.*, 1994). Microinjection of antibodies against the constitutive mammalian Hsp73 inhibited import of human serum albumin conjugated to a PTS1 signal, SKL (HSA–SKL), and Hsp73 could substitute for the cytosolic requirement for peroxisomal protein import in permeabilized cells. Proteins of the Hsp70 class have been shown to be associated with peroxisome membranes from rat liver (Walton *et al.*, 1994) and cucumber (Corpas & Trelease, 1997). An Hsp70 has also been detected within the matrix of watermelon glyoxysomes and shown to be encoded by a gene whose product is targeted to both plastids and glyoxysomes by use of alternate transcription sites (Wimmer *et al.*, 1997). Hsp70 proteins perform their functions in conjunction with a co-chaperone of the DnaJ or GrpE family and it is these partner proteins which are thought to determine the specific function of a particular Hsp70 (Rassow *et al.*, 1995). A protein with similarity to DnaJ has been detected in cucumber glyoxysome membranes (Preisig-Müller *et al.*, 1994) and the yeast gene *DJP1* encodes a cytosolic protein, Djp1p, with homology to DnaJ-like proteins that is required for the efficient import of peroxisomal matrix proteins and the proper maturation of the organelle (Hettema *et al.*, 1998). While there are undoubtedly Hsp70- and DnaJ-like proteins in or associated with peroxisomes, their role in protein import is unclear. In other organelle protein translocation systems, Hsp70s function on the *cis* side of the membrane to retain the protein in an import-competent (non-native) conformation and provide the driving force, coupled to ATP hydrolysis, to render translocation of the polypeptide chain unidirectional by binding to the unfolded translocating chain on the *trans* side of the membrane as it exits the translocation complex. Such a mode of action is not easily reconcilable with the import of folded proteins as is the case for at least several peroxisome matrix proteins (Fig. 4).

Important insights into the mechanisms of protein translocation into organelles have been gained by the ability to analyse trapped translocation intermediates *in vitro*, map their interactions with components of the import machinery and dissect the import process into a series of ordered steps. Studies from our laboratory using an *in vitro* import system for plant glyoxysomes (Behari & Baker, 1993; Horng *et al.*, 1995) have shown that a recombinant purified fusion protein comprising the immunoglobulin G binding domains of *Staphylococcus aureus* protein A fused at the amino-terminus of glycolate oxidase binds to the surface of glyoxysomes in the absence of ATP and competes with the binding of authentic glyoxysomal/peroxisomal proteins (Pool *et al.*, 1998). Upon ATP hydrolysis the protein is chased to a transmembrane state but is incompletely translocated. These two species can now be used to study the interaction of a translocating protein with components of the peroxisome import machinery.

MECHANISMS OF PROTEIN IMPORT

Although there is now ample evidence to implicate specific peroxins in peroxisomal protein import the actual mechanism remains elusive. Any proposed mechanism must be consistent with the known functions of peroxins and permit the uptake of folded and oligomeric proteins without

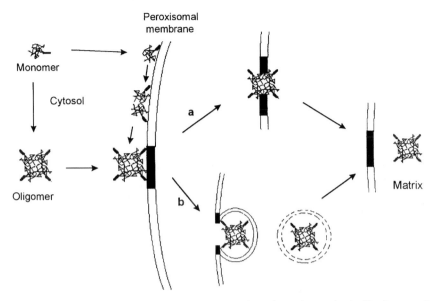

Fig. 8. Models for peroxisomal protein import. Peroxisomal proteins synthesized in the cytosol can oligomerize in either the cytosol or the peroxisomal membrane. Proteins may be imported by a regulated pore that opens when the protein docks at the membrane (a) or by an endocytic-like event (b) which involves further degradation of the vesicle-forming membrane.

loss of organelle integrity. The ER/chloroplast/mitochondrial paradigm where proteins are maintained or rendered unfolded on the cytoplasmic side of the membrane, translocated in an extended conformation and refolded on the *trans* side of the membrane (Fig. 4) does not fit with the translocation of folded proteins, although it is not known whether *all* peroxisomal proteins are *obligatorily* transported in their native conformation. This will be an important question to answer. McNew & Goodman (1996) proposed two possible models: uptake of oligomeric proteins through a regulated pore and import by an endocytic-like event (Fig. 8). At present there is no real evidence to support or refute either model. A thiolase dimer measures 80 Å × 50 Å (Mathieu *et al.*, 1994). For comparison nuclear pores have a passive diffusion limit of about 9 nm (90 Å) and can open to accommodate substrates of up to 26 nm during active nuclear transport (Alberts *et al.*, 1994). No obvious pores akin to nuclear pores have been seen in peroxisome membranes by electron microscopy. However, it seems likely that the opening of such a channel or pore would be regulated or gated to prevent leakage of small molecules (see 'The peroxisomal membrane' section), or could even be assembled upon docking of import receptors and their cargo at the peroxisome membrane. Equally, no real evidence for internalized membrane vesicles within peroxisomes has been obtained. However, if, as seems likely, some membrane proteins traffic to peroxisomes via the ER (see 'The endoplasmic reticulum and peroxisome biogenesis: back to the future?' section), vesicular fusion may play an important role in biogenesis of the peroxisome membrane, and perhaps indirectly of matrix proteins. As import is a transitory event, the ability to trap translocation intermediates (Pool *et al.*, 1998) may be crucial in differentiating between these models.

Paradigms for the import of folded proteins

In the past few years it has become apparent that peroxisomal protein import is not a unique case of translocation of folded proteins. Proteins which are secreted through the outer membrane of Gram-negative bacteria transit the membrane in a folded state as do the subunits of pili and indeed filamentous phage (Pugsley, 1993). A number of gene products are known to be required for this process, but only one, the D protein, is located in the outer membrane and could conceivably form a channel (Pugsley *et al.*, 1997). Likewise, certain cofactor-requiring proteins of bacteria may be exported across the cytoplasmic membrane to the periplasm fully folded with the cofactor attached (Berks, 1996). These proteins have a characteristic amino-terminal signal sequence with the consensus (S/T)-R-R-x-F-L-K, and there are even examples of periplasmic metalloproteins which do not have such a signal but are known to form complexes with another polypeptide which has this signal. For one class of these proteins, the periplasmic [Fe] hydrogenases, absence of one subunit prevents the export of the other subunit (van Dongen *et al.*, 1988). This is highly reminiscent of the 'piggyback' import seen in peroxisomes (see 'Import of folded proteins' section; McNew & Goodman, 1994; Glover *et al.*, 1994; Lee *et al.*, 1997). A further similarity is that several peroxisomal proteins bind FAD (e.g. acyl-CoA oxidase, urate oxidase) and several of the bacterial proteins exported by the twin arginine pathway are FAD-binding. Although the location of cofactor addition for peroxisomal proteins during biogenesis is unknown, in many cases it can take place in the cytoplasm as proteins such as acyl-CoA oxidase (FAD), catalase (haem) and alanine:glyoxylate aminotransferase (pyridoxal phosphate) are active in the cytoplasm if they fail to be imported. However, the signal sequences used by the bacterial periplasmic proteins are unlike known PTSs so these may be analogous rather than homologous pathways. In contrast, the bacterial twin arginine transfer peptides are highly reminiscent of signals which direct proteins into chloroplast thylakoids by the ΔpH pathway, and can in fact substitute for an authentic thylakoid transfer peptide (Wexler *et al.*, 1998). The thylakoid ΔpH pathway probably also transports folded proteins (Creighton *et al.*, 1995).

Another example in which a folded protein is taken up from the cytoplasm into a eukaryotic organelle in a folded and oligomeric conformation is aminopeptidase I in *S. cerevisiae*, which is transported into the vacuole as a dodecamer (Kim *et al.*, 1997). API is targeted by an amino-terminal amphipathic helix. Mutants defective in this cytoplasm to vacuole pathway (*cvt* mutants) overlap with autophagy mutants (Scott *et al.*, 1996). During autophagy, in response to nitrogen starvation, cytoplasmic proteins and organelles become non-selectively enveloped in large (500 nm) double-membraned vesicles which then fuse with the vacuole, resulting in the delivery of a still-intact membrane-bound autophagic body which is then degraded. As all autophagy mutants which were tested were also defective in the biosynthetic delivery of API, many of the same components would seem to be involved (Scott *et al.*, 1996).

Future progress in unravelling peroxisomal protein import and these other pathways should enable us to understand how many mechanisms cells have for the transport of fully folded proteins across biological membranes and whether they are mechanistically or evolutionarily related.

THE ER AND PEROXISOME BIOGENESIS: BACK TO THE FUTURE?

Early models of peroxisome biogenesis proposed in the 1960s and 70s (Lazarow & Fujiki, 1985; Trelease, 1984; and references within) envisaged that peroxisomes were formed by budding from the ER (Fig. 9). This view was based primarily on electron micrographs which showed peroxisomes in close proximity to ER in many cell types, and on subcellular fractionation experiments in which peroxisomal proteins were detected in ER fractions and glycoproteins were detected in peroxisomes. However, by the late 70s and early 80s a number of results which were inconsistent with this model demanded a re-evaluation (see Trelease, 1984; Lazarow & Fujiki, 1985; and references within). Principal amongst these were the findings that peroxisomal proteins, including membrane proteins, were synthesized on free rather than membrane-bound polysomes (see 'Peroxisomal protein targeting' section) as would have been expected if they were inserted co-translationally into the ER. Some peroxisomal proteins could be demonstrated to be imported from cytosolic pools without involvement of the ER. More careful subcellular fractionation experiments showed that the protein and lipid compositions of rat liver peroxisome and ER membranes were distinct (see 'The peroxisomal membrane' section; Fujiki & Lazarow, 1982) and that earlier experiments suggesting the presence of glycoproteins in the peroxisomes and peroxisomal proteins in the ER were probably artefacts attributable to cross-contamination of fractions or aggregation of proteins coupled with very sensitive detection methods. Although peroxisomes are frequently seen in close proximity to ER, micrographs showing unequivocal lumenal connections are probably non-existent, whereas numerous micrographs showing division of peroxisomes (especially in yeasts) have been published.

With the discovery of PTS1 and PTS2 signals in the late 1980s and increasingly detailed knowledge of the components of the peroxisomal and ER protein translocation machineries it became apparent that the two

1970s. Peroxisomes are derived from the ER

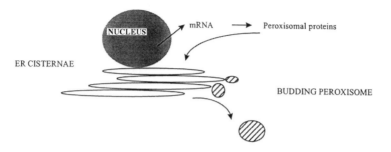

1980s Peroxisomes are derived from pre-existing peroxisomes

1990s ?

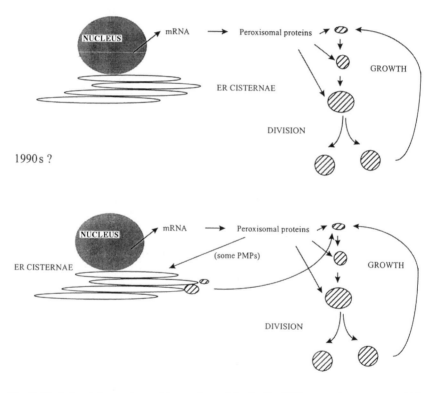

Fig. 9. Evolution of peroxisome biogenesis models. In the 1970s peroxisomes were widely believed to be derived from the ER by budding. During the 1980s this view changed due to the accumulation of evidence which was incompatible with the ER vesiculation model, and it was generally accepted that peroxisomes arose from pre-existing peroxisomes via import of proteins and division of the organelle. In the past couple of years evidence has begun to accumulate which suggests that some peroxisomal membrane proteins do traffic to peroxisomes via the ER, although the import of matrix proteins and some membrane proteins is post-translational. These observations have led to a hybrid model which can now be tested.

protein translocation processes were completely distinct. The data were most consistent with a model where peroxisomes arose by growth and division of existing peroxisomes (Lazarow & Fujiki, 1985; Fig. 9). It was envisaged that all cells had at least one rudimentary peroxisome which, under appropriate conditions, could grow through post-translational import of proteins across its membrane.

While it is clear that probably most (if not all) matrix proteins and at least some membrane proteins are imported post-translationally from the cytosol directly into peroxisomes, several recent results have provided circumstantial evidence that the ER may be involved in the sorting of at least some peroxisomal membrane proteins.

1. Most *pex* mutants have residual membranous structures which contain some peroxisome membrane proteins ('ghosts'), but ghosts are not detectable in *pex3* null mutants. Nevertheless, when *pex3* null mutants are retransformed with the *PEX3* gene they regain peroxisomes (Baerends *et al.*, 1996; Wiemer *et al.*, 1996). This suggests that peroxisomes are not obligatorily derived from pre-existing peroxisomes.

2. In *S. cerevisiae*, overexpression of the peroxisomal membrane protein Pex15p causes proliferation of the ER membrane. These membranes contain Pex15p and Pex3p (Elgersma *et al.*, 1998b).

3. Pex15p is *O*-mannosylated, an ER-specific modification in *S. cerevisiae* (Elgersma *et al.*, 1998b).

4. Pex15p can target cytosolic invertase to the ER; invertase becomes *N*-glycosylated (an ER-specific modification). The transmembrane domain of Pex15p is required for ER targeting (Elgersma *et al.*, 1998b).

5. The amino-terminal 16 amino acids of *H. polymorpha* Pex3p target a passenger protein to the ER (Baerends *et al.*, 1996).

6. Brefeldin A (BFA) interferes with the formation of COP coated vesicles required for vesicular traffic within the secretory pathway. BFA treatment results in the detection of some peroxisomal proteins in the ER (Salomons *et al.*, 1997).

7. RnPex11p has a cytosolic C-terminal tail with the consensus ER retention sequence KXKXX. RnPex11p is a peroxisomal protein, and highly purified rat liver peroxisomes were shown to bind coatomer subunits and ARF in a GTPγS-dependent manner (Passreiter *et al.*, 1998).

8. Two peroxins, Pex1p and Pex6p, are members of the AAA family of ATPases. On the basis of sequence similarity they are close relatives of NSF and p97, two proteins which are involved in vesicle fusion events within the endo membrane system.

9. Pex1p and Pex6p are found on small vesicles independent of peroxisomes (Faber *et al.*, 1998). *pex1* and *pex6* mutants have small peroxisomes with greatly reduced matrix content.

10. *pex1* and *pex6* mutants result in the accumulation of some peroxisomal proteins in the ER in *Y. lipolytica* (Titorenko & Rachubinski, 1998)

None of these observations provides direct proof of the involvement of the ER in peroxisome biogenesis; several are circumstantial and each can be subject to alternative interpretation. For example, overexpression of proteins can lead to mis-sorting and have non-specific effects on cell morphology. Inhibitors like BFA can have non-specific toxic effects. Pex1p and Pex6p are also related in sequence to proteosome subunits which are not involved in membrane fusion. Protein coats may be required for peroxisome division rather than the recruitment of ER-derived vesicles. However, collectively these results do indicate that there may be a biogenetic connection between the ER and peroxisomes after all. It is possible that some peroxisomal membrane proteins (Pex15p and Pex3p would be strong candidates) are delivered first to the ER (although probably not via the SRP/Sec61 pathway) and travel, via (coated?) vesicles containing Pex1p and Pex6p, to fuse with existing peroxisomes. This would provide a mechanism for enlarging the peroxisome membrane and allowing the organelle to grow by post-translational import of matrix and other membrane proteins (Fig. 9). The challenge is now to test this model and modify or reject it as appropriate.

REFERENCES

Albertini, M., Rehling, P., Erdmann, R., Girzalsky, W., Kiel, J. A. K. W., Veehuis, M. & Kunau, W.-H. (1997). Pex14p, a peroxisomal membrane protein binding both receptors of the two PTS-dependent import pathways. *Cell* **89**, 83–92.

Alberts, B., Bray, D., Lewis, J., Raff, M., Roberts, K. & Watson, J. D. (1994). *Molecular Biology of the Cell*, 3rd edn, pp. 563–564. New York & London: Garland Publishing.

Baerends, R. J. S., Rasmussen, S. W., Hilbrands, R. E. & 7 other authors (1996). The *Hansenula polymorpha* PER9 gene encodes a peroxisomal membrane protein essential for peroxisome assembly and integrity. *Journal of Biological Chemistry* **271**, 8887–8894.

Behari, R. & Baker, A. (1993). The carboxyl terminus of isocitrate lyase is not essential for import into glyoxysomes in an in vitro system. *Journal of Biological Chemistry* **268**, 7315–7322.

de Bellis, L. & Nishimura, M. (1991). Development of enzymes of the glyoxylate cycle during senescence of pumpkin cotyledons. *Plant Cell and Physiology* **23**, 555–561.

de Bellis, L., Picciarelli, P., Pistelli, L. & Alpi, A. (1990). Localisation of glyoxylate cycle marker enzymes in peroxisomes of senescent leaves and green cotyledons. *Planta* **180**, 435–439.

Berks, B. C. (1996). A common export pathway for proteins binding complex redox factors? *Molecular Microbiology* **22**, 393–404.

van den Bosch, H., Schutgens, R. B. H., Tager, J. M. & Wanders, J. A. (1992). Biochemistry of peroxisomes. *Annual Review of Biochemistry* **61**, 157–197.

Braverman, N., Steel, G., Obie, C., Moser, A., Moser, H., Gould, S. J. & Valle, D. (1997). Human PEX7 encodes the peroxisomal PTS2 receptor and is responsible for rhizomelic chondrodysplasia punctata. *Nature Genetics* **15**, 369–376.

Brickner, D. G. & Olsen, L. (1998). Nucleotide triphosphates are required for the transport of glycolate oxidase into peroxisomes. *Plant Physiology* **116**, 309–317.

Brocard, C., Lametschwandtner, G., Koudelka, K. & Hartig, A. (1997). Pex14p is a member of the protein linkage map of Pex5p. *EMBO Journal* **16**, 5491–5500.

Chang, C. C., Lee, W. H., Moser, H., Valle, D. & Gould, S. J. (1997). Isolation of the human PEX12 gene, mutated in group 3 of the peroxisome biogenesis disorders. *Nature Genetics* **15**, 385–388.

Chapman, K. D. & Trelease, R. N. (1991). Acquisition of membrane lipids by differentiating glyoxysomes: role of lipid bodies. *Journal of Cell Biology* **115**, 995–1007.

Chapman, K. D. & Trelease, R. N. (1992). Characterization of membrane proteins in enlarging cottonseed glyoxysomes. *Plant Physiology and Biochemistry* **30**, 1–10.

Corpas, F. J. & Trelease, R. N. (1997). The plant 73 kDa peroxisomal membrane protein (PMP73) is immunorelated to molecular chaperones. *European Journal of Cell Biology* **73**, 49–57.

Corpas, F. J., Bunkelmann, J. & Trelease, R. N. (1994). Identification and immuno-chemical characterization of a family of peroxisome membrane proteins (PMPs) in oilseed glyoxysomes. *European Journal of Cell Biology* **65**, 280–290.

Cregg, J. M., Vanklei, I. J., Sulter, G. J., Veenhuis, M. & Harder, W. (1989). Peroxisome deficient mutants of *Hansenula polymorpha*. *Yeast* **6**, 87–97.

Creighton, A. M., Hulford, A., Mant, A., Robinson, D. & Robinson, C. (1995). A monomeric, tightly folded stromal intermediate on the ΔpH-dependent thylakoid protein import pathway. *Journal of Biological Chemistry* **270**, 1663–1669.

Diestelkotter, P. & Just, W. W. (1993). In vitro insertion of the 22-kD peroxisomal membrane protein into isolated rat liver peroxisomes. *Journal of Cell Biology* **123**, 1717–1725.

Distel, B., Erdmann, R., Gould, S. J. & 22 other authors (1996). A unified nomenclature for peroxisome biogenesis factors. *Journal of Cell Biology* **135**, 1–3.

Dodt, G. & Gould, S. J. (1996). Multiple PEX genes are required for proper subcellular distribution and stability of Pex5p, the PTS1 receptor: evidence that PTS1 protein import is mediated by a cycling receptor. *Journal of Cell Biology* **135**, 1763–1774.

Dodt, G., Braverman, N., Wong, C., Moser, A., Moser, H. W., Watkins, P., Valle, D. & Gould, S. J. (1995). Mutations in the PTS1 receptor gene PXR-1 define complementation group 2 of the peroxisome biogenesis disorders. *Nature Genetics* **9**, 115–125.

Donaldson, R. P., Tully, R. E., Young, O. A. & Beevers, H. (1981). Organelle membranes from germinating castor bean endosperm. II. Enzymes, cytochromes, and permeability of the glyoxysomal membrane. *Plant Physiology* **67**, 21–25.

van Dongen, W., Hagen, W., van den Berg, W. & Veeger, C. (1988). Evidence for an unusual mechanism of membrane translocation of the periplasmic hydrogenase of *Desulfovibrio vulgaris* (Hildenborough) as derived from expression in *Escherichia coli*. *FEMS Letters* **50**, 5–9.

de Duve, C. (1996). The peroxisome in retrospect. *Annals of the New York Academy of Sciences* **804**, 1–10.

Dyer, J. M., McNew, J. A. & Goodman, J. M. (1996). The sorting sequence of the peroxisomal integral membrane protein PMP47 is contained within a short hydrophilic loop. *Journal of Cell Biology* **133**, 269–280.

Elgersma, Y., van den Berg, M., Tabak, H. F. & Distel, B. (1993). An efficient positive selection procedure for the isolation of peroxisomal import and peroxisome assembly mutants of *Saccharomyces cerevisiae*. *Genetics* **135**, 731–740.

Elgersma, Y., Van Roermund, C. W. T., Wanders, R. J. A. & Tabak, H. F. (1995).

Peroxisomal and mitochondrial acetyl carnitine transferases of *Saccharomyces cerevisiae* are encoded by a single gene. *EMBO Journal* **14**, 3472–3479.

Elgersma, Y., Vos, A., van den Berg, M., van Roermund, C. W. T., van der Sluijs, P., Distel, B. & Tabak, H. F. (1996a). Analysis of the carboxy-terminal peroxisomal targeting signal 1 in a homologous context in *Saccharomyces cerevisiae*. *Journal of Biological Chemistry* **271**, 26375–26382.

Elgersma, Y., Kwast, L., Klein, A., Voorn-Brouwer, T., van den Berg, M., Metzig, B., America, T., Tabak, H. F. & Distel, B. (1996b). The SH3 domain of the *Saccharomyces cerevisiae* peroxisomal membrane protein Pex13p functions as a docking site for Pex5p, a mobile receptor for the import of PTS-1 containing proteins. *Journal of Cell Biology* **135**, 97–109.

Elgersma, Y., Elgersma-Hooisma, M., Wenzel, T., McCaffery, J. M., Farquhar, M. G. & Subramani, S. (1998a). A mobile PTS2 receptor for peroxisomal protein import in *Pichia pastoris*. *Journal of Cell Biology* **140**, 807–820.

Elgersma, Y., Kwast, L., van den Berg, M., Snyder, W. B., Distel, B., Subramani, S. & Tabak, H. F. (1998b). Overexpression of Pex15p, a phosphorylated peroxisomal integral membrane protein required for peroxisome assembly in *S. cerevisiae*, causes proliferation of the endoplasmic reticulum. *EMBO Journal* **16**, 7326–7341.

Erdmann, R. & Blobel, G. (1995). Giant peroxisomes in oleic acid-induced *Saccharomyces cerevisiae* lacking the peroxisomal membrane protein Pmp27p. *Journal of Cell Biology* **128**, 509–523.

Erdmann, R. & Blobel, G. (1996). Identification of Pex13p, a peroxisomal membrane receptor for the PTS1 recognition factor. *Journal of Cell Biology* **135**, 111–121.

Erdmann, R. & Kunau, W.-H. (1992). A genetic approach to the biogenesis of peroxisomes in the yeast *Saccharomyces cerevisiae*. *Cell Biochemistry and Function* **10**, 167–174.

Erdmann, R., Veenhuis, M., Mertens, D. & Kunau, W.-H. (1989). Isolation of peroxisome deficient mutants of *Saccharomyces cerevisiae*. *Proceedings of the National Academy of Sciences, USA* **86**, 5419–5423.

Erdmann, R., Veenhuis, M. & Kunau, W. H. (1997). Peroxisomes: organelles at the crossroads. *Trends in Cell Biology* **7**, 400–407.

Faber, K. N., Heyman, J. A. & Subramani, S. (1998). Two AAA family peroxins, PpPex1p and PpPex6p, interact with each other in an ATP-dependent manner and are associated with different subcellular membranous structures distinct from peroxisomes. *Molecular Cell Biology* **18**, 936–943.

Fahimi, H. D., Baumgart, E. & Volkl, A. (1993). Ultrastructural aspects of the biogenesis of peroxisomes in rat liver. *Biochimie* **75**, 201–208.

Fransen, M., Brees, C., Baumgart, E., Vanhooren, J. C. T., Baes, M., Mannaerts, G. P. & Van Veldhoven, P. P. (1995). Identification and characterization of the putative human peroxisomal C-terminal targeting signal import receptor. *Journal of Biological Chemistry* **270**, 7731–7736.

Fransen, M., Terlecky, S. R. & Subramani, S. (1998). Identification of a human PTS1 receptor docking protein directly required for peroxisomal protein import. *Proceedings of the National Academy of Sciences, USA* **95**, 8087–8092.

Fujiki, Y. & Lazarow, P. B. (1982). Posttranslational uptake of an in vitro synthesized peroxisomal polypeptide of rat-liver. *Journal of Cell Biology* **95**, A398.

Fujiki, Y., Fowler, S., Shio, H., Hubbard, A. L. & Lazarow, P. B. (1982). Polypeptide and phospholipid composition of the membrane of rat liver peroxisomes. Comparison with endoplasmic reticulum and mitochondrial membranes. *Journal of Cell Biology* **93**, 103–110.

Glover, J. R., Andrews, D. W. & Rachubinski, R. A. (1994). *Saccharomyces cerevisiae* peroxisomal thiolase is imported as a dimer. *Proceedings of the National Academy of Sciences, USA* **91**, 10541–10545.

Goodman, J. M., Trapp, S. B., Hwang, H. & Veenhuis, M. (1990). Peroxisomes induced in *Candida boidinii* by methanol, oleic acid and D-alanine vary in metabolic function but share common integral membrane proteins. *Journal of Cell Science* **97**, 193–204.

Götte, K., Girzalsky, W., Linkert, M., Baumgart, E., Kammerer, S., Kunau, W.-H. & Erdmann, R. (1998). Pex19p, a farnesylated protein essential for peroxisome biogenesis. *Molecular Cell Biology* **18**, 616–628.

Gould, S. J., Keller, G. & Subramani, S. (1987). Identification of a peroxisomal targeting signal at the carboxyterminus of firefly luciferase. *Journal of Cell Biology* **105**, 2923–2931.

Gould, S. J., Keller, G. A., Hosken, N., Wilkinson, N. & Subramani, S. (1989). A conserved tripeptide sorts proteins to peroxisomes. *Journal of Cell Biology* **108**, 1657–1664.

Gould, S. J., McCollum, D., Spong, A. P., Heyman, J. A. & Subramani, S. (1992). Development of the yeast *Pichia pastoris* as a model organism for a genetic and molecular analysis of peroxisome assembly. *Yeast* **8**, 613–628.

Gould, S. J., Kalish, J. E., Morrell, J. C., Bjorkman, J., Urquhart, A. J. & Crane, D. I. (1996). Pex13p is an SH3 protein of the peroxisome membrane and is a docking protein for the predominantly cytoplasmic PTS1 receptor. *Journal of Cell Biology* **135**, 85–95.

Gut, H. & Matile, P. (1988). Apparent induction of key enzymes of the glyoxylic acid cycle in senescent barley leaves. *Planta* **176**, 548–550.

Hausler, T. Y., Stierhof, Y., Wirtz, E. & Clayton, C. (1996). Import of DHFR hybrid protein into glycosomes in vivo is not inhibited by the folate-analogue aminopterin. *Journal of Cell Biology* **132**, 311–324.

Hettema, E. H., van Roermund, C. W. T., Distel, B., van den Berg, M., Vilela, C., Rodrigues-Pousada, C., Wanders, R. J. A. & Tabak, H. F. (1996). The ABC transporter proteins Pat1p and Pat2p are required for import of long chain fatty acids into peroxisomes of *Saccharomyces cerevisiae*. *EMBO Journal* **15**, 3813–3822.

Hettema, E. H., Ruigrok, C. C. M., Koerkamp, M. G., van den Berg, M., Tabak, H. F., Distel, B. & Braakman, I. (1998). The cytosolic DnaJ-like protein Djp1p is involved specifically in peroxisomal protein import. *Journal of Cell Biology* **142**, 421–434.

de Hoop, M. J. & Ab, G. (1992). Import of proteins into peroxisomes and other microbodies. *Biochemical Journal* **286**, 657–669.

Horng, J. T., Behari, R., Burke, L. E. C. A. & Baker, A. (1995). Investigation of the energy requirement and targeting signal for the import of glycolate oxidase into glyoxysomes. *European Journal of Biochemistry* **230**, 157–163.

Huang, A. H. C., Trelease, R. N. & Moore, T. S. (1983). *Plant Peroxisomes*. New York: Academic Press.

Hühse, B., Rehling, P., Albertini, M., Blank, L., Meller, K. & Kunau, W.-H. (1998). Pex17p of *Saccharomyces cerevisiae* is a novel peroxin and component of the peroxisomal protein translocation machinery. *Journal of Cell Biology* **140**, 49–60.

Imanaka, T., Small, G. M. & Lazarow, P. B. (1987). Translocation of acyl-CoA oxidase into peroxisomes requires ATP hydrolysis but not a membrane potential. *Journal of Cell Biology* **105**, 2915–2922.

Kalish, J. E., Theda, C., Morrel, J. C., Berg, J. M. & Gould, S. J. (1995). Formation of the peroxisome lumen is abolished by loss of *Pichia pastoris* Pas7p, a zinc binding integral membrane protein of the peroxisome. *Molecular Cell Biology* **15**, 6406–6419.

Kalish, J. E., Keller, G.-A., Morrel, J. C., Mihalik, S. J., Smith, B., Cregg, J. M. & Gould, S. J. (1996). Characterisation of a novel component of the peroxisomal import apparatus using fluorescent peroxisomal proteins. *EMBO Journal* **15**, 3275–3285.

Kamiryo, T., Sakasegawa, Y. & Tan, H. (1989). Expression and transport of *Candida tropicalis* peroxisomal acyl-coenzyme A oxidase in the yeast *Candida maltosa*. *Agricultural and Biological Chemistry* **53**, 179–186.

Keller, G.-A., Krisans, S., Gould, S. J., Sommer, J. M., Wang, C. C., Schliebs, W., Kunau, W., Brody, S. & Subramani, S. (1991). Evolutionary conservation of a microbody targeting signal that targets proteins to peroxisomes, glyoxysomes and glycosomes. *Journal of Cell Biology* **114**, 893–904.

Kim, J., Scott, S. V., Oda, M. N. & Klionsky, D. J. (1997). Transport of a large oligomeric protein by the cytoplasm to vacuole protein targeting pathway. *Journal of Cell Biology* **137**, 609–618.

van der Klei, I. J., Hilbrands, R. E., Swaving, G. J., Waterham, H. R., Vrieling, E. G., Titorenko, V. I., Cregg, J. M., Harder, W. & Veenhuis, M. (1995). The *Hansenula polymorpha* PER3 gene is essential for import of PTS1 proteins into the peroxisomal matrix. *Journal of Biological Chemistry* **270**, 17229–17236.

van der Klei, I. J., Hilbrands, R. E., Kiel, J. A. K. W., Rasmussen, S. W., Cregg, J. M. & Veenhuis, M. (1998). The ubiquitin-conjugating enzyme Pex4p of *Hansenula polymorpha* is required for efficient functioning of the PTS1 import machinery. *EMBO Journal* **17**, 3608–3618.

Komori, M., Rasmussen, S. W., Kiel, J. A. K. W., Baerends, R. J. S., Cregg, J. M., van der Klei, I. J. & Veenhuis, M. (1997). The *Hansenula polymorpha* PEX14 gene encodes a novel peroxisomal membrane protein essential for peroxisome biogenesis. *EMBO Journal* **16**, 44–53.

Kragler, F., Langeder, A., Raupachova, J., Binder, M. & Harting, A. (1993). Two independent peroxisomal targeting signals in catalase A of *Saccharomyces cerevisiae*. *Journal of Cell Biology* **129**, 665–673.

Lazarow, P. B. & Fujiki, Y. (1985). Biogenesis of peroxisomes. *Annual Review of Cell Biology* **1**, 489–530.

Lee, M. S., Mullen, R. T. & Trelease, R. N. (1997). Oilseed isocitrate lyases lacking their essential type 1 peroxisomal targeting signal are piggybacked to glyoxysomes. *Plant Cell* **9**, 185–197.

Lemmens, M., Verheyden, K., Van Veldhoven, P., Vereecke, J., Mannaerts, G. P. & Carmeliet, E. (1989). Single-channel analysis of a large conductance channel in peroxisomes from rat liver. *Biochimica et Biophysica Acta* **984**, 351–359.

Lopez-Huertas, E., Sandalio, L. M. & del Río, L. A. (1995). Integral membrane polypeptides of pea leaf peroxisomes: characterization and response to plant stress. *Plant Physiology and Biochemistry* **33**, 295–302.

Luster, D. G., Bowditch, M. G., Eldridge, K. M. & Donaldson, R. P. (1988). Characterization of membrane-bound electron transport enzymes from castor bean glyoxysomes and endoplasmic reticulum. *Archives of Biochemistry and Biophysics* **265**, 50–61.

McCammon, M. T., McNew, J. A., Willy, P. J. & Goodman, J. M. (1994). An internal region of the peroxisomal membrane protein PMP47 is essential for sorting to peroxisomes. *Journal of Cell Biology* **124**, 915–925.

McCollum, E., Monosov, E. & Subramani, S. (1993). The Pas8 mutant of *Pichia pastoris* exhibits the peroxisomal protein import deficiencies of Zellweger syndrome cells – the PAS8 protein binds to the COOH-terminal tripeptide peroxisomal targeting signal, and is a member of the TPR protein family. *Journal of Cell Biology* **121**, 761–774.

McNew, J. A. & Goodman, J. M. (1994). An oligomeric protein is imported into peroxisomes in vivo. *Journal of Cell Biology* **127**, 1245–1257.

McNew, J. A. & Goodman, J. M. (1996). The targeting and assembly of peroxisomal proteins: some old rules do not apply. *Trends in Biochemical Sciences* **21**, 54–58.

Marshall, P. A., Dyer, J. M., Quick, M. E. & Goodman, J. M. (1996). Redox-sensitive homodimerization of Pex11p: a proposed mechanism to regulate peroxisomal division. *Journal of Cell Biology* **35**, 123–137.

Marzioch, M. R., Erdmann, R., Veenhuis, M. & Kunau, W.-H. (1994). PAS7 encodes a novel yeast member of the WD-40 protein family essential for import of 3-oxoacyl-CoA thiolase, a PTS2-containing protein, into peroxisomes. *EMBO Journal* **13**, 4908–4918.

Mathieu, M., Zeelen, J. P., Pauptit, R. A., Erdmann, R., Kunau, W.-H. & Wierenga, R. K. (1994). The 2.8 Å crystal structure of peroxisomal 3-ketoacyl-CoA thiolase of *Saccharomyces cerevisiae*: a five layered $\alpha\beta\alpha\beta\alpha$ structure constructed from 2 core domains of identical topology. *Structure* **2**, 797–808.

Motley, A., Hettema, E., Distel, B. & Tabak, H. (1994). Differential protein import deficiencies in human peroxisome assembly disorders. *Journal of Cell Biology* **125**, 755–767.

Motley, A. M., Hettema, E. H., Hogenhout, E. M. & 8 other authors (1997). Rhizomelic chondrodysplasia punctata is a peroxisomal protein targeting disease caused by a non-functional PTS2 receptor. *Nature Genetics* **15**, 377–380.

Nicolay, K., Veenhuis, M., Douma, A. C. & Harder, W. (1987). A 31P NMR assay study of the internal pH of yeast peroxisomes. *Archives of Microbiology* **147**, 37–41.

Nishimura, M., Yamaguchi, J., Mori, H., Akazawa, T. & Yokota, S. (1986). Immunocytochemical analysis shows that glyoxysomes are directly transformed to leaf peroxisomes during greening of pumpkin cotyledons. *Plant Physiology* **81**, 313–316.

Nuttley, W. M., Brade, A. M., Gaillardin, C., Eitzen, G. A. & Glover, J. R. (1993). Rapid identification and characterisation of peroxisome assembly mutants in *Yarrowia lipolytica*. *Yeast* **9**, 507–517.

Okumoto, K., Itoh, R., Shimozawa, N., Suzuki, Y., Tamura, S., Kondo, N. & Fujiki, Y. (1998a). Mutations in PEX10 is the cause of Zellweger peroxisome deficiency syndrome of complementation group B. *Human Molecular Genetics* **7**, 1399–1405.

Okumoto, K., Shimozawa, N., Kawai, A. & 8 other authors (1998b). PEX12, the pathogenic gene of group III Zellweger Syndrome: cDNA cloning by functional complementation of a CHO cell mutant, patient analysis and characterisation of Pex12p. *Molecular Cell Biology* **18**, 4324–4336.

Olsen, L. (1998). The surprising complexity of peroxisome biogenesis. *Plant Molecular Biology* **38**, 163–189.

Otera, H., Okumoto, K., Tateishi, K. & 8 other authors (1998). Peroxisomal targeting signal type 1 receptor is involved in the import of both PTS1 and PTS2; studies with PEX5 deficient CHO cell mutants. *Molecular Cell Biology* **18**, 388–399.

Palma, J. M., Garrido, M., Rodriguez-Garcia, M. I. & del Río, L. A. (1991). Peroxisome proliferation and oxidative stress mediated by activated oxygen species in plant peroxisomes. *Archives of Biochemistry and Biophysics* **287**, 68–74.

Passreiter, M., Anton, M., Lay, D., Frank, R., Harter, C., Wieland, F. T., Gorgas, K. & Just, W. W. (1998). Peroxisome biogenesis: involvement of ARF and Coatomer. *Journal of Cell Biology* **141**, 373–383.

Patel, S. & Latterich, M. (1998). The AAA team: related ATPases with diverse functions. *Trends in Cell Biology* **8**, 65–71.

Pause, B., Diestelkotter, P., Heid, H. & Just, W. W. (1997). Cytosolic factors mediate protein insertion into the peroxisomal membrane. *FEBS Letters* **414**, 95–98.

Pool, M. R., Lopez-Huertas, E. & Baker, A. (1998). Characterization of intermediates in the process of plant peroxisomal protein import. *EMBO Journal* **17**, 6854–6862.

Portsteffen, H., Beyer, A., Becker, E., Epplen, C., Pawlak, A., Kunau, W.-H. & Dodt, G. (1997). Human PEX1 is mutated in complementation group 1 of the peroxisome biogenesis disorders. *Nature Genetics* **17**, 449–452.

Preisig-Müller, R., Muster, G. & Kindl, H. (1994). Heat shock enhances the amount of prenylated DnaJ protein at membranes of glyoxysomes. *European Journal of Biochemistry* **219**, 57–63.

Pugsley, A. P. (1993). The complete general secretory pathway in gram negative bacteria. *Microbiological Reviews* **57**, 50–108.

Pugsley, A. P., Francetic, O., Possot, O. M., Sauvonnet, N. & Hardie, K. R. (1997). Recent progress and future directions in studies of the main terminal branch of the general secretory pathway in Gram negative bacteria – a review. *Gene* **192**, 12–19.

Purdue, P. E., Zhang, J. W., Skoneczny, M. & Lazarow, P. B. (1997). Rhizomelic chondrodysplasia punctata is caused by deficiency of human PEX7, a homologue of the yeast PTS2 receptor. *Nature Genetics* **15**, 381–384.

Rachubinski, R. A. & Subramani, S. (1995). How proteins penetrate peroxisomes. *Cell* **83**, 525–528.

Rapp, S., Soto, U. & Just, W. W. (1993). Import of firefly luciferase into peroxisomes of permeabilized Chinese hamster ovary cells: a model system to study peroxisomal protein import in vivo. *Experimental Cell Research* **205**, 59–65.

Rassow, J., Voos, W. & Pfanner, N. (1995). Partner proteins determine multiple functions of Hsp70. *Trends in Cell Biology* **5**, 207–212.

Rehling, P., Marzioch, M., Niesen, F., Wittke, E., Veenhuis, M. & Kunau, W.-H. (1996). The import receptor for the peroxisomal targeting signal 2 (PTS2) in *Saccharomyces cerevisiae* is encoded by the PAS7 gene. *EMBO Journal* **15**, 2901–2913.

Reuber, B. E., Germain-Lee, E., Collins, C. S., Morrell, J. C., Ameritunga, R., Moser, H. W., Valle, D. & Gould, S. J. (1997). Mutations in PEX1 are the most common cause of peroxisome biogenesis disorders. *Nature Genetics* **17**, 445–448.

Reumann, S., Maier, E., Benz, R. & Heldt, H. W. (1995). The membrane of leaf peroxisomes contains a porin-like channel. *Journal of Biological Chemistry* **270**, 17559–17565.

Reumann, S., Bettermann, M., Bent, R. & Heldt, H. W. (1997). Evidence for the presence of a porin in the membrane of glyoxysomes of *Ricinus communis* L. *Plant Physiology* **115**, 891–899.

Reumann, S., Bettermann, M., Bent, R. & Heldt, H. W. (1998). Permeability properties of the porin of spinach leaf peroxisomes. *European Journal of Biochemistry* **251**, 359–366.

del Río, L. A., Sandalio, L. M., Palma, J. M., Bueno, P. & Corpas, F. J. (1992). Metabolism of oxygen radicals in peroxisomes and subcellular implications. *Free Radicals in Biology and Medicine* **13**, 557–580.

del Río, L. A., Pastori, G. M., Palma, J. M., Sandalio, L. M., Sevilla, F., Corpas, F. J., Jimenez, A., Lopez-Huertas, E. & Hernandez, J. A. (1998). The activated oxygen role of peroxisomes in senescence. *Plant Physiology* **116**, 1195–1200.

van Roermund, C. W. T., Elgersma, Y., Singh, N., Wanders, R. J. A. & Tabak, H. F. (1995). The membrane of peroxisomes in *Saccharomyces cerevisiae* is impermeable to NAD(H) and acetyl-CoA under in vivo conditions. *EMBO Journal* **14**, 3480–3486.

Sakai, Y., Marshall, P. A., Saiganji, A., Takabe, K., Saiki, H., Kato, N. & Goodman, J. (1995). The *Candida boidinii* peroxisomal membrane protein Pmp30 has a role in peroxisomal proliferation and is functionally homologous to Pmp27 from *Saccharomyces cerevisiae*. *Journal of Bacteriology* **177**, 6773–6781.

Salomons, F. A., van der Klei, I. J., Kram, A. M., Harder, W. & Veenhuis, M. (1997). Brefeldin A interferes with peroxisomal protein sorting in the yeast *Hansenula polymorpha*. *FEBS Letters* **411**, 133–139.

Scott, S. V., Hefner-Gravink, A., Morano, K. A., Noda, T., Ohsumi, Y. & Klionsky, D. J. (1996). Cytoplasm-to-vacuole targeting and autophagy employ the same machinery to deliver proteins to the yeast vacuole. *Proceedings of the National Academy of Sciences, USA* **93**, 12304–12308.

Small, G. M., Szabo, L. J. & Lazarow, P. B. (1988). Acyl-CoA oxidase contains two targeting sequences each of which can mediate import into peroxisomes. *EMBO Journal* **7**, 1167–1173.

Subramani, S. (1997). Pex genes on the rise. *Nature Genetics* **15**, 331–333.

Subramani, S. (1998). Components involved in peroxisome import, biogenesis, proliferation, turnover and movement. *Physiological Reviews* **78**, 171–188.

Sulter, G. J., Verheyden, K., Mannaerts, G., Harder, W. & Veenhuis, M. (1993). The in vitro permeability of yeast peroxisomal membranes is caused by a 31 kDa integral membrane protein. *Yeast* **9**, 733–742.

Swinkels, B. W., Gould, S. J., Bodnar, A. G., Rachubinski, R. A. & Subramani, S. (1991). A novel cleavable peroxisomal targeting signal at the aminoterminus of the rat 3-ketoacyl-CoA thiolase. *EMBO Journal* **10**, 3255–3262.

Terlecky, S., Nuttley, W. M., McCollum, D., Sock, E. & Subramani, S. (1995). The *Pichia pastoris* peroxisomal protein Pas8p is the receptor for the C-terminal tripeptide peroxisomal targeting signal. *EMBO Journal* **14**, 3627–3634.

Titorenko, V. I. & Rachubinski, R. A. (1998). Mutants of the yeast *Yarrowia lipolytica* defective in protein exit from the endoplasmic reticulum are also defective in peroxisome biogenesis. *Molecular Cell Biology* **18**, 2789–2803.

Titorenko, V. I., Smith, J. J., Szilard, R. K. & Rachubinski, R. A. (1998). Pex20p of the yeast *Yarrowia lipolytica* is required for the oligomerization of thiolase in the cytosol and its targeting to the peroxisome. *Journal of Cell Biology* **142**, 403–420.

Titus, D. E. & Becker, W. M. (1985). Investigation of the glyoxysome-peroxisome transition in germinating cucumber cotyledons using double label immunoelectron microscopy. *Journal of Cell Biology* **101**, 1288–1299.

Trelease, R. N. (1984). Biogenesis of glyoxysomes. *Annual Review of Plant Physiology* **35**, 321–347.

Tsukamoto, T., Miura, S. & Fujiki, Y. (1991). Restoration by a 35K membrane protein of peroxisome assembly in a peroxisome-deficient mammalian cell mutant. *Nature* **350**, 77–81.

Van der Leij, I., van den Berg, M., Boot, R., Franse, M., Distel, B. & Tabak, H. F. (1992). Isolation of peroxisome assembly mutants from *Saccharomyces cerevisiae* with different morphologies using a novel positive selection procedure. *Journal of Cell Biology* **119**, 153–162.

Van Veldhoven, P. P., Just, W. W. & Mannaerts, G. P. (1987). Permeability of the peroxisomal membrane to cofactors of beta-oxidation: evidence for the presence of a pore forming protein. *Journal of Biological Chemistry* **262**, 4310–4318.

Verheyden, K., Fransen, M., Van Veldhoven, P. P. & Mannaerts, G. P. (1992). Presence of small GTP-binding proteins in the peroxisomal membrane. *Biochimica et Biophysica Acta* **1109**, 48–54.

Volokita, M. (1991). The carboxy-terminal end of glycolate oxidase directs a foreign protein into tobacco leaf peroxisomes. *Plant Journal* **1**, 361–366.

Walton, P. A., Wendland, M., Subramani, S., Rachubinsky, R. A. & Welch, W. J. (1994). Involvement of 70-kD heat shock proteins in peroxisomal import. *Journal of Cell Biology* **125**, 1037–1046.

Walton, P. A., Hill, P. E. & Subramani, S. (1995). Import of stably folded proteins into peroxisomes. *Molecular Biology of the Cell* **6**, 675–683.

Waterham, H. R., Keizer-Gunnick, I., Goodman, J. M., Harder, W. & Veenhuis, M. (1990). Immunocytochemical evidence for the acidic nature of peroxisomes in methylotropic yeasts. *FEBS Letters* **262**, 17–19.

Waterham, H. R., Russell, K. A., deVries, Y. & Cregg, J. M. (1997). Peroxisomal targeting, import and assembly of alcohol oxidase in *Pichia pastoris*. *Journal of Cell Biology* **139**, 1419–1431.

Wendland, M. & Subramani, S. (1993). Cytosol-dependent peroxisomal protein import in a permeabilized cell system. *Journal of Cell Biology* **120**, 675–685.

Wexler, M., Bogsch, E. G., Klosgen, R. B., Palmer, T., Robinson, C. & Berks, B. C. (1998). Targeting signals for a bacterial Sec-independent export system direct plant thylakoid import by the ΔpH pathway. *FEBS Letters* **431**, 339–342.

Whitney, A. B. & Bellion, E. (1991). ATPase activities in peroxisome-proliferating yeast. *Biochimica et Biophysica Acta* **1058**, 345–355.

Wiemer, E. A. C., Luers, G. H., Faber, K. N., Wenzel, T., Veenhuis, M. & Subramani, S. (1996). Isolation and characterization of Pas2p, a peroxisomal membrane protein essential for peroxisome biogenesis in the methylotrophic yeast *Pichia pastoris*. *Journal of Biological Chemistry* **271**, 18973–18980.

Wimmer, B., Lottspeich, F., van der Klei, I., Veenhuis, M. & Gietl, C. (1997). The glyoxysomal and plastid molecular chaperones (70 kda heat shock protein) of watermelon cotyledons are encoded by a single gene. *Proceedings of the National Academy of Sciences, USA* **94**, 13624–13629.

Yahraus, T., Braverman, N., Dodt, G., Kalish, J. E., Morrell, J. C., Moser, H. W., Valle, D. & Gould, S. J. (1996). The peroxisome biogenesis disorder group 4 gene PXAAA1 encodes a cytoplasmic ATPase required for the stability of the PTS1 receptor. *EMBO Journal* **15**, 2914–2923.

Zhang, J. W. & Lazarow, P. B. (1995). PEB1 (PAS7) in *Saccharomyces cerevisiae* encodes a hydrophilic intraperoxisomal protein which is a member of the WD40 repeat family and is essential for the import of thiolase into peroxisomes. *Journal of Cell Biology* **129**, 65–80.

Zhang, J. W. & Lazarow, P. B. (1996). Peb1p (Pas7p) is an intraperoxisomal receptor for the NH₂-terminal, type 2, peroxisomal targeting sequence of thiolase: Peb1p itself is targeted to peroxisomes by an NH₂-terminal peptide. *Journal of Cell Biology* **132**, 325–334.

Zhang, W. J., Han, Y. & Lazarow, P. B. (1993). Novel peroxisome clustering mutants and peroxisome biogenesis mutants of *Saccharomyces cerevisiae*. *Journal of Cell Biology* **123**, 1133–1147.

TRANSPORT OF PROTEINS INTO AND ACROSS THE THYLAKOID MEMBRANE

COLIN ROBINSON, WAYNE EDWARDS, PETER J. HYNDS AND CHRISTOPHE TISSIER

Department of Biological Sciences, University of Warwick, Coventry CV4 7AL, UK

INTRODUCTION

The chloroplast carries out the crucial processes of light capture, photosynthetic electron transport, ATP synthesis and carbon dioxide fixation, as well as a variety of other important metabolic functions. These functions require the presence of numerous proteins – upwards of 1000 – amongst which are the most abundant proteins in the biosphere. It is now widely accepted that chloroplasts arose from endosymbiotic cyanobacteria, and they still contain their own prokaryotic-type genetic system. However, most of the genes have been transferred to the nucleus and higher plant chloroplasts now synthesize only about 90 of their resident polypeptides. As a consequence, chloroplast biogenesis requires the import of proteins from the cytosol on a massive scale. This process has been intensively studied over the last two decades but another facet of chloroplast biogenesis has attracted at least as much attention: the problem of *intraorganellar protein sorting*. This is a major issue because chloroplasts are highly complex in terms of structure, containing six distinct compartments: the outer and inner envelope membranes, the intermembrane space enclosed by these membranes (about which very little is known), the soluble stromal phase (site of most of the metabolic pathways), internal thylakoid membrane, and finally the lumenal space enclosed by the thylakoid membrane. Cytosolically synthesized proteins are imported and directed to all of these compartments, and studies on chloroplast protein biogenesis have thus addressed two basic questions: how are proteins imported into this organelle and how are they 'sorted' so that they arrive at the correct destination.

The import of thylakoid proteins has attracted particular interest because of the complexity of the pathway. These studies are moreover facilitated by the abundance of this membrane and its component proteins; the thylakoid membrane accounts for up to 99% of chloroplast lipid and it is literally packed with protein. Thylakoid membrane proteins have to be transported across both envelope membranes and the stromal phase in order to reach their destination; numerous proteins are resident in the thylakoid membrane and these proteins must be additionally transported across the thylakoid membrane before their targeting pathway is completed. *In vitro* assays have been used to dissect the steps involved and it is now clear that multiple mechanisms are used for the targeting of thylakoid proteins. In this article we review recent progress in this area which has resulted in a much-improved understanding of the rationale behind this targeting complexity. We also describe how studies on the targeting of certain thylakoid lumen proteins have resulted in the discovery of a novel type of translocation system that has now been found to be present in a wide range of bacteria.

A TWO-STEP TARGETING PATHWAY FOR IMPORTED THYLAKOID LUMEN PROTEINS

The thylakoid lumen contains several well-characterized photosynthetic proteins such as plastocyanin and the 33, 23 and 16 kDa proteins (33K, 23K, 16K) of the oxygen-evolving complex. All are synthesized in the cytosol with bipartite presequences in which two distinct domains are present: an envelope transfer domain followed by a thylakoid-targeting domain. The envelope transit domain directs import of the protein into the stroma and is structurally and functionally equivalent to the targeting signal presequences of imported stromal proteins (Ko & Cashmore, 1989; Hageman *et al.*, 1990). These signals are hydrophilic, basic and enriched in hydroxylated amino acids (von Heijne *et al.*, 1989; Robinson *et al.*, 1998), and are usually removed in the stroma. The thylakoid-targeting domains, on the other hand, can be divided into three distinct domains that very much resemble those of classical 'signal peptides' that direct the export of proteins by the secretory (Sec) pathway in bacteria. These peptides are characterized by the presence of a positively charged amino-terminal domain (N-region), hydrophobic core region (H-region) and a more polar carboxy terminal (C-) region ending with short-chain residues (usually Ala) at the -3 and -1 positions, relative to the terminal processing site. Bacterial signal peptides have long been known to direct the export of periplasmic and outer-membrane proteins by the 'general secretory pathway' using the Sec apparatus to achieve translocation across the inner membrane (reviewed by Pugsley, 1993; Izard & Kendall, 1994). Hence, these observations suggested at first that thylakoid lumen proteins must be transported by a Sec-type system in the thylakoid membrane – a reasonable proposition since chloroplasts are widely believed to have originated from endosymbiotic cyanobacteria. Moreover, these peptides are cleaved after entry into the thylakoid lumen by a signal-type peptidase that is clearly homologous to bacterial signal peptidases (Halpin *et al.*, 1989; Chaal *et al.*, 1998). It is only within the last few years that a far more complex story has emerged.

TWO DISTINCT MECHANISMS FOR PROTEIN TRANSLOCATION ACROSS THE THYLAKOID MEMBRANE

Although all known lumenal proteins are synthesized with superficially similar targeting signals, the advent of assays for the import of proteins into isolated plant thylakoids revealed the surprising finding that lumenal proteins fell into two clear groups in terms of their requirements for translocation across the thylakoid membrane. Plastocyanin and 33K are dependent on stromal protein factors and ATP for import (Hulford et al., 1994), while a transthylakoid proton gradient is not required although the ΔpH does enhance translocation to a degree. These findings were consistent with transport by a Sec pathway, and the stromal factor has since been identified as SecA, the translocation ATPase that drives the insertion and transport of the pre-protein (Yuan et al., 1994; Nakai et al., 1994). However, other lumenal proteins, including 23K and 16K, required neither stromal factors nor any form of nucleoside triphosphate (NTP) for translocation, and import was instead totally reliant on the thylakoidal ΔpH (Mould & Robinson, 1991; Cline et al., 1992). Further studies revealed that these proteins are in fact translocated by two entirely separate pathways. Competition experiments (Cline et al., 1993) showed that 33K and plastocyanin compete with each other for transport across the thylakoid membrane, as do 23K and 16K, but the two groups of proteins do not compete with each other. Finally, it was found that replacement of the plastocyanin thylakoid-targeting signal with that of 23K led to a quantitative rerouting of the protein from the Sec pathway to the ΔpH pathway (Robinson et al., 1994; Henry et al., 1994). These studies proved conclusively that the two pathways are different in terms of both their operating requirements and the targeting signals they recognize. These studies have been followed by others in which further lumenal proteins were assigned to one or other pathway (reviewed by Robinson et al., 1998).

The discovery of parallel targeting pathways for lumen proteins was unexpected because, as mentioned above, the targeting signals for lumenal proteins all resemble typical bacterial signal peptides to a large extent. All contain broadly similar hydrophobic core regions and an Ala-X-Ala consensus motif which specifies cleavage by the thylakoidal processing peptidase (Robinson et al., 1998). However, signals for the ΔpH system contain a common feature: a twin-arginine motif just before the start of the H-region. This motif is crucial for targeting by the ΔpH pathway, and alteration of either residue, even to Lys, almost completely blocks translocation across the thylakoid membrane (Chaddock et al., 1995). This in itself is not sufficient to dictate ΔpH-dependent targeting, however, and it has been shown more recently that a second important determinant is the presence of a highly hydrophobic residue two or three residues after the twin-Arg (Brink et al., 1998).

The available evidence indicates that the known lumenal proteins are transported very specifically by only a single pathway, and there is as yet no evidence for cross-talk between pathways. It is perhaps not surprising that Sec substrates are not recognized by the ΔpH pathway because none has so far been found to contain the required twin-Arg motif. However, it is interesting that ΔpH substrates fail to be transported by the Sec pathway when, for example, transport is totally blocked by uncouplers or removal of the twin-Arg motif (Cline *et al*., 1992; Chaddock *et al*., 1995). It turns out that two factors prevent this. Most signals for the ΔpH pathway contain a basic residue in the C-region, as well as a twin-Arg in the N-region. Mutagenesis studies on the pre-plastocyanin targeting signal showed that the presence of *either* feature is no impediment to Sec-dependent translocation, but the *combined* presence of these features totally blocks transport (Bogsch *et al*., 1997). These results suggest that this combination of charges may represent a 'Sec-avoidance' signal, although further studies are required to determine whether this holds true for other ΔpH targeting signals.

The second feature concerns the mature proteins themselves. Studies using chimeric proteins have shown quite clearly that the mature proteins targeted by the ΔpH pathway are inherently difficult for the Sec system to translocate, even when authentic signal peptides are attached (Clausmeyer *et al*., 1989; Henry *et al*., 1997; Bogsch *et al*., 1997). Some proteins, for example 23K, can not be transported at all by the Sec pathway. Thus, some feature of these proteins makes them inherently 'Sec-incompatible'. The underlying reasons have not yet been established but one possibility is that the ΔpH-dependent proteins may fold too tightly for the Sec system to handle (see below). Whatever the reason, these results are of interest because they provide a rationale for the presence of a second, Sec-independent pathway in chloroplasts.

THE ΔpH-DRIVEN TRANSLOCASE TRANSPORTS FOLDED PROTEINS

The operational requirements of the ΔpH-dependent system are unique because no other form of protein translocase requires neither soluble proteins nor NTPs, but recent studies have revealed that this system is indeed very different from all other known translocation systems. Two studies have shown that the system is able to transport fully folded proteins – the first demonstration that this can occur across tightly sealed membranes. Clark & Theg (1997) coupled the targeting signal of a ΔpH-dependent protein to bovine pancreatic trypsin inhibitor (BPTI) and showed that the construct was efficiently targeted into the lumen even when the BPTI domain was locked in a folded position by internal cross-linking. Similar findings were made by Hynds *et al*. (1998) using a chimera comprising pre-23K linked to dihydrofolate reductase (DHFR). DHFR has been a favourite tool in protein targeting studies because it is stabilized to a high degree by the binding of the folate analogue methotrexate. Under these conditions, it can not be trans-

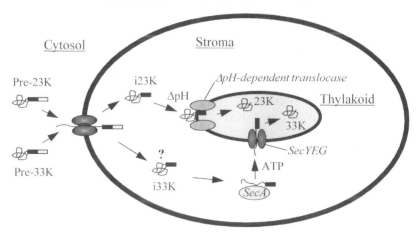

Fig. 1. Pathways for the targeting of thylakoid lumen proteins in chloroplasts. The figure illustrates the differing characteristics of the ΔpH- and Sec-dependent pathways. Both types of protein are synthesized with bipartite presequences containing envelope transit signals and thylakoid-targeting signals in tandem. The former are recognized by a protein transport system(s) in the envelope membranes, which facilitates translocation into the stroma. The envelope transit signals are usually removed at this point and the resultant intermediate forms are directed along two distinct routes. Proteins such as 23K and 16K are believed to refold in the stroma before being transported in a folded form by a ΔpH-driven translocase; other lumenal proteins, such as 33K and plastocyanin, are transported by the Sec route. It is presently unclear whether these proteins fold in the stroma; the later stages involve SecA driving translocation of the intermediate in an unfolded state through the membrane-bound Sec apparatus. After translocation, substrates on both pathways are processed to the mature forms by the thylakoidal processing peptidase.

ported across the mitochondrial envelope, the endoplasmic reticulum or across the bacterial plasma membrane by the Sec pathway (e.g. Eilers & Schatz, 1986). This is because these systems all transport folded proteins in an unfolded state, and the same holds true for the chloroplast envelope although the unfolding action of this translocase is sufficiently powerful to enable translocation even of DHFR with pre-bound methotrexate (Guéra *et al.*, 1993; America *et al.*, 1994). This has been considered a logical evolutionary step because all of these membranes have to avoid leakage of ions or protons and this would appear to be easier if proteins are 'threaded' across the bilayer. However, the thylakoidal ΔpH-dependent system is able to transport DHFR with methotrexate bound in the active site (Hynds *et al.*, 1998), effectively demonstrating the translocation of a fully folded protein. The transport of folded proteins has been demonstrated across the peroxisome membrane and the outer membranes of some bacteria (Pugsley, 1993; Walton *et al.*, 1995) but the above studies provided the first direct demonstration that folded globular proteins could be transported across energy-transducing membranes that are designed to be impermeable to protons. Fig. 1 depicts the key features of the ΔpH- and Sec-dependent pathways for the targeting of thylakoid lumen proteins.

A RELATED SYSTEM IN BACTERIA

Until relatively recently, the origins of the ΔpH-dependent system were a complete mystery. However, elegant work by Voelker & Barkan (1995) succeeded in isolating the first plant mutants defective in this pathway and one of the mutants, designated *hcf106*, was shown to be specifically affected in the ΔpH pathway. Because the maize mutant was created by transposon insertion, this facilitated the isolation of the disrupted gene and the sequence of the Hcf106 protein was published (Settles *et al.*, 1997). The Hcf106 protein is indeed located in the thylakoid membrane and the sequence data predict a single-span protein with a soluble domain exposed on the stromal face. Most importantly of all, however, homologues of Hcf106 were found to be encoded by unassigned reading frames in numerous bacteria, suggesting for the first time that prokaryotes possess a second, Sec-independent protein transport system.

This scenario was predicted to some extent by Berks (1996), who analysed the signal peptides of bacterial periplasmic proteins binding any of a range of redox cofactors such as FeS or molybdopterin centres; most of the proteins are involved in anaerobic respiration pathways. He speculated that these proteins may be exported by a Sec-independent process because the cofactors are almost certainly inserted in the cytoplasm, necessitating the export of a folded protein (supposedly impossible by the Sec pathway). He furthermore noted that all of these periplasmic proteins are synthesized with signal peptides containing twin-Arg motifs and speculated that they might be exported by a system similar to the thylakoidal ΔpH-dependent one. This scenario has proved to be correct.

The *Escherichia coli* genome encodes two *hcf106* homologues, one of which is monocistronic whereas the other forms the first gene in an apparent four-gene operon. Disruption of either gene affects the export of several proteins bearing the above types of redox factors (Sargent *et al.*, 1998) and these genes have been designated *tat* genes (for twin-arginine translocase). A double disruption leads to a complete block in this export pathway. Other studies have shown that the TatB and TatC proteins are also important in this process (Weiner *et al.*, 1998; Bogsch *et al.*, 1998) and it is now clear from database searches that this pathway operates in a wide range of prokaryotes. The two contrasting pathways for protein export in *E. coli* are illustrated in Fig. 2.

TARGETING OF THYLAKOID MEMBRANE PROTEINS IN CHLOROPLASTS

The thylakoid membrane contains numerous integral membrane proteins, most of which are imported from the cytosol. Studies on the insertion of these proteins have revealed at least two further pathways, and the emerging data

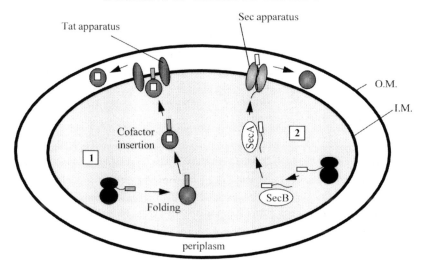

Fig. 2. Protein export pathways in bacteria. [1] The Tat pathway. Most of the substrates for this pathway bind redox cofactors and function in the periplasm. After synthesis in the cytoplasm, the substrates are predicted to fold and bind the cofactor (white square), after which the pre-protein is transferred through the Tat apparatus in a folded form. [2] The Sec route. Soon after emerging from the ribosome, the pre-protein is usually bound by the SecB protein, which prevents folding. The substrate is then transferred to SecA, which drives ATP-dependent translocation through the SecYEG complex. Additional SecD and SecF proteins modulate translocation but are omitted for simplicity. O.M, I.M., outer and inner membranes.

suggest an unexpected diversity of requirements. For a long time, only one protein was analysed in any detail: the light-harvesting chlorophyll-binding protein of photosystem II (Lhcb1). This protein spans the thylakoid membrane three times and is synthesized in the cytosol as a larger precursor. Unlike the presequences of lumenal proteins, the Lhcb1 presequence represents a single targeting domain that specifies translocation only as far as the stroma; additional information in the mature protein specifies insertion into the thylakoid membrane (Lamppa, 1988; Viitanen et al., 1988). The Lhcb1 insertion process has been reconstituted using isolated thylakoids and shown to require a stromal homologue of bacterial signal recognition particle (SRP) together with GTP (Hoffman & Franklin, 1994; Li et al., 1995). In addition, protein transport apparatus in the thylakoid membrane is essential, since proteolysis of thylakoids destroys their ability to integrate Lhcb1 (Robinson et al., 1996). These findings strongly suggest that this pathway, like the Sec- and ΔpH-dependent pathways, was inherited from the cyanobacterial-like progenitors of chloroplasts. There is now clear evidence that many inner membrane proteins use an SRP-dependent route for insertion in bacteria, and many of these proteins are predicted to require the membrane-bound Sec machinery for the final insertion step (de Gier et al., 1997; Ulbrandt et al., 1997). It appears that SRP binds preferentially to highly hydrophobic molecules; stromal SRP has been shown to interact strongly with Lhcb1 but not with the more hydrophilic signal peptides of the signal peptides within lumenal protein precursors (High *et al.*, 1997). There is also evidence for the involvement of an additional targeting factor, FtsY, in the targeting of bacterial membrane proteins (reviewed by de Gier *et al.*, 1997) and it seems likely that a homologue is likewise involved in the targeting of SRP-dependent thylakoid membrane proteins.

A NOVEL PATHWAY FOR THE INSERTION OF THYLAKOID MEMBRANE PROTEINS

While the Sec-, ΔpH- and SRP-dependent pathways described above can be traced back to the initial prokaryotic endosymbionts that gave rise to chloroplasts, there is clear evidence that chloroplasts possess a fourth protein targeting pathway that has few, if any, parallels in other types of organism or organelle. A series of single-span membrane proteins, namely subunit II of the CF_o complex (CF_oII) and subunits X and W of photosystem II (PsbX, PsbW), are synthesized with bipartite presequences that very much resemble those of imported lumenal proteins, and which are similarly removed by the thylakoidal processing peptidase on the lumenal side of the membrane. Surprisingly, however, their insertion into the thylakoid membrane does not require any of the factors or targeting machinery characterized for the other known pathways. Efficient insertion is observed in the absence of SecA, SRP, NTPs or the thylakoidal ΔpH (Michl *et al.*, 1994; Lorkovic *et al.*, 1995; Kim *et al.*, 1998), and proteolysis of thylakoids blocks the other pathways yet has no effect on the insertion of these three single-span proteins (Robinson *et al.*, 1996). Because the known transport machinery is not involved (other than cleavage by the thylakoidal processing peptidase), it is highly likely that these proteins insert spontaneously into the thylakoid membrane, although it can not yet be ruled out that other, as yet uncharacterized, thylakoid proteins may assist in the insertion process.

The CF_oII-type 'spontaneous' insertion mechanism is highly unusual because cleavable signal peptides almost always interact with proteinaceous targeting apparatus (SRP and the Sec61p complex in the ER membrane, or the Sec or Tat apparatus in bacteria/chloroplasts). Indeed, only one protein has previously been shown to be synthesized with a signal peptide yet inserted by an SRP/Sec-independent mechanism: the coat protein of bacteriophage M13. This single-span protein is synthesized as a precursor protein (procoat) in *E. coli* and elegant studies have shown that insertion takes place by an apparently spontaneous mechanism in which the two hydrophobic regions (one in the signal peptide, the other in the mature protein) insert cooperatively into the bilayer and flip the intervening region across. Cleavage by signal peptidase then yields the mature protein with the N-terminus in the

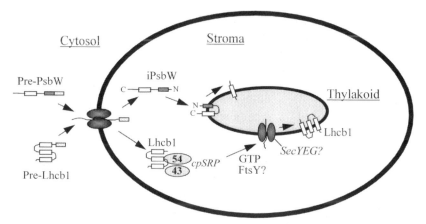

Fig. 3. Distinct routes for the insertion of thylakoid membrane proteins. Most multispanning proteins, such as the major light-harvesting chlorophyll-binding protein (Lhcb1), are synthesized only with stroma-targeting signals (shown as hatched rectangles); a series of single-span proteins such as PsbW are synthesized with bipartite presequences containing a stroma-targeting signal followed by a hydrophobic, cleavable signal peptide. Signal peptides are represented by grey rectangles and membrane-spanning regions of the mature proteins by white rectangles. Pre-Lhcb1 is imported, processed to the mature size in the stroma and bound by signal recognition particle (SRP), a complex of 54 kDa and 43 kDa subunits. This factor then mediates the GTP-dependent insertion into the thylakoid membrane, probably via the SecYEG complex used for a subset of lumenal proteins. By analogy with bacterial systems, an additional factor, FtsY, probably functions as a soluble signal peptide receptor mediating transfer to SecYEG. Pre-PsbW, on the other hand, is imported into the stroma and processed to an intermediate form, after which the protein inserts into the thylakoid membrane by an apparently spontaneous mechanism. In this process, the hydrophobic regions in the signal peptide and mature protein insert coordinately to drive translocation of the intervening region; cleavage by the thylakoidal processing peptidase then yields the mature protein.

periplasm (reviewed by Kuhn, 1995). A similar loop intermediate is formed during the insertion of at least one of the thylakoid proteins (PsbW), suggesting that, as in procoat insertion, the primary role of the signal peptide is to provide an additional hydrophobic region to assist in driving the translocation of the N-terminus of the mature protein (Thompson *et al.*, 1998). Fig. 3 shows the basic features of the SRP-dependent and spontaneous pathways for the insertion of thylakoid membrane proteins.

CONCLUDING REMARKS

A great deal of information has emerged in the last few years on the targeting of thylakoid proteins, and it is now clear that at least four distinct pathways operate for the targeting of nuclear-encoded proteins to this membrane system. Because only a small fraction of thylakoid proteins have been analysed (and *chloroplast*-encoded thylakoid proteins have hardly been studied at all) there is every chance that yet further mechanisms and pathways will emerge in the future. This pathway complexity was totally unexpected and one of the most notable points to emerge from these studies is the remarkable capabilities of the ΔpH-driven system. The presence of related systems in so many bacterial species strongly suggests that the translocation of folded proteins is far more widespread than previously thought and it will be very interesting indeed to determine the structure of this type of system and unravel its mode of action. To date, this is the only mainstream mechanism identified for the transport of folded proteins across tightly sealed biological membranes.

The biogenesis of membrane proteins is also better understood, largely due to the use of relatively efficient *in vitro* assays for the insertion of proteins into isolated thylakoid membranes. In a sense, the favoured approaches for the study of membrane proteins are diametrically opposite in the bacterial and chloroplast fields – a simple reflection of the advantages and limitations of the systems. Bacteria, and *E. coli* in particular, offer the advantage of rapid and efficient genetic analysis, and components of the translocation machinery have often been identified through the isolation of mutant strains. Once identified, the function of the gene product is usually studied by its selective depletion. However, this mode of analysis is often rather indirect and a recurring problem in the bacterial field is the difficulty encountered in the analysis *in vitro* of membrane protein insertion. Thylakoids, on the other hand, have turned out to be excellent systems for *in vitro* analysis whereas genetic approaches are far more difficult. Nevertheless, this combination of systems is now yielding a great deal of information on the pathways adopted by membrane proteins in both systems. Future studies should help to understand these processes in real detail and the prospects for structural and biophysical studies look good.

REFERENCES

America, T., Hageman, J., Guéra, A., Rook, F., Archer, K., Keegstra, K. & Weisbeek, P. (1994). Methotrexate does not block import of a DHFR fusion protein into chloroplasts. *Plant Molecular Biology* **24**, 283–294.

Berks, B. C. (1996). A common export pathway for proteins binding complex redox cofactors? *Molecular Microbiology* **22**, 393–404.

Bogsch, E., Brink, S. & Robinson, C. (1997). Pathway specificity for a ΔpH-dependent precursor thylakoid lumen protein is governed by a "Sec-avoidance" motif in the transfer peptide and a "Sec-incompatible" mature protein. *EMBO Journal* **16**, 3851–3858.

Bogsch, E. G., Sargent, F., Stanley, N. R., Berks, B. C., Robinson, C. & Palmer, T. (1998). An essential component of a novel bacterial protein export system with homologues in plastids and mitochondria. *Journal of Biological Chemistry* **273**, 18003–18006.

Brink, S., Bogsch, E. G., Edwards, W. R., Hynds, P. J. & Robinson, C. (1998). Targeting of thylakoid proteins by the ΔpH-driven twin-arginine translocation pathway requires a specific signal in the hydrophobic domain in conjunction with the twin-arginine motif. *FEBS Letters* **434**, 425–430.

Chaal, B. K., Mould, R. M., Barbrook, A. C., Gray, J. C. & Howe, C. J. (1998). Characterization of a cDNA encoding the thylakoidal processing peptidase from *Arabidopsis thaliana*. Implications for the origin and catalytic mechanism of the enzyme. *Journal of Biological Chemistry* **273**, 689–692.

Chaddock, A. M., Mant, A., Karnauchov, I., Brink, S., Herrmann, R. G., Klösgen, R. B. & Robinson, C. (1995). A new type of signal peptide: central role of a twin-arginine motif in transfer signals for the ΔpH-dependent thylakoidal protein translocase. *EMBO Journal* **14**, 2715–2722.

Clark, S. A. & Theg, S. M. (1997). A folded protein can be transported across the chloroplast envelope and thylakoid membranes. *Molecular Biology of the Cell* **8**, 923–934.

Clausmeyer, S., Klösgen, R. B. & Herrmann, R. G. (1989). Protein import into chloroplasts. The hydrophilic lumenal proteins exhibit unexpected import and sorting specificities in spite of structurally conserved transit peptides. *Journal of Biological Chemistry* **268**, 13869–13876.

Cline, K., Ettinger, W. F. & Theg, S. M. (1992). Protein-specific energy requirements for protein transport across or into thylakoid membranes. Two lumenal proteins are transported in the absence of ATP. *Journal of Biological Chemistry* **267**, 2688–2696.

Cline, K., Henry, R., Li, C. & Yuan, J. (1993). Multiple pathways for protein transport into or across the thylakoid membrane. *EMBO Journal* **12**, 4105–4114.

Eilers, M. & Schatz, G. (1986). Binding of a specific ligand inhibits import of a purified precursor protein into isolated mitochondria. *Nature* **322**, 228–232.

de Gier, J.-W. L., Valent, Q. A., von Heijne, G. & Luirink, J. (1997). The *E. coli* SRP: preferences of a targeting factor. *FEBS Letters* **408**, 1–4.

Guéra, A., America, T., van Waas, M. & Weisbeek, P. J. (1993). A strong protein unfolding activity is associated with the binding of precursor chloroplast proteins to chloroplast envelopes. *Plant Molecular Biology* **23**, 309–324.

Hageman, J., Baecke, C., Ebskamp, M., Pilon, R., Smeekens, S. & Weisbeek, P. (1990). Protein import into and sorting inside the chloroplast are independent processes. *Plant Cell* **2**, 479–494.

Halpin, C., Elderfield, P. D., James, H. E., Zimmermann, R., Dunbar, B. & Robinson, C. (1989). The reaction specificities of the thylakoidal processing peptidase and *Escherichia coli* leader peptidase are identical. *EMBO Journal* **8**, 3917–3921.

von Heijne, G., Steppuhn, J. & Herrmann, R. G. (1989). Domain structure of mitochondrial and chloroplast targeting peptides. *European Journal of Biochemistry* **180**, 535–545.

Henry, R., Kapazoglou, A., McCaffery, M. & Cline, K. (1994). Differences between lumen targeting domains of chloroplast transit peptides determine pathway specificity for thylakoid transport. *Journal of Biological Chemistry* **269**, 10189–10192.

Henry, R., Carrigan, M., McCaffrey, M., Ma, X. & Cline, K. (1997). Targeting determinants and proposed evolutionary basis for the Sec and the delta pH protein transport systems in chloroplast thylakoid membranes. *Journal of Cell Biology* **136**, 823–832.

High, S., Henry, R., Mould, R. M., Valent, Q. A., Meacock, S., Cline, K., Gray, J. C. & Luirink, J. (1997). Chloroplast SRP54 interacts with a specific subset of thylakoid precursor proteins. *Journal of Biological Chemistry* **272**, 11622–11628.

Hoffman, N. E. & Franklin, A. E. (1994). Evidence for a stromal GTP requirement for the integration of a chlorophyll *a/b* binding polypeptide into thylakoid membranes. *Plant Physiology* **105**, 295–304.

Hulford, A., Hazell, L., Mould, R. M. & Robinson, C. (1994). Two distinct mechanisms for the translocation of proteins across the thylakoid membrane, one requiring the presence of a stromal protein factor and nucleotide triphosphates. *Journal of Biological Chemistry* **269**, 3251–3256.

Hynds, P. J., Robinson, D. & Robinson, C. (1998). The Sec-independent twin-arginine translocation system can transport both tightly folded and malfolded proteins across the thylakoid membrane. *Journal of Biological Chemistry* **273**, 34868–34874.

Izard, J. W. & Kendall, D. A. (1994). Signal peptides: exquisitely designed transport promoters. *Molecular Microbiology* **13**, 765–773.

Kim, S. J., Robinson, C. & Mant, A. (1998). Sec/SRP-independent insertion of two thylakoid membrane proteins bearing cleavable signal peptides. *FEBS Letters* **424**, 105–108.

Ko, K. & Cashmore, A. R. (1989). Targeting of proteins to the thylakoid lumen by the bipartite transit peptide of the 33kDa oxygen-evolving protein. *EMBO Journal* **8**, 3187–3194.

Kuhn, A. (1995). Major coat proteins of bacteriophage Pf3 and M13 as model systems for Sec-independent protein transport. *FEMS Microbiology Reviews* **17**, 185–190.

Lamppa, G. K. (1988). The chlorophyll a/b-binding protein inserts into the thylakoids independent of its cognate transit peptide. *Journal of Biological Chemistry* **263**, 14996–14999.

Li, X., Henry, R., Yuan, J., Cline, K. & Hoffman, N. E. (1995). A chloroplast homologue of the signal recognition particle subunit SRP54 is involved in the post-translational integration of a protein into thylakoid membranes. *Proceedings of the National Academy of Sciences, USA* **92**, 3789–3793.

Lorkovic, Z. J., Schröder, W. P., Pakrasi, H. B., Irrgang, K.-D., Herrmann, R. G. & Oelmüller, R. (1995). Molecular characterization of PSII-W, the only nuclear-encoded component of the photosystem II reaction center. *Proceedings of the National Academy of Sciences, USA* **92**, 8930–8934.

Michl, D., Robinson, C., Shackleton, J. B., Herrmann, R. G. & Klösgen, R. B. (1994). Targeting of proteins to thylakoids by bipartite presequences: CFoII is imported by a novel, third pathway. *EMBO Journal* **13**, 1310–1317.

Mould, R. M. & Robinson, C. (1991). A proton gradient is required for the transport of two lumenal oxygen-evolving proteins across the thylakoid membrane. *Journal of Biological Chemistry* **266**, 12189–12193.

Nakai, M., Goto, A., Nohara, T., Sugita, D. & Endo, T. (1994). Identification of the SecA protein homolog in pea chloroplasts and its possible involvement in thylakoidal protein transport. *Journal of Biological Chemistry* **269**, 31338–31341.

Pugsley, A. P. (1993). The complete general secretory pathway in Gram-negative bacteria. *Microbiological Reviews* **57**, 50–108.

Robinson, C., Cai, D., Hulford, A., Brock, I. W., Michl, D., Hazell, L., Schmidt, I., Herrmann, R. G. & Klösgen, R. B. (1994). The presequence of a chimeric construct dictates which of two mechanisms is utilised for translocation across the thylakoid membrane: evidence for the existence of two distinct translocation systems. *EMBO Journal* **13**, 279–285.

Robinson, C., Hynds, P. J., Robinson, D. & Mant, A. (1998). Multiple pathways for the targeting of thylakoid proteins in chloroplasts. *Plant Molecular Biology* **38**, 209–221.

Robinson, D., Karnauchov, I., Herrmann, R. G., Klösgen, R. B. & Robinson, C. (1996). Protease-sensitive thylakoidal import machinery for the Sec-, ΔpH- and signal recognition particle-dependent protein targeting pathways, but not for CF_oII integration. *Plant Journal* **10**, 149–155.

Sargent, F., Bogsch, E. G., Stanley, N. R., Wexler, M., Robinson, C., Berks, B. C. & Palmer, T. (1998). Overlapping functions of components of a bacterial Sec-independent export pathway. *EMBO Journal* **17**, 3640–3650.

Settles, M. A., Yonetani, A., Baron, A., Bush, D. R., Cline, K. & Martienssen, R. (1997). Sec-independent protein translocation by the maize Hcf106 protein. *Science* **278**, 1467–1470.

Thompson, S. J., Kim, S. J. & Robinson, C. (1998). Sec-independent insertion of thylakoid membrane proteins: analysis of insertion forces and identification of a loop intermediate. *Journal of Biological Chemistry* **273**, 18979–18983.

Ulbrandt, N. D., Newitt, J. A. & Bernstein, H. D. (1997). The *E. coli* signal recognition particle is required for the insertion of a subset of inner membrane proteins. *Cell* **88**, 187–196.

Viitanen, P. V., Doran, E. R. & Dunsmuir, P. (1988). What is the role of the transit peptide in thylakoid integration of the light-harvesting chlorophyll a/b protein? *Journal of Biological Chemistry* **263**, 15000–15007.

Voelker, R. & Barkan, A. (1995). Two nuclear mutations disrupt distinct pathways for targeting proteins to the chloroplast thylakoid. *EMBO Journal* **14**, 3905–3914.

Walton, P. A., Hill, P. E. & Subramani, S. (1995). Import of stably folded proteins into peroxisomes. *Molecular Biology of the Cell* **3**, 186–190.

Weiner, J. H., Bilous, P. T., Shaw, G. M., Lubitz, S. P., Frost, L., Thomas, G. H., Cole, J. A. & Turner, R. J. (1998). A novel and ubiquitous system for membrane targeting and secretion of cofactor-containing proteins. *Cell* **93**, 93–101.

Yuan, J., Henry, R., McCaffery, M. & Cline, K. (1994). SecA homolog in protein transport within chloroplasts: evidence for endosymbiont-derived sorting. *Science* **266**, 796–798.

EVOLUTIONARY ORIGINS OF TRANSMEMBRANE TRANSPORT SYSTEMS

MILTON H. SAIER, JR AND TSAI-TIEN TSENG

Department of Biology, University of California at San Diego, La Jolla, CA 92093-0116, USA

INTRODUCTION

The primary structural element of most biological membranes, the phospholipid bilayer, exhibits a hydrophobic interior bordered by strongly hydrophilic surfaces. Although many functions have been ascribed to this structure, the primary function of virtually all cellular and organellar membranes is that of a barrier. Membranes prevent the free flow of molecules between external and internal environments and allow organisms to maintain cytoplasmic homeostatic conditions consistent with life in the face of a constantly varying environment. The maintenance and regulation of cytoplasmic inorganic, organic and macromolecular constituent concentrations is the function of membrane-embedded transporters and their accessory proteins. These transporters are essential for virtually all aspects of life as we know it on Earth.

Over the past decade, our laboratory and others have been concerned with molecular archaeological studies aimed at revealing the origins and evolutionary histories of permeases. These studies have revealed that several different families, defined on the basis of sequence similarities, arose independently of each other, at different times in evolutionary history, following different routes.

When complete microbial genomes first became available for analysis, we adapted pre-existing software and designed new programs that allowed us to quickly identify probable transmembrane proteins, estimate their topologies and determine the likelihood that they function in transport. This work allowed us to expand previously recognized families and to identify dozens of new families.

All of this work then led us to attempt to design a rational but comprehensive classification system that would be applicable to the complete complement of transport systems found in all living organisms. The classification system that we have devised is based primarily on mode of transport and energy coupling mechanism, secondarily on molecular phylogeny, and lastly on the substrate specificities of the individual permeases.

In this symposium article we shall utilize the knowledge that has accrued from the classification and characterization of transport systems and their protein constituents in order to gain insight into the evolutionary origins of transporters found in living organisms. This information is based in part on our genome analyses. Websites are available describing the genome analysis work (Website #1: http://www-biology.ucsd.edu/~ipaulsen/transport/titlepage.html) (see also Paulsen et al., 1998a, b) as well as the classification system (Website #2: http://www-biology.ucsd.edu/~msaier/transport/titlepage.html) (see also Saier, 1998, 1999a, b). The latter site includes (1) names, (2) abbreviations, (3) transport commission numbers (TC #), (4) descriptions of the families, (5) primary references and (6) representative well-characterized family members including their (a) names, (b) substrate specificities, (c) organismal source and (d) database accession numbers.

INDEPENDENT ORIGINS OF DIFFERENT PERMEASE FAMILIES

Our current classification system includes over 200 different families of transporters (see our Website #2). These families are defined on the basis of the degrees of sequence similarity exhibited by their members. Homology of two proteins is considered to be proven if comparable regions of these proteins (60 residues or more) exhibit a comparison score in excess of 9 standard deviations (Devereux et al., 1984). This value corresponds to a probability that the observed degree of similarity occurred by chance is less than 10^{-19} (Dayhoff et al., 1983). If two proteins do not exhibit significant sequence similarity, they are assigned to different families. Two families may have arisen independently of each other, or they may have arisen from a common ancestor by extensive sequence diversion. In many cases it is difficult, if not impossible, to distinguish between these two possibilities. However, in some cases, evidence for the latter possibility can be obtained by conducting matrix/motif analyses using programs such as PSI–BLAST (Altschul et al., 1997), or MEME and MAST (Bailey & Gribskov, 1998a, b). In other cases, specifically for proteins that have arisen by intragenic multiplication events, it is sometimes possible to establish independent origins.

We have previously discussed the origins of three families for which compelling evidence of independent origin exists (Saier, 1994, 1996). These families are (1) the mitochondrial carrier (MC) family of anion exchangers (TC #2.29; Kuan & Saier, 1993), (2) the major intrinsic protein (MIP) family of aquaporins and glycerol channels (TC #1.1; Park & Saier, 1996) and (3) the major facilitator (MF) superfamily of secondary carriers (TC #2.1; Pao et al., 1998). The MC family arose in eukaryotic organelles by an intragenic triplication event in which a 2 transmembrane spanning segment (TMS) gave rise to a 6 TMS segment. The MIP family probably arose in prokaryotes before the three major kingdoms of life (bacteria, archaea and eukarya) diverged from each other. In this case, an intragenic duplication event occurred in which a 3-TMS-encoding genetic element gave rise to a 6-TMS-encoding element. Finally, the MFS probably arose very early, either soon after life on Earth became established or before life came to Earth. In the MFS, an intragenic duplication event gave rise to a 12-TMS-encoding element from a 6 TMS unit.

TRANSPORTER FAMILIES IN WHICH MEMBERS EXHIBIT REPEAT UNITS

Table 1 lists families of transporters in which tandem intragenic duplication, triplication or quadruplication events have apparently given rise to currently existing permeases containing repeat elements. Several families clearly arose by intragenic duplication events. Five of these families include members that have variable numbers of repeat units, between one and four. These five families will be discussed in this section.

The voltage-gated ion channel (VIC) family

The voltage-gated ion channel (VIC) family (TC #1.5) includes tetrameric K^+ channels in which the basic channel-forming peptide unit spans the membrane twice. The 2 TMS channel-forming element may be present once or twice in a single polypeptide chain depending on the system. Superimposed on and preceding this 2 TMS element, in all 6 TMS polypeptide chains, is a 4 TMS module present in all voltage-gated channels. The 4 TMS elements serve as a leader sequence (TMS 1) and as the voltage sensor (TMSs

Table 1. *Permease families for which members exhibit easily recognizable internal repeat sequences*

Family	TC #	No. of repeats per polypeptide chain	No. of TMS per polypeptide chain[a]
MIP	1.1	2	6 (3 + 3)
VIC	1.5	1, 2 or 4	2 or 4 (2 + 2); 6 or 12 (6 + 6), or 24 (6 + 6 + 6 + 6)
MF	2.1	2	12 (6 + 6) or 14 (6 + 2 + 6)
RND	2.6	1 or 2	6 or 12; variable
CaCA	2.19	2	12
PiT	2.20	2	12
MC	2.29	3	6 (2 + 2 + 2)
PNaS	2.58	2	12
ABC	3.1	1 or 2	6 or 12; variable
F-ATPase	3.2, subunit c	1 or 2	2 or 4 (2 + 2); variable
ArsB	3.4	2	12
Chr	99.7	1 or 2	6 or 10 (4 + 6)

[a] The usual numbers of putative or established TMSs per polypeptide chain are indicated. The numbers per repeat unit are provided in parentheses. When more than one topological possibility is observed for different family members, these possibilities are indicated; 'variable' indicates that additional TMSs may be tacked on to the basic structure in some of the family members or that one or two of these TMSs have been deleted in others.

2–4). In addition to the 6 TMS K^+-transporting polypeptide chains, Ca^{2+} channels, with either 12 or 24 TMSs (two or four repeat units, respectively) per polypeptide chain, and Na^+ channels with 24 TMSs (four repeat units) comprise distinct subfamilies within the VIC family (Hille, 1992). K^+ channels, found ubiquitously (in bacteria, archaea and eukaryotes), are believed to have been the primordial channels from which Ca^{2+} channels were derived, and Na^+ channels may have arisen from Ca^{2+} channels (Hille, 1992). The X-ray crystallographic structure of a bacterial VIC family K^+ channel homologue, with just 2 TMSs per polypeptide chain, corresponding in sequence to TMSs 5 and 6 in full-length 6 TMS proteins, suggests the pathway for the evolution of the latter proteins (Doyle *et al.*, 1998; Sansom, 1998).

As noted above, TMS 1 in the 6 TMS channels may function in protein insertion, serving as a 'leader peptide'; TMSs 2–4 probably serve as the voltage sensor, conferring on the channel regulatory properties, and TMSs 5 and 6 actually comprise the channel. The four amphipathic TMSs 6 in the tetrameric protein line the channel while the four hydrophobic TMSs 5 reinforce the channel and interact with lipids (Armstrong & Hille, 1998; Doyle *et al.*, 1998; Sansom, 1998). A modular origin for VIC family members is therefore suggested (R. D. Nelson and others, unpublished).

The dimeric and tetrameric Ca^{2+} channels arose, respectively, by one and two internal gene duplications, with the full-length 6 TMS units found in tetrameric K^+ channel proteins serving as the repeat element (Hille, 1992; Armstrong & Hille, 1998). Interestingly, both the 2 and 6 TMS polypeptides of the VIC family are found ubiquitously, but the 12 and 24 TMS polypeptide channels have so far been identified only in eukaryotes (Salkoff & Jegla, 1995). Na^+ channels of the VIC family all exhibit four repeat units with a total of 24 TMSs. Because no such proteins have yet been identified in prokaryotes, they probably arose in eukaryotes after eukaryotes diverged from prokaryotes. One can infer that little horizontal transfer of genetic material encoding these channels has occurred between prokaryotes and eukaryotes.

The resistance–nodulation–division (RND) superfamily

Most members of the resistance–nodulation–division (RND) superfamily (TC #2.6; Table 2) arose by internal duplicative events (Saier *et al.*, 1994), but some members of the family have just one repeat unit per polypeptide chain. Proteins with just one repeat unit, though rare, have been clearly identified both in archaea and eukaryotes, and they probably exist in bacteria as well (Tseng *et al.*, 1999). Sequence analyses of these divergent proteins have led to the suggestion that the internal gene duplication event that gave rise to the full-length proteins occurred not once but several times during the evolution of the family. This family of proteins is unusual in that most if not all

Table 2. *Families included within the resistance–nodulation–division (RND) superfamily (TC #2.6)*[a]

2.6.1	Heavy metal efflux (HME) family
2.6.2	Largely Gram-negative bacterial hydrophobe/amphiphile efflux (HAE1) family
2.6.3	Putative nodulation factor exporter (NFE) family
2.6.4	SecDF (SecDF) family
2.6.5	Putative Gram-positive bacterial hydrophobe/amphiphile (HAE2) family
2.6.6	Putative eukaryotic sterol homeostasis (ESH) family
2.6.7	Putative archaeal hydrophobe/amphiphile transporter (HAE3) family

[a] Transport substrates for families 2.6.1 and 2.6.2 are established. The nodulation factor-lipooligosaccharide substrates of one member of family 2.6.3 and the actinorhodin (drug) substrate of one member of family 2.6.5 are reasonably well-established. SecDF proteins may play an auxiliary role in protein secretion, but their biochemical substrates have not been identified (Bolhuis *et al.*, 1998). The involvement of several members of family 2.6.7 in sterol (cholesterol) homeostasis and/or metabolism has been demonstrated (Loftus *et al.*, 1997; see our Website #2 for details and original references).

members of the family exhibit large extracytoplasmic hydrophilic domains of 200–400 residues between TMSs 1 and 2, and again between TMSs 7 and 8. Large hydrophilic external domains are particularly rare in prokaryotic transport proteins. The functions of these domains are unknown, but they unexpectedly exhibit sequences that suggest the presence of nucleotide-binding folds (Tseng *et al.*, 1999). It is possible that these unusual domains serve regulatory functions.

ATP-binding cassette (ABC) superfamily

ABC permeases (TC #3.1) include prokaryotic uptake systems which function with extracytoplasmic solute-binding receptors, and ubiquitous efflux systems, some of which act on macromolecules as substrate (Table 3). The efflux porters function exclusively by receptor-independent processes. The uptake permeases can consist of up to five distinct polypeptide chains, but the efflux systems may have as few as one. In the latter case, at least four domains are fused into a single protein. The minimal requirement for the function of an ABC permease is apparently two integral membrane proteins or domains (M) plus two ATP-hydrolysing proteins or domains (A). The two integral membrane constituents of ABC transporters sometimes have the two ATP-hydrolysing domains fused to them (Linton & Higgins, 1998). As for the RND superfamily, sequence analyses reveal that fusion of ABC family proteins probably occurred multiple times during the evolution of the family. Thus, some duplicated ABC family membrane proteins contain only the M constituents of the system (domain order MM), and the A constituents are separate polypeptide chains (Higgins, 1992; Saurin *et al.*, 1999). In others, the A domains are part of the same polypeptide chains which incorporate the integral membrane constituents. In these latter cases,

Table 3. *Families included within the ATP-binding cassette (ABC) superfamily (TC #3.1)*[a]

ABC-type uptake permeases [all from prokaryotes (bacteria and archaea)]	
3.1.1	Carbohydrate uptake transporter-1 (CUT1) family
3.1.2	Carbohydrate uptake transporter-2 (CUT2) family
3.1.3	Polar amino acid uptake transporter (PAAT) family
3.1.4	Hydrophobic amino acid uptake transporter (HAAT) family
3.1.5	Peptide/opine/nickel uptake transporter (PepT) family
3.1.6	Sulfate uptake transporter (SulT) family
3.1.7	Phosphate uptake transporter (PhoT) family
3.1.8	Molybdate uptake transporter (MolT) family
3.1.9	Phosphonate uptake transporter (PhnT) family
3.1.10	Ferric iron uptake transporter (FeT) family
3.1.11	Polyamine/opine/phosphonate uptake transporter (POPT) family
3.1.12	Quaternary amine uptake transporter (QAT) family
3.1.13	Vitamin B_{12} uptake transporter ($VB_{12}T$) family
3.1.14	Iron chelates uptake transporter (FeCT) family
3.1.15	Manganese/zinc/iron chelate uptake transporter (MZT) family
3.1.16	Nitrate/nitrite/cyanate uptake transporter (NitT) family
3.1.17	Taurine uptake transporter (TauT) family
3.1.18	Putative cobalt uptake transporter (CoT) family
ABC-type efflux permeases (bacterial)	
3.1.31	Capsular polysaccharide exporter (CPSE) family
3.1.32	Lipooligosaccharide exporter (LOSE) family
3.1.33	Lipopolysaccharide exporter (LPSE) family
3.1.34	Teichoic acid exporter (TAE) family
3.1.35	Drug exporter (DrugE) family
3.1.36	Putative lipid A exporter (LipidE) family
3.1.37	Putative haem exporter (HemeE) family
3.1.38	β-Glucan exporter (GlucanE) family
3.1.39	Protein-1 exporter (Prot1E) family
3.1.40	Protein-2 exporter (Prot2E) family
3.1.41	Peptide-1 exporter (Pep1E) family
3.1.42	Peptide-2 exporter (Pep2E) family
3.1.43	Peptide-3 exporter (Pep3E) family
3.1.44	Probable glycolipid exporter (DevE) family
3.1.45	Na^+ exporter (NatE) family
3.1.46	Microcin B17 exporter (McbE) family
3.1.47	Multidrug exporter (MDE) family
ABC-type efflux permeases (mostly eukaryotic)	
3.1.61	Multidrug resistance exporter (MDR) family
3.1.62	Cystic fibrous transmembrane conductance exporter (CFTR) family
3.1.63	Peroxisomal fatty acyl-CoA transporter (FAT) family
3.1.64	Eye pigment precursor transporter (EPP) family
3.1.65	Pleiotropic drug resistance (PDR) family
3.1.66	α-Factor sex pheromone exporter (STE) family
3.1.67	Conjugate transporter-1 (CT1) family
3.1.68	Conjugate transporter-2 (CT2) family
3.1.69	MHC peptide transporter (TAP) family

[a] Families 3.1.1–3.1.18 are all prokaryotic uptake systems; families 3.1.31–3.1.47 are all prokaryotic efflux systems, and families 3.1.61–3.1.69 are efflux systems, mostly from eukaryotes. The uptake systems, but not the efflux systems, function with extracytoplasmic solute-binding receptors. Uptake systems are phylogenetically distinct from the efflux systems. Each family is identified on the basis of its phylogenetic properties, but it also exhibits characteristic motifs and a limited range of related substrates. Arguments have been presented suggesting that sequence similarity (phylogeny) primarily reflects the evolutionary process and not constraints imposed upon the process of sequence divergence due to substrate recognition and binding (Kuan et al., 1995).

the domain order can be AM, MA, A-M-A-M or M-A-M-A. While most M domains exhibit a 5 or 6 TMS topology, variations on this pattern are observed (Table 1).

Together with E. Dassa, W. Saurin and M. Hofnung at the Pasteur Institute in Paris, we have analysed the proteins of ABC uptake permeases (Tam & Saier, 1993; Saurin & Dassa, 1994; Kuan et al., 1995). Phylogenetic analyses have provided evidence that the three protein constituents of ABC uptake permeases, the A, M and R (receptor) domains, evolved in parallel from a common evolutionary precursor without apparent shuffling of constituents between systems. It seems that although constituents of closely related ABC permeases may be functionally substitutable, at least those of distantly related proteins are not. This lack of functional complementarity may have prevented the shuffling of constituents between ABC permeases of dissimilar structure and function throughout the long and colourful evolutionary history of this large superfamily.

The F-type ATPase family

The proteolipid subunits c of H^+- or Na^+-transporting F-ATPase complexes (TC #3.2) exhibit 2 TMSs in F-type ATPases, but often 4 in the homologous V-type ATPases of eukaryotes. The four TMS c-subunits arose by an internal duplication event, and these duplicated c-subunits are present in half the number of copies observed for the unduplicated c-subunits (Blair et al., 1996). Variations on this theme of 2 versus 4 TMSs have been noted, however, as subunits c sometimes display 2, 3, 4 or 5 TMSs per polypeptide chain (Hilario & Gogarten, 1998). The 2 TMS c-subunits of F-type ATPases are generally about 75 amino acyl residues long while the 4 TMS c-subunits of V-type ATPases are about twice this length, as expected. However, additional variations probably resulted from insertions and deletions in the evolving genes that code for these subunits.

Three integral membrane polypeptides of F-type ATPases comprise the membrane sector F_0 of F-type ATPases or V_0 of V-type ATPases. The stoichiometry of these subunits is believed to be ab_2c_{12} for F-type ATPases (Jones et al., 1998), and ab_2c_6 for V-type ATPases, due to the internal duplication within the c-subunit as noted above. Interestingly, cyanobacteria and chloroplasts possess F-type ATPases that exhibit an $abb'c_{12}$ subunit composition. In this last-mentioned example, a late-occurring extragenic duplication event evidently gave rise to two distinct subunits that presumably serve related but different functions. Thus, in most F_0 domains, b forms a homodimer in which both b-subunits presumably serve the same function, but in the cyanobacterial and chloroplast F_0 domains, the two b-subunits have differentiated into a functionally more refined heterodimer.

While the b-subunits of F-type ATPases have only 1 TMS, the a-subunits, which occur only once per complex, have been predicted to have 6 TMSs based on hydropathy plots. However, Long *et al.* (1998) as well as Valiya-veetil & Fillingame (1998) have recently provided strong evidence for a 5 TMS topology in the *Escherichia coli* subunit a. One of the six putative TMSs may actually be localized to the membrane surface. It is possible that this 5 TMS topology arose from a more conventional 6 TMS protein during evolutionary history. Documentation of a similar change in topology is provided in the next section.

The chromate (Chr) transporter family

The bacterial-specific chromate ion transporter (Chr) family (TC #99.7; Nies *et al.*, 1998) provides another example of a transporter family in which the integral membrane proteins which comprise the family exhibit either one or two repeat units. All functionally characterized members of the Chr family exhibit two repeat units, but sequencing of the *Bacillus subtilis* genome revealed two adjacent genes, each of which encodes a single unit Chr family protein with 6 putative TMSs. No evidence for a sequencing error giving rise to artificial splicing could be found. Moreover, the two encoded proteins are both more similar in sequence to each other than to other Chr family members (Nies *et al.*, 1998). This observation led to the prediction that the two *B. subtilis* proteins are truly distinct, and that they arose by an extragenic duplication event that occurred more recently than the intragenic duplication event that gave rise to other members of the family. The evolutionary pathway that was responsible for the appearance of this duplex of tandem genes is not clear, but the situation is reminiscent of other families of transporters where variable numbers of repeat units are found within the polypeptide chains that comprise these proteins.

CHANGES IN NUMBERS OF TMSs DURING PERMEASE FAMILY EVOLUTION

The chromate (Chr) transporter family

Although full-length Chr family members arose by an intragenic duplication event, as discussed in the previous section, the two halves of at least one member of this family, the ChrA protein of *Alcaligenes eutrophus*, do not exhibit the same numbers of TMSs (Nies *et al.*, 1998). Gene fusion analyses revealed that the first half of the ChrA protein of *A. eutrophus* spans the membrane four times while the second half spans the membrane six times. It seems likely that the primordial protein unit contained 6 TMSs, but that the first two TMSs in the N-terminal unit of the internally duplicated protein were lost. This may have been due to amino acid substitutions that occurred late in the evolutionary process, causing the two N-terminal spanners to lose their hydrophobic character and assume a more hydrophilic nature. This

N-terminal segment of the *A. eutrophus* ChrA protein appears to be localized to the cytoplasmic side of the membrane (Nies *et al.*, 1998).

The major facilitator (MF) superfamily

A contrasting situation is observed for some proteins within the major facilitator superfamily (MFS; TC #2.1; Pao *et al.*, 1998). Of the 24 currently established families within the MFS (see Table 4 and our Website #2), 21 have members with 12 putative or established TMSs, while three have members with 14. A case in point is the drug:H^+ antiporter-1 (DHA1) family (TC #2.1.2) and the DHA2 family (TC #2.1.3), which have 12 and 14 established TMSs, respectively (Paulsen *et al.*, 1996). Sequence comparisons

Table 4. *Families included within the major facilitator (MF) superfamily (TC #2.1)*[a]

2.1.1	Sugar porter (SP) family
2.1.2	Drug:H^+ antiporter-1 (12 spanner) (DHA1) family
2.1.3	Drug:H^+ antiporter-2 (14 spanner) (DHA2) family
2.1.4	Organophosphate:P_i antiporter (OPA) family
2.1.5	Oligosaccharide:H^+ symporter (OHS) family
2.1.6	Metabolite:H^+ symporter (MHS) family
2.1.7	Fucose:H^+ symporter (FHS) family
2.1.8	Nitrate/nitrite porter (NNP) family
2.1.9	Phosphate:H^+ symporter (PHS) family
2.1.10	Nucleoside:H^+ symporter (NHS) family
2.1.11	Oxalate:formate antiporter (OFA) family
2.1.12	Sialate:H^+ symporter (SHS) family
2.1.13	Monocarboxylate porter (MCP) family
2.1.14	Anion:cation symporter (ACS) family
2.1.15	Aromatic acid:H^+ symporter (AAHS) family
2.1.16	Unknown major facilitator-1 (UMF1) family
2.1.17	Cyanate permease (CP) family
2.1.18	Polyol permease (PP) family
2.1.19	Organic cation transporter (OCT) family
2.1.20	Sugar efflux transporter (SET) family
2.1.21	Drug:H^+ antiporter-3 (12 spanner) (DHA3) family
2.1.22	Vesicular neurotransmitter transporter (VNT) family
2.1.23	Conjugated bile salt transporter (BST) family
2.1.24	Unknown major facilitator-2 (UMF2) family
2.2	Glycoside-pentoside-hexuronide (GPH):cation symporter family
2.17	Proton-dependent oligopeptide transporter (POT) family
2.60	Organo anion transporter (OAT) family
100.7	Putative bacteriochlorophyll delivery (BCD) family

[a] Each established family in the MFS exhibits (1) distinct phylogenetic clustering, (2) characteristic motifs that overlap to some extent with those of other MFS families and (3) a limited and characteristic range of related substrates. The four families presented at the bottom of the table (GPH, POT, OAT and BCD) share motif and sequence matrix properties with established MFS families and are therefore probably distant constituents of the MFS. However, this fact cannot be established by standard statistical analyses of the primary sequences of their members.

of proteins within these two closely related families reveal that TMSs 1–6 and 7–12 in proteins of the DHA1 family are homologous to TMSs 1–6 and 9–14 in the DHA2 family. The two extra TMSs evidently arose by amino acid substitution of residues in the large cytoplasmic loop between TMSs 6 and 7 in a primordial protein exhibiting the topology of the DHA1 family proteins followed by (or concomitant with) insertion of this loop into the membrane. This interpretation is supported by the fact that the loop between TMSs 6 and 7 in the DHA1 family members is about as long as the total region between TMSs 6 and 9 in the DHA2 family members (Paulsen *et al.*, 1996).

The LysE family

Still another example of a probable change in number of TMSs during the evolution of a family is the LysE family (TC #2.65). The LysE lysine/arginine exporter of *Corynebacterium glutamicum* exhibits 6 putative TMSs as do all members of the family, but experimental examination of LysE surprisingly revealed that it has only 5 TMSs. Putative TMS 2 in the topological model that results from prediction based on hydropathy analyses appears not to span the membrane (M. Vrljic and others, unpublished). This is particularly surprising because this putative TMS exhibits the greatest degree of hydrophobicity of the 6 putative TMSs when the sequences for all family members are averaged. Assuming the experimental data to be correct, one must propose that putative TMS 2 is not transmembrane but is either surface-localized or merely dips into the membrane.

The VIC and MIP families of channel proteins

A well-documented example of regions of channel proteins that dip into but do not traverse the membrane are the K^+ channels of the VIC family (TC #1.5), discussed in a previous section. In this case, the loop between TMSs 5 and 6 clearly dips into the membrane to serve as a channel specificity determinant. This fact was established by the X-ray crystallographic data, which revealed the three-dimensional structure of the 2 TMS protein of *Streptomyces lividans* (Doyle *et al.*, 1998; Sansom, 1998).

Another example where an inter TMS loop probably dips into the membrane without traversing it is to be found in representative aquaporins and glycerol facilitators of the MIP family (TC #1.1). In these proteins, the highly conserved loop regions between TMSs 2 and 3, and again between the homologous TMSs 5 and 6 are believed to dip into the membrane, possibly to constrict the channel in a pseudosymmetrical way (Reizer *et al.*, 1993; Jung *et al.*, 1994; Park & Saier, 1996; Shukla & Chrispeels, 1998). Because the two halves of these proteins exhibit reverse topology (from in to out for the first halves, and from out to in for the second halves), the channel can be expected to be constricted from both sides of the membrane and to exhibit the unusual property of pseudosymmetry.

TOPOLOGICAL CHANGES IN PERMEASE PROTEINS ARISING DURING EVOLUTION

When a tandem internal gene duplication event gives rise to a protein of twice the original size, the two repeat units usually exhibit very similar three-dimensional structures, particularly if recognizable sequence similarity is still observable between the two halves. However, probable exceptions have been observed both for soluble enzymes and for integral membrane proteins (Godzik, 1996; Persson *et al.*, 1998; Wang *et al.*, 1994). HIV reverse transcriptase provides an interesting example (Wang *et al.*, 1994). This enzyme is a homodimer with very different structures for the two subunits. This structural asymmetry is apparently triggered by proteolytic cleavage of one subunit which causes this subunit (but not the other one) to drastically change structure. Thus, although the two 'halves' of the protein have essentially the same amino acid sequence, they have very different conformations. Conformational asymmetry for proteins with two or more identical subunits is often a reflection of negative cooperativity (Saier, 1987).

Recently, Persson *et al.* (1998) have examined two phosphate permeases of the yeast *Saccharomyces cerevisiae*. One is Pho84, a member of the phosphate uptake transporter (PhoP) family (TC #2.1.7) within the major facilitator superfamily (TC #2.1). As also documented by Pao *et al.* (1998), this family exhibits clear sequence and motif similarity between the two 6 TMS halves of these proteins. As expected, conserved motifs occur in corresponding positions in the two halves of the protein in the topological model presented by Persson *et al.* (1998). This model is reproduced in Fig. 1, top.

A second yeast phosphate permease protein, Pho89, is a member of the PiT family (TC #2.20; see also Table 1). Like Pho84, Pho89 exhibits an apparent 12 TMS topology, based on predictions derived using the TopPred program (Claros & von Heijne, 1994), and it exhibits a sufficient degree of sequence similarity between the two halves to allow the establishment of homology (unpublished results). However, in this case, the TopPred program predicts completely different topologies for the two homologous portions of Pho89. Thus, in the model presented in Fig. 1, bottom, the homologous regions begin at the ends of TMSs 1 and 9 and end at the ends of TMSs 5 and 12, respectively. The fact that the homologous regions encompass 5 putative TMSs in the first half and only 4 putative TMSs in the second half reflects the presence of much larger inter-TMS loop regions in the second half of the Pho89 model. Visual sequence gazing clearly suggests that the TopPred-based model presented by Persson *et al.* (1998) is correct. This prediction will be a particularly interesting one to confirm experimentally.

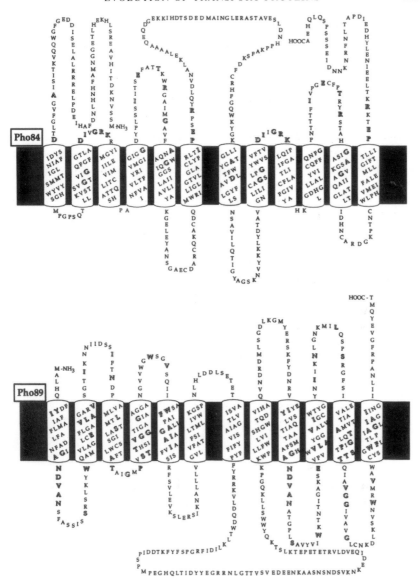

Fig. 1. Secondary topological models of Pho84 (top) and Pho89 (bottom) derived using the TopPred algorithm (Claros & von Heijne, 1994). The one-letter amino acid code is used, and putative transmembrane segments are shown as cylinders. Highlighted amino acid residues represent the residues that are conserved between the two duplicated halves of the Pho84 protein or the Pho89 protein. Reprinted from *Biochimica et Biophysica Acta*, **1365**, Persson, B. L., Berhe, A., Fristedt, U., Martinez, P., Pattison, J., Petersson, J. & Weinander, R., Phosphate permeases of *Saccharomyces cerevisiae*, pp. 23–30, copyright 1998, with permission from Elsevier Science.

PERMEASES WITH DIFFERENT TRANSPORT MODES WITHIN A SINGLE SUPERFAMILY

Use of the PSI–BLAST program has provided evidence that several transporter families, the members of which do not exhibit sufficient sequence similarity to allow the establishment of homology between them, are nonetheless distantly related. These families, which comprise the newly recognized ion transporter (IT) superfamily (Table 5), include the divalent anion sodium symporter (DASS) family (TC #2.47) (Pos et al., 1998), the arsenical (Ars) efflux family (TC #3.4) (Silver et al., 1993) and the tripartite ATP-independent periplasmic transporter (TRAP) family (TC #2.56) (Forward et al., 1997; Rabus et al., 1999). Characterized members of these three families can function (a) as periplasmic binding-receptor-independent secondary sodium symport carriers (DASS), (b) as periplasmic binding-receptor-dependent secondary proton symport carriers (TRAP) or (c) as either ATP-dependent or ATP-independent efflux uniporters (Ars). The integral membrane ArsB constituent of the Ars family alone catalyses electrogenic uniport of arsenite and antimonite, but coexpression of ArsB with ArsA, an ATPase that superficially resembles those of ABC transporters but shows no significant sequence similarity with them, results in the generation of an ATP-driven arsenite efflux pump that functions with much greater efficiency than does the ArsB uniporter alone (Kuroda et al., 1997). It seems that proteins within these three mechanistically divergent families of the IT superfamily have diverged functionally to accommodate different energy-coupling mechanisms and to utilize extracytoplasmic receptors. The incorporation of high-affinity receptors may confer upon the permeases a higher degree of substrate specificity as well as increased substrate affinity.

MOSAIC ORIGINS OF SOME PERMEASE SUPERFAMILIES

The phosphoenolpyruvate-dependent sugar group translocating phosphotransferase system (PTS) superfamily consists of six families of mosaic

Table 5. *Families tentatively included within the ion transporter (IT) superfamily (Rabus et al., 1999)*

TC#	Family
2.8	Gluconate: H^+ symporter (GntP) family
2.11	Citrate-Mg^{2+} : H^+ – citrate: H^+ symporter (CitMHS) family
2.13	C_4-dicarboxylate uptake (Dcu) family
2.14	Lactate permease (LctP) family
2.34	NhaB Na^+ : H^+ antiporter (NhaB) family
2.35	NhaC Na^+ : H^+ antiporter (NhaC) family
2.47	Divalent anion: Na^+ symporter (DASS) family
2.56	Tripartite ATP-independent periplasmic transporter (TRAP-T) family
2.61	C_4-dicarboxylate uptake C (DcuC) family
3.4	Arsenical (Ars) efflux family

permease enzymes (TC #4.1–4.6; Table 6). Each sugar-specific permease or Enzyme II complex of the PTS possesses three distinct proteins or protein domains designated IIA, IIB and IIC. One of these domains (IIC) spans the membrane six or more times and catalyses sugar transport as well as sugar phosphorylation. The other two (IIA and IIB) serve as sugar-specific, water-soluble, energy-coupling domains or proteins (Saier & Reizer, 1992). Additional proteins, called Enzyme I and HPr (TC #98.7 and 98.8, respectively), serve as general energy-coupling proteins of the PTS, catalysing phosphoryl transfer to the sugar-specific proteins via the following phosphoryl transfer chain:

$$\text{PEP} \longrightarrow \text{EI} \sim \text{P} \longrightarrow \text{HPr} \sim \text{P} \longrightarrow \text{IIA} \sim \text{P} \longrightarrow \text{IIB} \sim \text{P} \xrightarrow{\text{IIC}} \text{sugar–P}$$

Both the general energy-coupling proteins and the sugar-specific constituents of the PTS have undergone extensive domain fusion, splicing and shuffling during evolution of the system. Genetic deletions and duplications have also given rise to loss or gain of specific domains as will be elaborated in the next section.

As noted above, PTS permeases generally consist of three domains, here abbreviated A, B and C (Saier & Reizer, 1992, 1994). Extensive sequence data have revealed that these domains may exist as separate polypeptide chains or as covalently linked domains connected to each other by flexible linkers (Reizer & Saier, 1997). Thus, A, B and C have been found as distinct polypeptide chains encoded by distinct genes. Additionally, the B and C domains have been found linked to each other in either the BC or the CB order, and A has been found linked to B in one case (Saier & Reizer, 1992; Stolz et al., 1993). A and C, however, have not yet been found covalently linked together in a polypeptide chain that lacks a B domain.

Three-dimensional structural analyses of the A and B protein domains of various PTS permeases by X-ray crystallography and multidimensional NMR have surprisingly revealed that not all A domains, and not all B domains are homologous (Saier & Reizer, 1994). Thus, three structurally

Table 6. *Families included within the bacterial phosphotransferase system (PTS) functional superfamily*[a]

4.1	PTS glucose-glucoside (Glc) family
4.2	PTS fructose-mannitol (Fru) family
4.3	PTS lactose-*N,N*′-diacetylchitobiose-*β*-glucoside (Lac) family
4.4	PTS glucitol (Gut) family
4.5	PTS galactitol (Gat) family
4.6	PTS mannose-fructose-sorbose (Man) family

[a] These permeases are mosaic multidomain systems in which some (but not other) of the domains may be homologous with those of other PTS families. Family 4.6 proteins consist of four recognized domains, IIA, B, C and D, all of which evolved independently of the IIA, B and C domains of families 4.1–4.5. The IIC domains of family 4.6 permeases may share homology with integral membrane proteins of ABC permeases (TC #3.1).

elucidated IIA protein domains (those present in PTS permeases specific for glucose, mannose and mannitol) are nonhomologous (Bordo *et al.*, 1998), while three structurally defined IIB protein domains (those present in PTS permeases specific for mannose, glucose and chitobiose) similarly are nonhomologous (van Montfort *et al.*, 1997). These observations are particularly surprising in view of the evidence that the IIC proteins/domains of the glucose, mannitol and chitobiose PTS permeases probably arose from a common evolutionary origin, although the IIC protein of the *E. coli* mannose PTS permease probably arose independently from an integral membrane protein related to those of ABC permeases (M. H. Saier, Jr, unpublished observations). These observations clearly suggest that the PTS is a mosaic superfamily which extracted its functionally equivalent domains from a variety of different sources. At least one of the PTS permease families, the Man family (TC #4.6), probably evolved completely independently of the other five families, and some of the domains within the other five families may have evolved independently of some of their functional counterparts (Table 6).

MODULAR CONSTRUCTION OF MULTIDOMAIN PERMEASE FAMILIES

Several permease types have apparently evolved as modifiable modules rather than as immutable systems. Thus, for example, the 6 TMS unit of voltage-gated ion channel family proteins arose by superimposition on the channel (formed from TMSs 5 and 6) of the voltage sensor (formed from TMSs 2–4) and a leader sequence concerned primarily with membrane insertion (TMS 1). Additionally, ABC permeases generally consist of integral membrane proteins and ATP-hydrolysing constituents which can be linked together in a variety of orders (Linton & Higgins, 1998), and these proteins can function in conjunction with extracytoplasmic solute-specific receptors.

In a related but different way, the modular nature of the PTS has been demonstrated. The three principal domains of all PTS permeases can exist in one or more polypeptide chains. Considering PTS permeases that consist of single polypeptide chains, the A, B and C domains could theoretically be linked to each other in six orders: ABC, ACB, BAC, BCA, CAB and CBA. Of these orders, only three have been identified. Thus as shown in Fig. 2, the order ABC occurs in the fructose (Fru) system of *Mycoplasma genitalium* and a fructose-like system (HrsA or Frx) of *E. coli* (Reizer *et al.*, 1996), the order BCA occurs in the β-glucoside (Bgl) system of *E. coli* and the sucrose system of *Streptococcus mutans*, and the order CBA occurs in the mannitol and *N*-acetylglucosamine systems of *E. coli* (Lee & Saier, 1983; Schnetz *et al.*, 1987; Peri & Waygood, 1988; Sato *et al.*, 1989). In the *E. coli* glucose permease, either the native domain order, CB, or the artificial domain order, BC, has been shown to be functional (Gutknecht *et al.*, 1998). So far, no PTS permease has been found in which the A domain is linked N-terminally to

(a)

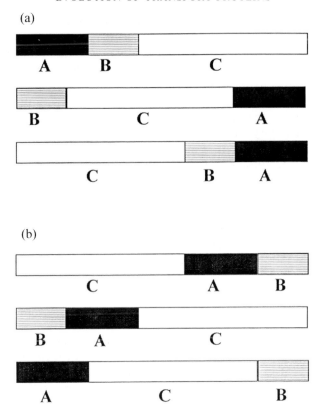

(b)

Fig. 2. Domain structures of (a) known covalent Enzyme II complexes of the PTS and (b) Enzyme II complexes that have not yet been identified. Only polypeptides containing the three PTS domains [IIA (A), IIB (B) and IIC (C)] are shown. The AC domain linkage may be prohibited since it has not yet been identified. The domain order ABC has so far been found only in the fructose-specific Enzyme II complex of *Mycoplasma genitalium* and the fructose-like Enzyme II complex HrsA of *E. coli*. The domain order BCA is found, for example, in the β-glucoside permease (Bgl) of *E. coli* and the sucrose permease (Scr) of *Streptococcus mutans*. The CBA arrangement is found, for example, in the mannitol and *N*-acetylglucosamine permeases of *E. coli*. These latter two arrangements (BCA and CBA) are found in almost all full-length, single polypeptide PTS permeases. Reprinted from *Current Opinion in Structural Biology*, 7, Reizer, J. & Saier, M. H., Jr, Modular multidomain phosphoryl transfer proteins of bacteria, pp. 407–415, copyright 1997, with permission from Elsevier Science.

the C domain (i.e. ACB or BAC), and the domain order CAB has also not been detected (Fig. 2) (Reizer & Saier, 1997).

In several *E. coli* Enzyme II complexes, specifically those for sucrose (Scr), trehalose (Tre), maltose (Mal) and β-glucosides (Asc), the IIA domains have been lost, and they have been functionally replaced by the glucose IIA protein. In this case, the IIA domain is not completely sugar-specific as it phosphorylates a subset of PTS IIB domains (Lengeler *et al.*, 1994). Furthermore, in several fructose systems (e.g. those of *Rhodobacter capsulatus*, *E. coli*, *Haemophilus influenzae* and *Xanthomonas campestris*), the IIB domains have been duplicated (Wu & Saier, 1990; De Crecy-Lagard *et al.*, 1995; Reizer *et al.*, 1995, 1996). These observations suggest that as has been demonstrated for the enzyme pyruvate:phosphate dikinase (Herzberg *et al.*, 1996) the domains are sufficiently flexibly attached to each other so as to permit the swivelling of their active sites, thereby allowing functional binding to two or more interactive enzymes. The fact that the A and C domains have not been observed to be linked to each other in the order AC may suggest that steric constraints render this domain order unfavourable.

Mao *et al.* (1995) have artificially constructed fusion proteins in which all requisite PTS phosphoryl transfer protein domains are linked together in a single polypeptide chain that is C-terminal to the permease domain with the domain order IIC^{Glc}-IIB^{Glc}-IIA^{Glc}-HPr-Enzyme I. The chosen domain order is the same as in the similar *N*-acetylglucosamine permease of *E. coli* (IIC-B-A^{NAG}) (Peri & Waygood, 1988) and in the three-domain phosphoryl transfer protein of *R. capsulatus* (MTP with the domain order IIA^{Fru}-HPr-Enzyme I) (Wu *et al.*, 1990). AP-rich linkers, which are found in the latter protein, were used by Mao *et al.* (1995) in the construction of the artificial PTS fusion protein. The phosphotransfer activity of the fusion protein proved to be substantially higher than that of an equimolar mixture of the isolated subunits in *in vitro* enzyme assays, and the complex was insensitive to competitive inhibition by other PTS Enzyme II complexes. The linking of functional domains, therefore, confers certain advantages to a specific PTS permease by rendering it more efficient, but it also confers disadvantages by rendering the energy-coupling proteins less effective for the phosphorylation of other PTS proteins, and of protein targets of regulation. The fusion protein constructed by Mao and others may be functionally equivalent to a multienzyme PTS complex in which each constituent of the PTS assumes a specific orientation in the complex relative to the other proteins of the system (Saier & Staley, 1977; Saier *et al.*, 1982).

CONCLUSIONS AND PERSPECTIVES

In the preceding discussion we have argued that families of transporters have arisen continuously over at least the past 4 billion years. Those that have arisen recently, i.e. since the three major domains of living organisms (bacteria, archaea and eukaryotes) separated, are in general found only in the domain in which that family arose. Many of the families that are restricted to only the bacterial, archaeal or eukaryotic domain undoubtedly arose within that domain, and little or no horizontal transmission of relevant genetic material caused the family to become dispersed throughout the living world (Saier, 1998, 1999a, b). However, it should be kept in mind that only in a few cases have independent origins been proven for different permease families, and two families, defined on the basis of primary sequence similarity, or the lack thereof, may share a common origin but have diverged from each other beyond recognition. Such extensive sequence divergence could have occurred because of early divergence of the host organisms, or because of functional differentiation of the proteins. Either possibility could have prevented exchange of genetic material between the two gene pools.

We have seen that many permease proteins arose by tandem intragenic duplication events. In a few cases, extragenic duplications have given rise to functional differentiation of constituents within the permease complexes. In other cases, independent origins of the constituents of modularly constructed permeases have been proposed and documented. It is clear that different families have evolved at different times in evolutionary history, from different sources, following different routes.

A common topological unit of 6 TMSs is found in many families of permeases, and it has been possible to show that the 6 TMS unit evolved independently in several distinct families. Variations in this 6 TMS unit have resulted from the documented gain or loss of TMSs by amino acid substitutions that increased or decreased the hydrophobicities of portions of a polypeptide chain. In other cases, it seems more likely that proteinaceous elements were incorporated into or deleted from a pre-existing 6 TMS unit, giving rise to the observed topological variability. Such events could have occurred at the gene level by DNA splicing, fusion and deletion. The fact that so many permeases of distinct origin exhibit units of 6 TMSs, and that so many primary and secondary carriers include two such units in either one or at most two polypeptide chains, is still without explanation. It seems, however, that channel function allows construction of the transmembrane hydrophilic pore from many polypeptide chains that may reversibly come together and separate, but that the maintenance of carrier function requires that the complete functional unit be found within just one, or at most two, polypeptide chains so that a well-defined transport pathway is formed. A requirement for an architecturally constrained solute transport pathway may be a prerequisite for stereospecific solute recognition and carrier-mediated transport.

We have seen that motif/matrix analysis of permease families has led to the tentative identification of superfamilies, the members of which may exhibit very little sequence similarity. Of great potential significance is the fact that any such well-conserved residues and sequence motifs are likely to be of structural and/or functional significance. The identification of these residues and motifs is likely to be the first step in providing detailed three-dimensional structures and mechanistic schemes for these permeases.

Finally, phylogenetic analyses based on primary protein structure provide a reliable guide to macromolecular structure, function and mechanism with very few exceptions. Any laboratory involved in structure/function studies must therefore be cognizant of the benefits and restrictions of the phylogenetic approach. A knowledge of the family members allows extrapolation of data from many proteins to all members of the family, and detailed information derived from just one family member will be applicable to all family members to degrees that in general will be directly related to their degrees of sequence similarity. Such extrapolations are not justified for nonhomologous proteins. It seems certain that when genome sequencing increases by several orders of magnitude the volume of molecular data available for analysis, the experimental biologist will become increasingly dependent on molecular archaeology to provide a guide through the wilderness (Saier, 1998, 1999a, b).

ACKNOWLEDGEMENTS

I am grateful to Mary Beth Hiller and Milda Simonaitis for their assistance in the preparation of this manuscript. Work in this laboratory was supported by USPHS grants 5RO1 AI21702 from the National Institutes of Allergy and Infectious Diseases and 9RO1 GM55434 from the National Institute of General Medical Sciences, as well as by the M. H. Saier, Sr memorial research fund.

REFERENCES

Altschul, S. F., Madden, T. L., Schäffer, A. A., Zhang, J., Zhang, Z., Miller, W. & Lipman, D. J. (1997). Gapped BLAST and PSI–BLAST: a new generation of protein database search programs. *Nucleic Acids Research* **25**, 3389–3402.

Armstrong, C. M. & Hille, B. (1998). Voltage-gated ion channels and electrical excitability. *Neuron* **20**, 371–380.

Bailey, T. L. & Gribskov, M. (1998a). Combining evidence using p-values: application to sequence homology searches. *Bioinformatics* **14**, 48–54.

Bailey, T. L. & Gribskov, M. (1998b). Methods and statistics for combining motif match scores. *Journal of Computational Biology* **5**, 211–221.

Blair, A., Ngo, L., Park, J., Paulsen, I. T. & Saier, M. H., Jr (1996). Phylogenetic analyses of the homologous transmembrane channel-forming proteins of the F_0F_1-ATPases of bacteria, chloroplasts and mitochondria. *Microbiology* **142**, 17–32.

Bolhuis, A., Broekhuizen, C. P., Sorokin, A., van Roosmalen, M. L., Venema, G., Bron, S., Quax, W. J. & van Dijl, J. M. (1998). SecDF of *Bacillus subtilis*, a molecular Siamese twin required for the efficient secretion of proteins. *Journal of Biological Chemistry* **273**, 21217–21224.

Bordo, D., van Montfort, R. L. M., Pijning, T., Kalk, K. H., Reizer, J., Saier, M. H., Jr & Dijkstra, B. W. (1998). The three-dimensional structure of the nitrogen regulatory protein IIA^{Ntr} from *Escherichia coli*. *Journal of Molecular Biology* **279**, 245–255.

Claros, M. G. & von Heijne, G. (1994). TopPred II: an improved software for membrane protein structure predictions. *Computer Applications in the Biosciences* **10**, 685–686.

Dayhoff, M. O., Barker, W. C. & Hunt, L. T. (1983). Establishing homologies in protein sequences. *Methods in Enzymology* **91**, 524–545.

De Crecy-Lagard, V., Binet, M. & Danchin, A. (1995). Fructose phosphotransferase system of *Xanthomonas campestris* pv. *campestris*: characterization of the *fruB* gene. *Microbiology* **141**, 2253–2260.

Devereux, J., Haeberli, P. & Smithies, O. (1984). A comprehensive set of sequence analysis programs for the VAX. *Nucleic Acids Research* **12**, 387–395.

Doyle, D. A., Cabral, J. M., Pfuetzner, R. A., Kuo, A., Gulbis, J. M., Cohen, S. L., Chait, B. T. & MacKinnon, R. (1998). The structure of the potassium channel: molecular basis of K$^+$ conduction and selectivity. *Science* **280**, 69–77.

Forward, J., Behrendt, M. C., Wyborn, N. R., Cross, R. & Kelly, D. J. (1997). TRAP transporters: a new family of periplasmic solute transport systems encoded by the *dctPQM* genes of *Rhodobacter capsulatus* and by homologs in diverse Gram-negative bacteria. *Journal of Bacteriology* **179**, 5482–5493.

Godzik, A. (1996). Structural diversity in a family of homologous proteins. *Journal of Molecular Biology* **258**, 349–366.

Gutknecht, R., Manni, M., Mao, Q. & Erni, B. (1998). The glucose transporter of *Escherichia coli* with circularly permuted domains is active *in vivo* and *in vitro*. *Journal of Biological Chemistry* **273**, 25745–25750.

Herzberg, O., Chen, C. C. H., Kapadia, G., McGuire, M., Carroll, L. J., Noh, S. J. & Dunaway-Mariano, D. (1996). Swiveling-domain mechanism for enzymatic phosphotransfer between remote reaction sites. *Proceedings of the National Academy of Sciences, USA* **93**, 2652–2657.

Higgins, C. F. (1992). ABC transporters: from microorganisms to man. *Annual Review of Cell Biology* **8**, 67–113.

Hilario, E. & Gogarten, J. P. (1998). The prokaryote-to-eukaryote transition reflected in the evolution of the V/F/A-ATPase catalytic and proteolipid subunits. *Journal of Molecular Evolution* **46**, 703–715.

Hille, B. (1992). *Ionic Channels of Excitable Membranes*, 2nd edn. Sunderland, MA: Sinauer Associates.

Jones, P. C., Jiang, W. & Fillingame, R. H. (1998). Arrangement of the multicopy H$^+$-translocating subunit c in the membrane sector of the *Escherichia coli* F$_1$F$_0$ ATP synthase. *Journal of Biological Chemistry* **273**, 17178–17185.

Jung, J. S., Preston, G. M., Smith, B. L., Guggino, W. B. & Agre, P. (1994). Molecular structure of the water channel through aquaporin CHIP. The hourglass model. *Journal of Biological Chemistry* **269**, 14648–14654.

Kuan, G., Dassa, E., Saurin, W., Hofnung, M. & Saier, M. H., Jr (1995). Phylogenetic analyses of the ATP-binding constituents of bacterial extracytoplasmic receptor-dependent ABC-type nutrient uptake permeases. *Research in Microbiology* **146**, 271–278.

Kuan, J. & Saier, M. H., Jr (1993). The mitochondrial carrier family of transport proteins: structural, functional and evolutionary relationships. *Critical Reviews in Biochemistry and Molecular Biology* **28**, 209–233.

Kuroda, M., Dey, S., Sanders, O. I. & Rosen, B. P. (1997). Alternate energy coupling of ArsB, the membrane subunit of the Ars anion-translocating ATPase. *Journal of Biological Chemistry* **272**, 326–331.

Lee, C. A. & Saier, M. H., Jr (1983). Mannitol-specific Enzyme II of the bacterial phosphotransferase system. III. The nucleotide sequence of the permease gene. *Journal of Biological Chemistry* **58**, 10761–10767.

Lengeler, J. W., Jahreis, K. & Wehmeier, U. F. (1994). Enzymes II of the phosphoenolpyruvate-dependent phosphotransferase systems: their structure and function in carbohydrate transport. *Biochimica et Biophysica Acta* **1188**, 1–28.

Linton, K. J. & Higgins, C. F. (1998). The *Escherichia coli* ATP-binding cassette (ABC) proteins. *Molecular Microbiology* **28**, 5–13.

Loftus, S. K., Morris, J. A., Carstea, E. D. & 9 other authors (1997). Murine model of Niemann-Pick C disease: mutation in a cholesterol homeostasis gene. *Science* **277**, 232–235.

Long, J. C., Wang, S. & Vik, S. B. (1998). Membrane topology of a subunit a of the

F_1F_0 ATP synthase as determined by labeling of unique cysteine residues. *Journal of Biological Chemistry* **273**, 16235–16240.

Mao, Q., Schunk, T., Gerber, B. & Erni, B. (1995). A string of enzymes: purification and characterization of a fusion protein comprising the four subunits of the glucose phosphotransferase system of *Escherichia coli*. *Journal of Biological Chemistry* **270**, 18295–18300.

van Montfort, R. L. M., Pijning, T., Kalk, K. H., Reizer, J., Saier, M. H., Jr, Thunnissen, M. M. G. M., Robillard, G. T. & Dijkstra, B. W. (1997). The structure of an energy-coupling protein from bacteria, IIB[cellobiose], reveals similarity to eukaryotic protein tyrosine phosphatases. *Structure* **5**, 217–225.

Nies, D. H., Koch, S., Wachi, S., Peitzsch, N. & Saier, M. H., Jr (1998). CHR, a novel family of prokaryotic proton motive force-driven transporters probably containing chromate/sulfate antiporters. *Journal of Bacteriology* **180**, 5799–5802.

Pao, S. S., Paulsen, I. T. & Saier, M. H., Jr (1998). Major facilitator superfamily. *Microbiology and Molecular Biology Reviews* **62**, 1–32.

Park, J. H. & Saier, M. H., Jr (1996). Phylogenetic characterization of the MIP family of transmembrane channel proteins. *Journal of Membrane Biology* **153**, 171–180.

Paulsen, I. T., Brown, M. H. & Skurray, R. A. (1996). Proton-dependent multidrug efflux systems. *Microbiological Reviews* **60**, 575–608.

Paulsen, I. T., Sliwinski, M. K., Nelissen, B., Goffeau, A. & Saier, M. H., Jr (1998a). Unified inventory of established and putative transporters encoded within the complete genome of *Saccharomyces cerevisiae*. *FEBS Letters* **430**, 116–125.

Paulsen, I. T., Sliwinski, M. K. & Saier, M. H., Jr (1998b). Microbial genome analyses: global comparisons of transport capabilities based on phylogenies, bioenergetics and substrate specificities. *Journal of Molecular Biology* **277**, 573–592.

Peri, K. G. & Waygood, E. B. (1988). Sequence of cloned Enzyme II[N-acetylglucosamine] of the phosphoenolpyruvate:*N*-acetylglucosamine phosphotransferase system of *Escherichia coli*. *Biochemistry* **27**, 6054–6061.

Persson, B. L., Berhe, A., Fristedt, U., Martinez, P., Pattison, J., Petersson, J. & Weinander, R. (1998). Phosphate permeases of *Saccharomyces cerevisiae*. *Biochimica et Biophysica Acta* **1365**, 23–30.

Pos, K. M., Dimroth, P. & Bott, M. (1998). The *Escherichia coli* citrate carrier CitT: a member of a novel eubacterial transporter family related to the 2-oxoglutarate/malate translocator from spinach chloroplasts. *Journal of Bacteriology* **180**, 4160–4165.

Rabus, R., Jack, D. L., Kelley, D. J. & Saier, M. H., Jr (1999). TRAP transporters: an ancient family of extracytoplasmic solute receptor-dependent secondary active transporters. *Microbiology* (in press).

Reizer, J. & Saier, M. H., Jr (1997). Modular multidomain phosphoryl transfer proteins of bacteria. *Current Opinion in Structural Biology* **7**, 407–415.

Reizer, J., Reizer, A. & Saier, M. H., Jr (1993). The MIP family of integral membrane channel proteins: sequence comparisons, evolutionary relationships, reconstructed pathway of evolution, and proposed functional differentiation of the two repeated halves of the proteins. *Critical Reviews in Biochemistry and Molecular Biology* **28**, 235–257.

Reizer, J., Reizer, A. & Saier, M. H., Jr (1995). Novel phosphotransferase system genes revealed by bacterial genome analysis – a gene cluster encoding a unique Enzyme I and a putative anaerobic fructose-like system. *Microbiology* **141**, 961–971.

Reizer, J., Paulsen, I. T., Reizer, A., Titgemeyer, F. & Saier, M. H., Jr (1996). Novel phosphotransferase system genes revealed by bacterial genome analysis: the complete complement of *pts* genes in *Mycoplasma genitalium*. *Microbial and Comparative Genomics* **1**, 151–164.

Saier, M. H., Jr (1987). *Enzymes and Coenzyme Action in Metabolic Pathways. A Comparative Study of Mechanism, Structure, Evolution, and Control.* New York: Harper & Row.

Saier, M. H., Jr (1994). Computer-aided analyses of transport protein sequences: gleaning evidence concerning function, structure, biogenesis, and evolution. *Microbiological Reviews* **58**, 71–93.

Saier, M. H., Jr (1996). Phylogenetic approaches to the identification and characterization of protein families and superfamilies. *Microbial and Comparative Genomics* **1**, 129–150.

Saier, M. H., Jr (1998). Molecular phylogeny as a basis for the classification of transport proteins from bacteria, archaea and eukarya. *Advances in Microbial Physiology* **40**, 81–136.

Saier, M. H., Jr (1999a). A proposal for classification of transmembrane transport proteins in living organisms. In *Biomembrane Transport*, pp. 265–276. Edited by L. VanWinkle. San Diego, CA: Academic Press.

Saier, M. H., Jr (1999b). Eukaryotic transmembrane solute transport systems. In *International Review of Cytology: a Survey of Cell Biology*, pp. 61–136. Edited by K. W. Jeon. San Diego, CA: Academic Press.

Saier, M. H., Jr & Reizer, J. (1992). Proposed uniform nomenclature for the proteins and protein domains of the bacterial phosphoenolpyruvate:sugar phosphotransferase system. *Journal of Bacteriology* **174**, 1433–1438.

Saier, M. H., Jr & Reizer, J. (1994). The bacterial phosphotransferase system: new frontiers 30 years later. *Molecular Microbiology* **13**, 755–764.

Saier, M. H., Jr & Staley, J. T. (1977). Phosphoenolpyruvate:sugar phosphotransferase system in *Ancalomicrobium adetum*. *Journal of Bacteriology* **131**, 716–718.

Saier, M. H., Jr, Cox, D. F., Feucht, B. U. & Novotny, M. J. (1982). Evidence for the functional association of Enzyme I and HPr of the phosphoenolpyruvate-sugar phosphotransferase system with the membrane in sealed vesicles of *Escherichia coli*. *Journal of Cellular Biochemistry* **18**, 231–238.

Saier, M. H., Jr, Tam, R., Reizer, A. & Reizer, J. (1994). Two novel families of bacterial membrane proteins concerned with nodulation, cell division and transport. *Molecular Microbiology* **11**, 841–847.

Salkoff, L. & Jegla, T. (1995). Surfing the DNA databases for K^+ channels nets yet more diversity. *Neuron* **15**, 489–492.

Sansom, M. S. P. (1998). Ion channels: a first view of K^+ channels in atomic glory. *Current Biology* **8**, R450–452.

Sato, Y., Poy, F., Jacobson, G. R. & Kuramitsu, H. K. (1989). Characterization and sequence analysis of the *scrA* gene encoding Enzyme II[Scr] of the *Streptococcus mutans* phosphoenolpyruvate-dependent sucrose phosphotransferase system. *Journal of Bacteriology* **171**, 263–271.

Saurin, W. & Dassa, E. (1994). Sequence relationships between integral inner membrane proteins of binding protein-dependent transport systems: evolution by recurrent gene duplications. *Protein Science* **3**, 325–344.

Saurin, W., Hofnung, M. & Dassa, E. (1999). Getting in or out. Early segregation between importers and exporters in the evolution of ATP-binding cassette (ABC) transporters. *Journal of Molecular Evolution* **48**, 22–41.

Schnetz, K., Toloczyki, C. & Rak, B. (1987). β-Glucoside (*bgl*) operon of *Escherichia coli* K-12: nucleotide sequence, genetic organization, and possible evolutionary relationship to regulatory components of two *Bacillus subtilis* genes. *Journal of Bacteriology* **169**, 2579–2590.

Shukla, V. K. & Chrispeels, M. J. (1998). Aquaporins: their role and regulation in cellular water movement. In *Cellular Integration of Signaling Pathways in Plant Development*, NATO-ASI Series, subseries H, pp. 11–22. Springer.

Silver, S., Ji, G., Bröer, S., Dey, S., Dou, D. & Rosen, B. P. (1993). Orphan enzyme or patriarch of a new tribe: the arsenic resistance ATPase of bacterial plasmids. *Molecular Microbiology* **8**, 637–642.

Stolz, B., Huber, M., Markovic-Housley, Z. & Erni, B. (1993). The mannose transporter of *Escherichia coli*. Structure and function of the IIAB^{Man} subunit. *Journal of Biological Chemistry* **268**, 27094–27099.

Tam, R. & Saier, M. H., Jr (1993). Structural, functional, and evolutionary relationships among extracellular solute-binding receptors of bacteria. *Microbiological Reviews* **57**, 320–346.

Tseng, T.-T., Gratwick, K. S., Kollman, J., Park, D., Nies, D. H., Goffeau, A. & Saier, M. H., Jr (1999). The RND Permease Superfamily: an ancient, ubiquitous and diverse family that includes human disease and development proteins. *Journal of Molecular Microbiology and Biotechnology* **1** (in press).

Valiyaveetil, F. I. & Fillingame, R. H. (1998). Transmembrane topography of subunit a in the *Escherichia coli* F_1F_0 ATP synthase. *Journal of Biological Chemistry* **273**, 16241–16247.

Wang, J., Smerdon, S. J., Jäger, J., Kohlstaedt, L. A., Rice, P. A., Friedman, J. M. & Steitz, T. A. (1994). Structural basis of asymmetry in the human immunodeficiency virus type 1 reverse transcriptase heterodimer. *Proceedings of the National Academy of Sciences, USA* **91**, 7242–7246.

Wu, L.-F. & Saier, M. H., Jr (1990). Nucleotide sequence of the *fruA* gene encoding the fructose permease of the *Rhodobacter capsulatus* phosphotransferase system, and analyses of the deduced protein sequence. *Journal of Bacteriology* **172**, 7167–7178.

Wu, L.-F., Tomich, J. M. & Saier, M. H., Jr (1990). Structure and evolution of a multi-domain, multiphosphoryl transfer protein: nucleotide sequence of the *fruB(HI)* gene in *Rhodobacter capsulatus* and comparisons with homologous genes from other organisms. *Journal of Molecular Biology* **213**, 687–703.

INDEX

References to tables/figures are shown in italics